Advanced Concepts of Bearing Technology

Rolling Bearing Analysis
FIFTH EDITION

Advanced Concepts of Bearing Technology

Tedric A. Harris
Michael N. Kotzalas

Taylor & Francis
Taylor & Francis Group
Boca Raton London New York

CRC is an imprint of the Taylor & Francis Group,
an informa business

CRC Press
Taylor & Francis Group
6000 Broken Sound Parkway NW, Suite 300
Boca Raton, FL 33487-2742

International Standard Book Number-13: 978-1-138-74768-5 (pbk)

First issued in paperback 2022

Visit the Taylor & Francis Web site at
http://www.taylorandfrancis.com

and the CRC Press Web site at
http://www.crcpress.com

Preface

The main purpose of the first volume of this handbook was to provide the reader with information on the use, design, and performance of ball and roller bearings in common and relatively noncomplex applications. Such applications generally involve slow-to-moderate speed, shaft, or bearing outer ring rotation; simple, statically applied, radial or thrust loading; bearing mounting that does not include misalignment of shaft and bearing outer-ring axes; and adequate lubrication. These applications are generally covered by the engineering information provided in the catalogs supplied by the bearing manufacturers. While catalog information is sufficient to enable the use of the manufacturer's product, it is always empirical in nature and rarely provides information on the geometrical and physical justifications of the engineering formulas cited. The first volume not only includes the underlying mathematical derivations of many of the catalog-contained formulas, but also provides means for the engineering comparison of rolling bearings of various types and from different manufacturers.

Many modern bearing applications, however, involve machinery operating at high speeds; very heavy combined radial, axial, and moment loadings; high or low temperatures; and otherwise extreme environments. While rolling bearings are capable of operating in such environments, to assure adequate endurance, it is necessary to conduct more sophisticated engineering analyses of their performance than can be achieved using the methods and formulas provided in the first volume of this handbook. This is the purpose of the present volume.

When compared with its earlier editions, this edition presents updated and more accurate information to estimate rolling contact friction shear stresses and their effects on bearing functional performance and endurance. Also, means are included to calculate the effects on fatigue endurance of all stresses associated with the bearing rolling and sliding contacts. These comprise stresses due to applied loading, bearing mounting, ring speeds, material processing, and particulate contamination.

The breadth of the material covered in this text, for credibility, can hardly be covered by the expertise of the two authors. Therefore, in the preparation of this text, information provided by various experts in the field of ball and roller bearing technology was utilized. Contributions from the following persons are hereby gratefully acknowledged:

Neal DesRuisseaux	bearing vibration and noise
John I. McCool	bearing statistical analysis
Frank R. Morrison	bearing testing
Joseph M. Perez	lubricants
John R. Rumierz	lubricants and materials
Donald R. Wensing	bearing materials

Finally, since its initial publication in 1967, *Rolling Bearing Analysis* has evolved into this 5th edition. We have endeavored to maintain the material presented in an up-to-date and useful format. We hope that the readers will find this edition as useful as its earlier editions.

Tedric A. Harris
Michael N. Kotzalas

Authors

Tedric A. Harris is a graduate in mechanical engineering from the Pennsylvania State University, who received a B.S. in 1953 and an M.S. in 1954. After graduation, he was employed as a development test engineer at the Hamilton Standard Division, United Aircraft Corporation, Windsor Locks, Connecticut, and later as an analytical design engineer at the Bettis Atomic Power Laboratory, Westinghouse Electric Corporation, Pittsburgh, Pennsylvania. In 1960, he joined SKF Industries, Inc. in Philadelphia, Pennsylvania as a staff engineer. At SKF, Harris held several key management positions: manager, analytical services; director, corporate data systems; general manager, specialty bearings division; vice president, product technology & quality; president, SKF Tribonetics; vice president, engineering & research, MRC Bearings (all in the United States); director for group information systems at SKF headquarters, Gothenburg, Sweden; and managing director of the engineering & research center in the Netherlands. He retired from SKF in 1991 and was appointed as a professor of mechanical engineering at the Pennsylvania State University at University Park. He taught courses in machine design and tribology and conducted research in the field of rolling contact tribology at the university until retirement in 2001. Currently, he is a practicing consulting engineer and, as adjunct professor in mechanical engineering, teaches courses in bearing technology to graduate engineers in the university's continuing education program.

Harris is the author of 67 technical publications, mostly on rolling bearings. Among these is the book *Rolling Bearing Analysis*, currently in its 5th edition. In 1965 and 1968, he received outstanding technical paper awards from the Society of Tribologists and Lubrication Engineers and in 2001 from the American Society of Mechanical Engineers (ASME) Tribology Division. In 2002, he received the outstanding research award from the ASME.

Harris has served actively in numerous technical organizations, including the Anti-Friction Bearing Manufacturers' Association, ASME Tribology Division, and ASME Research Committee on Lubrication. He was elected ASME Fellow Member in 1973. He has served as chair of the ASME Tribology Division and as chair of the Tribology Division's Nominations and Oversight Committee. He holds three U.S. patents.

Michael N. Kotzalas graduated from the Pennsylvania State University with a B.S. in 1994, M.S. in 1997, and Ph.D. in 1999, all in mechanical engineering. During this time, the focus of his study and research was on the analysis of rolling bearing technology, including quasidynamic modeling of ball and cylindrical roller bearings for high-acceleration applications and spall progression testing and modeling for use in condition-based maintenance algorithms.

Since graduation, Dr. Kotzalas has been employed by The Timken Company in research and development and most recently in the industrial bearing business. His current responsibilities include advanced product design and application support for industrial bearing customers, while the previous job profile in research and development included new product and analysis algorithm development. From these studies, Dr. Kotzalas has received two U.S. patents for cylindrical roller bearing designs.

Outside of work, Dr. Kotzalas is also an active member of many industrial societies. As a member of the ASME, he currently serves as the chair of the publications committee and as a member of the rolling element bearing technical committee. He is a member of the awards committee in the Society of Tribologists and Lubrication Engineers (STLE). Dr. Kotzalas has also published ten articles in peer-reviewed journals and one conference proceeding. Some of his publications were honored with the ASME Tribology Division's Best Paper Award in 2001 and STLE's Hodson Award in 2003 and 2006. Also, working with the American Bearing Manufacturer's Association (ABMA), Dr. Kotzalas is one of the many instructors for the short course "Advanced Concepts of Bearing Technology".

Table of Contents

1 Distribution of Internal Loading in Statically Loaded Bearings: Combined Radial, Axial, and Moment Loadings—Flexible Support of Bearing Rings

LIST OF SYMBOLS

Symbol	Description	Units
A	Distance between raceway groove curvature centers	mm (in.)
B	$f_i + f_o - 1$	
c	Crown drop at end of roller or raceway effective length or crown gap at other locations	mm (in.)
C	Influence coefficient	mm/N (in./lb)
D	Ball or roller diameter	mm (in.)
D_{ij}	Influence coefficient to calculate nonideal roller–raceway contact deformations	
d_m	Bearing pitch diameter	mm (in.)
e	Eccentricity of loading	mm (in.)
E	Modulus of elasticity	MPa (psi)
f	r/D	
F	Applied load	N (lb)
F_a	Friction force due to roller end–ring flange sliding motions	N (lb)
h	Roller thrust couple moment arm	mm (in.)
I	Ring section moment of inertia	mm^4 ($in.^4$)
k	Number of laminae	
K	Load–deflection factor, axial load–deflection factor	N/mm^n ($lb/in.^n$)
l	Roller length	mm (in.)
M	Moment	$N \times mm$ ($lb \times in.$)
n	Load–deflection exponent	
P_d	Diametral clearance	mm (in.)
q	Load per unit length	N/mm (lb/in.)
Q	Ball or roller–raceway normal load	N (lb)
Q_a	Roller end–ring flange load in cylindrical roller bearing	N (lb)
Q_f	Roller end–ring flange load in tapered roller bearing	N (lb)

r	Raceway groove curvature radius	mm (in.)
r	Radius to raceway contact in tapered roller bearing	mm (in.)
r_f	Radius from inner-ring axis to roller end–flange contact in tapered roller bearing	mm (in.)
R_f	Radius from tapered roller axis to roller end–flange contact	mm (in.)
$\Re$	Ring radius to neutral axis	mm (in.)
$\Re$	Radius of locus of raceway groove curvature centers	mm (in.)
s	Distance between loci of inner and outer raceway groove curvature centers	mm (in.)
u	Ring radial deflection	mm (in.)
U	Strain energy	N × mm (lb × in.)
Z	Number of balls or rollers per bearing row	
α	Mounted contact angle	rad, °
α°	Free contact angle	rad, °
β	$\tan^{-1} l/(d_m - D)$	rad, °
γ	$D \cos \alpha / d_m$	
δ	Deflection or contact deformation	mm (in.)
δ_1	Distance between inner and outer rings	mm (in.)
Δ	Contact deformation due to ideal normal loading	mm (in.)
$\Delta \psi$	Angular spacing between rolling elements	rad, °
ζ	Roller tilt angle	rad, °
η	$\tan^{-1} l/D$	rad, °
θ	Bearing misalignment angle	rad, °
λ	Lamina position	
μ	Coefficient of sliding friction between roller end and ring flange	
σ	Normal contact stress or pressure	MPa (psi)
ξ	Poisson's ratio	
ξ	Roller skewing angle	rad, °

1.1 GENERAL

In most bearing applications, only applied radial, axial, or combined radial and axial loadings are considered. However, under very heavy applied loading or if shafting is hollow to minimize weight, the shaft on which the bearing is mounted may bend, causing a significant moment load on the bearing. Also, the bearing housing may be nonrigid due to design targeted at minimizing both size and weight, causing it to bend while accommodating moment loading. Such combined radial, axial, and moment loadings result in altered distribution of load among the bearing's rolling element complement. This may cause significant changes in bearing deflections, contact stresses, and fatigue endurance compared to these operating parameters associated with the simpler load distributions considered in Chapter 7 of the first volume of this handbook.

In cylindrical and tapered roller bearings, the moment loading caused by bending of the shaft results in nonuniform load per unit length along the roller–raceway contacts. Misalignment of the bearing inner ring on the shaft or outer ring in the housing also generates moment loading in the bearing, causing a nonuniform load per unit length along the roller–raceway contacts. Thus, the maximum roller–raceway contact stresses will be greater than those occurring if the contacts are loaded uniformly along their lengths. Moreover, when bearing rings are misaligned, thrust loading is induced in the rollers, causing the rollers to tilt, further exacerbating the nonuniform roller–raceway contact loading. As seen in Chapter 11 in the first volume of this

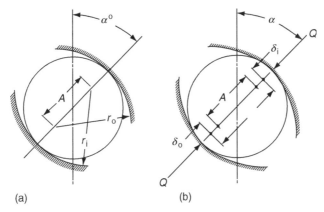

FIGURE 1.1 (a) Ball–raceway contact before applying load; (b) ball–raceway contact after load is applied.

handbook, fatigue life is inversely proportional to approximately the ninth power of contact stress. Hence, a nonuniform roller–raceway contact loading can result in significant reduction in bearing endurance.

In this chapter, methods to determine the distribution of applied loading among the rolling elements will be established considering each of the aforementioned effects.

1.2 BALL BEARINGS UNDER COMBINED RADIAL, THRUST, AND MOMENT LOADS

When a ball is compressed by load Q, since the centers of curvature of the raceway grooves are fixed with respect to the corresponding raceways, the distance between the centers is increased by the amount of the normal approach between the raceways. From Figure 1.1, it can be seen that

$$s = A + \delta_i + \delta_o \tag{1.1}$$

$$\delta_n = \delta_i + \delta_o = s - A \tag{1.2}$$

If a ball bearing that has a number of balls situated symmetrically about a pitch circle is subjected to a combination of radial, thrust (axial), and moment loads, the following relative displacements of inner and outer raceways may be defined:

δ_a Relative axial displacement
δ_r Relative radial displacement
θ Relative angular displacement

These relative displacements are shown in Figure 1.2.

Consider a rolling bearing before the application of a load. Figure 1.3 shows the positions of the loci of the centers of the inner and outer raceway groove curvature radii. It can be determined from Figure 1.4 that the locus of the centers of the inner-ring raceway groove curvature radii is expressed by

$$\Re_i = \frac{d_m}{2} + \left(r_i - \frac{D}{2}\right)\cos\alpha^o \tag{1.3}$$

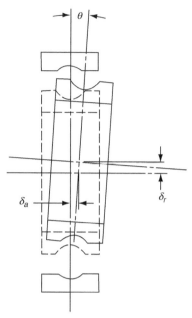

FIGURE 1.2 Displacements of an inner ring (outer ring fixed) due to application of combined radial, axial, and moment loadings.

where α° is the free contact angle determined by bearing diametral clearance. From Figure 1.3 then

$$\Re_o = \Re_i - A \cos \alpha^\circ \tag{1.4}$$

$$\Re_i - \Re_o = A \cos \alpha^\circ \tag{1.5}$$

In Figure 1.3, ψ is the angle between the most heavily loaded rolling element and any other rolling element. Because of symmetry $0 \le \psi \le \pi$.

    If the outer ring of the bearing is considered fixed in space as the load is applied to the bearing, then the inner ring will be displaced and the locus of inner-ring raceway groove radii centers will also be displaced as shown in Figure 1.5. From Figure 1.5 it can be determined that s, the distance between the centers of curvature of the inner- and outer-ring raceway grooves at any rolling element position ψ, is given by

$$s = [(A \sin \alpha^\circ + \delta_a + \Re_i\ \theta \cos \psi)^2 + (A \cos \alpha^\circ + \delta_r \cos \psi)^2]^{1/2} \tag{1.6}$$

or

$$s = A\left[\left(\sin \alpha^\circ + \bar{\delta}_a + \Re_i\ \bar{\theta} \cos \psi\right)^2 + \left(\cos \alpha^\circ + \bar{\delta}_r \cos \psi\right)^2\right]^{1/2} \tag{1.7}$$

where

$$\bar{\delta}_a = \frac{\delta_a}{A} \tag{1.8}$$

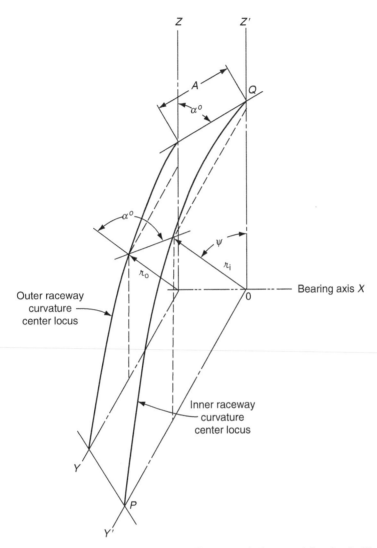

FIGURE 1.3 Loci of raceway groove curvature radii centers before applying load. (From Jones, A., *Analysis of Stresses and Deflections*, New Departure Engineering Data, Bristol, CT, 1946.)

$$\bar{\delta}_r = \frac{\delta_r}{A} \tag{1.9}$$

$$\bar{\theta} = \frac{\theta}{A} \tag{1.10}$$

Substituting Equation 1.7 into Equation 1.2 yields

$$\delta_n = A\left\{\left[\left(\sin\alpha^o + \bar{\delta}_a + \Re_i\bar{\theta}\cos\psi\right)^2 + \left(\cos\alpha^o + \bar{\delta}_r\cos\psi\right)^2\right]^{1/2} - 1\right\} \tag{1.11}$$

From Chapter 7 of the first volume of this book, the load vs. deformation relationship for a rolling element–raceway contact is given by

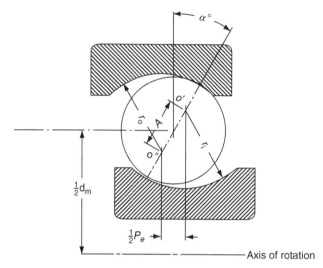

FIGURE 1.4 Radial ball bearing showing ball–raceway contact due to axial shift of inner and outer rings.

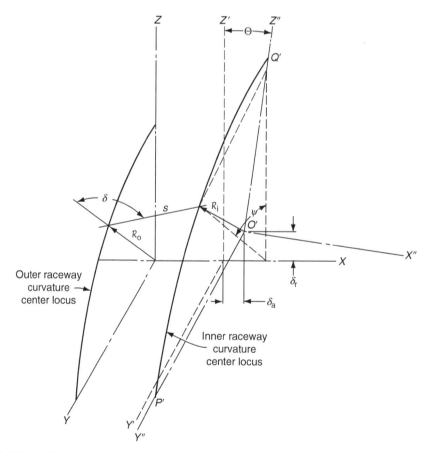

FIGURE 1.5 Loci of raceway groove curvature radii centers after displacement (From Jones, A., *Analysis of Stresses and Deflections*, New Departure Engineering Data, Bristol, CT, 1946.)

$$Q = K_n \delta^n \tag{1.12}$$

In Equation 1.12, exponent $n = 3/2$ for ball bearings and 10/9 for roller bearings. Substitution of Equation 1.11 into Equation 1.12 and using the former exponent gives

$$Q = K_n A^{1.5} \left\{ \left[\left(\sin \alpha^\circ + \overline{\delta}_a + \Re_i \overline{\theta} \cos \psi \right)^2 + \left(\cos \alpha^\circ + \overline{\delta}_r \cos \psi \right)^2 \right]^{1/2} - 1 \right\}^{1.5} \tag{1.13}$$

At any ball azimuth position ψ, the operating contact angle is α. This angle can be determined from

$$\sin \alpha = \frac{\sin \alpha^\circ + \overline{\delta}_a + \Re_i \overline{\theta} \cos \psi}{\left[\left(\sin \alpha^\circ + \overline{\delta}_a + \Re_i \overline{\theta} \cos \psi \right)^2 + \left(\cos \alpha^\circ + \overline{\delta}_r \cos \psi \right)^2 \right]^{1/2}} \tag{1.14}$$

or

$$\cos \alpha = \frac{\cos \alpha^\circ + \overline{\delta}_r \cos \psi}{\left[\left(\sin \alpha^\circ + \overline{\delta}_a + \Re_i \overline{\theta} \cos \psi \right)^2 + \left(\cos \alpha^\circ + \overline{\delta}_r \cos \psi \right)^2 \right]^{1/2}} \tag{1.15}$$

Equation 1.12 describes the normal load on the raceway acting through the contact angle. This normal load may be resolved into axial and radial components as follows:

$$Q_a = Q \sin \alpha \tag{1.16}$$

$$Q_r = Q \cos \psi \cos \alpha \tag{1.17}$$

If the radial and thrust loads applied to the bearing are F_r and F_a, respectively, then for static equilibrium to exist

$$F_a = \sum_{\psi = 0}^{\psi = \pm \pi} Q_\psi \sin \alpha \tag{1.18}$$

$$F_r = \sum_{\psi = 0}^{\psi = \pm \pi} Q_\psi \cos \psi \cos \alpha \tag{1.19}$$

Additionally, each of the thrust components produce a moment about the Y-axis such that

$$M_\psi = \frac{d_m}{2} Q_\psi \cos \psi \sin \alpha \tag{1.20}$$

For static equilibrium, the applied moment M about the Y-axis must equal the sum of the moments of each rolling element about the Y-axis (in the case of load symmetry, rolling element thrust component moments about the Z-axis are self-equilibrating).

$$M = \frac{d_m}{2} \sum_{\psi = 0}^{\psi = \pm \pi} Q_\psi \cos \psi \sin \alpha \tag{1.21}$$

Combining Equation 1.13, Equation 1.16, and Equation 1.18 yields

$$
F_a - K_n A^{1.5} \sum_{\psi=0}^{\psi=\pm\pi} \frac{\left\{ \left[\left(\sin\alpha^\circ + \bar\delta_a + \Re_i \bar\theta \cos\psi \right)^2 + \left(\cos\alpha^\circ + \bar\delta_r \cos\psi \right)^2 \right]^{1/2} - 1 \right\}^{1.5} \left(\sin\alpha^\circ + \bar\delta_a + \Re_i \bar\theta \cos\psi \right)}{\left[\left(\sin\alpha^\circ + \bar\delta_a + \Re_i \bar\theta \cos\psi \right)^2 + \left(\cos\alpha^\circ + \bar\delta_r \cos\psi \right)^2 \right]^{1/2}} = 0
$$

(1.22)

$$
F_r - K_n A^{1.5} \sum_{\psi=0}^{\psi=\pm\pi} \frac{\left\{ \left[\left(\sin\alpha^\circ + \bar\delta_a + \Re_i \bar\theta \cos\psi \right)^2 + \left(\cos\alpha^\circ + \bar\delta_r \cos\psi \right)^2 \right]^{1/2} - 1 \right\}^{1.5} \left(\cos\alpha^\circ + \bar\delta_r \cos\psi \right) \cos\psi}{\left[\left(\sin\alpha^\circ + \bar\delta_a + \Re_i \bar\theta \cos\psi \right)^2 + \left(\cos\alpha^\circ + \bar\delta_r \cos\psi \right)^2 \right]^{1/2}} = 0
$$

(1.23)

$$
M - \frac{d_m}{2} K_n A^{1.5} \sum_{\psi=0}^{\psi=\pm\pi} \frac{\left\{ \left[\left(\sin\alpha^\circ + \bar\delta_a + \Re_i \bar\theta \cos\psi \right)^2 + \left(\cos\alpha^\circ + \bar\delta_r \cos\psi \right)^2 \right]^{1/2} - 1 \right\}^{1.5} \left(\sin\alpha^\circ + \bar\delta_a + \Re_i \bar\theta \cos\psi \right) \cos\psi}{\left[\left(\sin\alpha^\circ + \bar\delta_a + \Re_i \bar\theta \cos\psi \right)^2 + \left(\cos\alpha^\circ + \bar\delta_r \cos\psi \right)^2 \right]^{1/2}} = 0
$$

(1.24)

These equations were developed by Jones [1].

Equation 1.22 through Equation 1.24 are simultaneous nonlinear equations with unknowns δ_a, δ_r, and θ. They may be solved by numerical methods; for example, the Newton–Raphson method. Having obtained δ_a, δ_r, and θ, the maximum ball load may be obtained from Equation 1.13 for $\psi = 0$.

$$
Q_{\max} = K_n A^{1.5} \left\{ \left[\left(\sin\alpha^\circ + \bar\delta_a + \Re_i \bar\theta \right)^2 + \left(\cos\alpha^\circ + \bar\delta_r \right)^2 \right]^{1/2} - 1 \right\}^{1.5}
$$

(1.25)

Solution of the indicated equations generally necessitates the use of a digital computer.

1.3 MISALIGNMENT OF RADIAL ROLLER BEARINGS

Although it is undesirable, radial cylindrical roller bearings and tapered roller bearings can support to a small extent the moment loading due to misalignment. The various types of misalignment are illustrated in Figure 1.6. Spherical roller bearings are designed to exclude moment loads from acting on the bearings and therefore are not included in this discussion. Figure 1.7 illustrates the misalignment of the inner ring of a cylindrical roller bearing relative to the outer ring.

To commence the analysis, it is assumed that any roller–raceway contact can be divided into a number of "slices" or laminae situated in planes parallel to the radial plane of the bearing. It is also assumed that shear effects between these laminae can be neglected owing to the small magnitudes of the contact deformations that develop. (Only contact deformations are considered.)

1.3.1 COMPONENTS OF DEFORMATION

In a misaligned cylindrical roller bearing subjected to radial load, at each lamina in a crowned roller–raceway contact, the deformation may be considered to be composed of three components: (1) Δ_{mj} due to the radial load at the roller azimuth location j, (2) c_λ due to the crown drop at lamina λ, and (3) the deformation due to bearing misalignment and

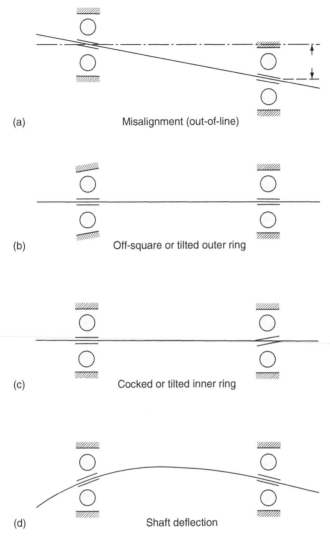

FIGURE 1.6 Types of misalignments.

roller tilt at the roller azimuth location j. These components are shown schematically in Figure 1.8.

The component due to radial load is the only contact deformation component considered in the simplified analytical methods presented in Chapter 7 of the first volume of this book. It needs no further explanation here.

1.3.1.1 Crowning

As stated previously, crowning of rollers and raceways is accomplished to avoid edge loading that can result in early fatigue failure of the rolling components. It may be accomplished in various forms. The simplest of these is the full circular profile crown illustrated in Figure 1.9. The rollers in most spherical roller bearings may be considered fully crowned whether of symmetrical contour (barrel-shaped) or of asymmetrical contour. In the latter case, the crown is offset from the roller mid-length point. Full crowning may also be applied to raceways as

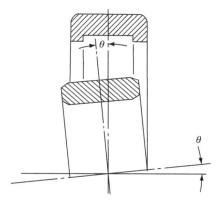

FIGURE 1.7 Misalignment of cylindrical roller bearing rings.

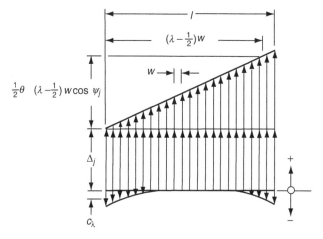

FIGURE 1.8 Components of roller–raceway contact deformation due to radial load, misalignment, and crowning.

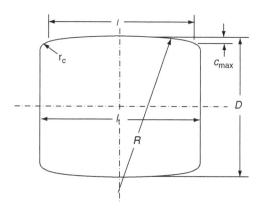

FIGURE 1.9 Schematic diagram of cylindrical roller with full circular profile crown.

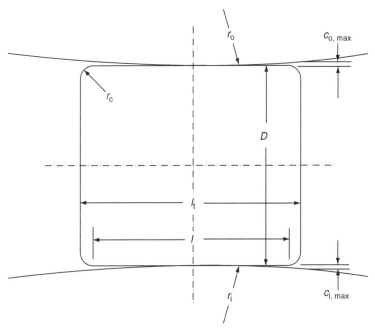

FIGURE 1.10 Schematic diagram of uncrowned (straight profile) cylindrical roller contacting inner and outer raceways, each with a full circular profile crown.

shown in Figure 1.10. This is commonly used in tapered roller bearings where often both the cone and cup raceways are crowned, and the rollers are not crowned.

Most cylindrical roller bearings employ rollers that are crowned only over a portion of the roller contour; the remaining portion is cylindrical (the contour is sometimes called flat or straight). A partially crowned cylindrical roller is illustrated in Figure 1.11.

From Figure 1.8, it can be seen that crown drop or crown gap c_λ at a selected lamina is considered as a negative deformation; that is, no roller–raceway loading can occur at a lamina until c_λ is overcome by the radial or the misalignment deformation. For both the fully crowned or partially crowned rollers that have circular profiles, Equation 1.26 defines c_λ in terms of the roller and crown dimensions, where $1 \leq \lambda \leq k$.

$$c_\lambda = \begin{cases} c_{max}\left[\dfrac{\left(\frac{2\lambda-1}{k}-1\right)^2-\left(\frac{l_s}{l}\right)^2}{1-\frac{l_s}{l}}\right] & \left(\dfrac{2\lambda-1}{k}-1\right)^2-\left(\dfrac{l_s}{l}\right)^2 > 0 \\ \\ 0 & \left(\dfrac{2\lambda-1}{k}-1\right)^2-\left(\dfrac{l_s}{l}\right)^2 \leq 0 \end{cases} \qquad (1.26)$$

For rollers with circular profile partial crowns, blending between the straight and crowned portions of the profile is necessary to minimize stress concentrations and the resulting reduced fatigue life. To avoid such stress concentrations, in lieu of a circular profile, a tangential profile might be used. In this case, the crown radius would be variable, and the crown gap at each lamina k would be calculated using

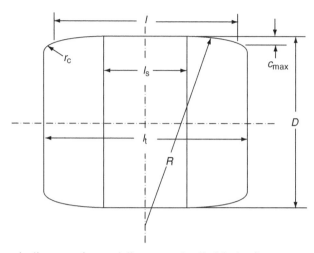

FIGURE 1.11 Schematic diagram of a partially crowned cylindrical roller.

$$c_\lambda = \left\{ c_{max} \left[\frac{\left| \frac{2\lambda-1}{k} - 1 \right| - \frac{l_s}{l}}{1 - \frac{l_s}{l}} \right]^2 \quad \left| \frac{2\lambda-1}{k} - 1 \right| - \frac{l_s}{l} > 0 \atop 0 \qquad\qquad\qquad \left| \frac{2\lambda-1}{k} - 1 \right| - \frac{l_s}{l} \le 0 \right\} \tag{1.27}$$

To minimize edge loading, Lundberg and sjövall [2] devised a fully crowned roller having a logarithmic profile. The crown gap at each lamina k is calculated using

$$c_\lambda = 0.2 \cdot ln \left[\frac{1}{1.0067 - \left(\frac{2\lambda-1}{k} - 1 \right)^2} \right] \tag{1.28}$$

Subsequently, Reussner [3] developed another logarithmic profile crown believed to be more effective. The crown gap at each lamina k for the Reussner crown profile is given by

$$c_\lambda = 2 \times 10^{-4} \Sigma \rho w^2 k^2 \cdot ln \left[\frac{1}{1 - \left(\frac{2\lambda-1}{k} - 1 \right)^6} \right] \tag{1.29}$$

It is possible to combine roller crowning and raceway crowning. In this case, the crown gap at each lamina k would be calculated as the sum of the crown gaps for the roller and raceway as follows:

$$c_{m\lambda} = c_{R\lambda} + c_{m\lambda} \tag{1.30}$$

In the above equation, subscript R refers to the roller and m to the raceway (m = i or m = o).

For the bearing misalignment θ shown in Figure 1.7, the effective misalignment at the azimuth location of the roller ψ_j is $\pm 1/2\theta \cos \psi_j$. The plus sign pertains to $0 \le \psi_j \le \pi/2$; the minus sign pertains to $\pi/2 \leq \psi_j \leq \pi$ (assuming symmetry of loading about the $0-\pi$ diameter). Therefore, the total roller–raceway deformation at roller location j and lamina λ is given by

$$\delta_{\lambda j} = \Delta_j \pm \frac{\theta}{2}\left(\lambda - \frac{1}{2}\right)w\cos\psi_j - c_\lambda \tag{1.31}$$

1.3.2 LOAD ON A ROLLER–RACEWAY CONTACT LAMINA

In Chapter 6 of the first volume of this book, the following equations were given to describe the deformation vs. load for a roller–raceway contact:

$$\delta = \frac{2Q(1-\xi^2)}{\pi El}\ln\left[\frac{\pi El^2}{Q(1-\xi^2)(1\mp\gamma)}\right] \tag{1.32}$$

$$\delta = 3.84 \times 10^{-5}\frac{Q^{0.9}}{l^{0.8}} \tag{1.33}$$

Equation 1.32 was developed by Lundberg and Sjövall [2] for an ideal line contact. In Equation 1.32, $\gamma = D\cos\alpha/d_m$, E is the modulus of elasticity, and ξ is Poisson's ratio. Equation 1.33 was developed empirically by Palmgren [4] from laboratory test data and pertains to the contact of a crowned roller on a raceway. While the load–deformation characteristic of an individual contact lamina may be described using either equation, the latter is applied here as the solution of a transcendental equation leads to force and moment equilibrium equations of greater complexity. Considering that the contact is divided into k laminae, each lamina of width w, the contact length is kw. Letting $q = Q/l$, Equation 1.33 becomes

$$\delta = 3.84 \times 10^{-5}q^{0.9}(kw)^{0.1} \tag{1.34}$$

Rearranging the above equation to define q yields

$$q = \frac{\delta^{1.11}}{1.24 \times 10^{-5}\,(kw)^{0.11}} \tag{1.35}$$

Equation 1.35 does not consider edge stresses; however, because these obtain only over very small areas, they can be neglected with little loss of accuracy when considering equilibrium of loading. Substitution of Equation 1.31 into Equation 1.35 gives

$$q_{\lambda j} = \frac{\left[\Delta_j \pm \theta\left(\lambda - \frac{1}{2}\right)w\cos\psi_j - c_\lambda\right]^{1.11}}{1.24 \times 10^{-5}(k_j w)^{0.11}} \tag{1.36}$$

Depending on the degree of loading and misalignment, all laminae in every contact may not be loaded; in Equation 1.36, k_j is the number of laminae under load at roller location j. Total roller loading is given by

$$Q_j = \frac{w^{0.89}}{1.24 \times 10^{-5}k_j^{0.11}}\sum_{\lambda=1}^{\lambda=k_j}\left[\Delta_j \pm \frac{1}{2}\theta\left(\lambda - \frac{1}{2}\right)w\cos\psi_j - c_\lambda\right]^{1.11} \tag{1.37}$$

1.3.3   EQUATIONS OF STATIC EQUILIBRIUM

To determine the individual roller loading, it is necessary to satisfy the requirements of static equilibrium. Hence, for an applied radial load,

$$\frac{F_\mathrm{r}}{2} - \sum_{j=1}^{j=Z/2+1} \tau_j Q_j \cos \psi_j = 0 \quad \begin{aligned} \tau_j = 0.5; & \quad \psi_j = 0, \pi \\ \tau_j = 1; & \quad \psi_j \neq 0, \pi \end{aligned} \tag{1.38}$$

Substituting Equation 1.37 into Equation 1.38 yields

$$\frac{0.62 \times 10^{-5} F_\mathrm{r}}{w^{0.89}} - \sum_{j=1}^{j=Z/2+1} \frac{\tau_j \cos \psi_j}{k_j^{0.11}} \sum_{\lambda=1}^{\lambda=k_j} \left[\Delta_j \pm \frac{1}{2} \theta \left(\lambda - \frac{1}{2} \right) w \cos \psi_j - c_\lambda \right]^{1.11} = 0 \tag{1.39}$$

For an applied coplanar misaligning moment load, the equilibrium condition to be satisfied is

$$\frac{\mathfrak{M}}{2} - \sum_{j=1}^{j=Z/2+1} \tau_j Q_j e_j \cos \psi_j = 0 \quad \begin{aligned} \tau_j = 0.5; & \quad \psi_j = 0, \pi \\ \tau_j = 1; & \quad \psi_j \neq 0, \pi \end{aligned} \tag{1.40}$$

where e_j is the eccentricity of loading at each roller location. e_j, which is illustrated in Figure 1.12, is given by

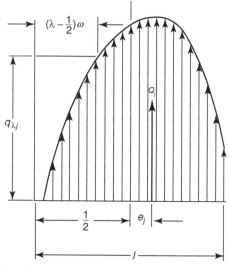

FIGURE 1.12 Load distribution for a misaligned crowned roller showing eccentricity of loading.

$$e_j = \frac{\sum\limits_{\lambda=1}^{\lambda=k_j} q_{\lambda j}\left(\lambda - \tfrac{1}{2}\right)w}{\sum\limits_{\lambda=1}^{\lambda=k_j} q_{\lambda j}} - \frac{l}{2} \quad j = 3, \quad \frac{Z}{2} + 3 \tag{1.41}$$

Hence,

$$\frac{0.62 \times 10^{-5}\mathfrak{M}}{w^{0.89}} - \sum_{j=1}^{j=Z/2+1} \frac{\tau_j \cos\psi_j}{k_j^{0.11}}$$

$$\times \left\{ \sum_{\lambda=1}^{\lambda=k_j} \left[\Delta_j \pm \frac{1}{2}\theta\left(\lambda - \frac{1}{2}\right)w\cos\psi_j - c_j \right]^{1.11} \left(\lambda - \frac{1}{2}\right)w \right. \tag{1.42}$$

$$\left. - \frac{l}{2} \sum_{\lambda=1}^{\lambda=k_j} \left[\Delta_j \pm \frac{1}{2}\theta\left(\lambda - \frac{1}{2}\right)w\cos\psi_j - c_\lambda \right]^{1.11} \right\} = 0$$

1.3.4 Deflection Equations

The remaining equations to be established are the radial deflection relationships. It is necessary here to determine the relative radial movement of the rings caused by the misalignment as well as that owing to radial loading. To assist in the first determination, Figure 1.13 shows schematically an inner ring–roller assembly misaligned with respect to the outer ring. From this sketch, it is evident that one half of the roller included angle is described by

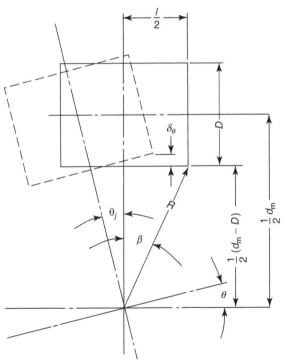

FIGURE 1.13 Schematic diagram of misaligned roller–inner ring assembly showing interference with outer ring.

$$\beta = \tan^{-1} \frac{l}{d_{\mathrm{m}} - D} \tag{1.43}$$

and

$$\sin \beta = \frac{l}{[(d_{\mathrm{m}} - D)^2 + l^2]^{1/2}} \tag{1.44}$$

The maximum radial interference between a roller and the outer ring owing to misalignment is given by

$$\delta_\theta = R\cos(\beta - \theta_j) - R\cos\beta \tag{1.45}$$

where

$$R = 0.5 \times [(d_{\mathrm{m}} - D)^2 + l^2]^{1/2} \tag{1.46}$$

In developing Equation 1.45 and Equation 1.46, the effect of crown drop was investigated and found to be negligible.

Expanding Equation 1.46 in terms of the trigonometric identity further yields

$$\delta_\theta = R(\cos\beta \cos\theta_j + \sin\beta \sin\theta_j - \cos\beta) \tag{1.47}$$

As θ_j is small, $\cos\theta_j \to 1$, and $\sin\theta_j \to \theta_j$. Moreover, $\theta_j = \pm\theta\cos\psi_j$ and $\sin\beta = l/2R$; therefore,

$$\delta_\theta = \pm\tfrac{1}{2}\, l\theta\cos\psi_j \tag{1.48}$$

The shift of the inner-ring center relative to the outer-ring center owing to radial loading and clearance, and the subsequent relative radial movement at any roller location are shown in Figure 1.14. The sum of the relative radial movement of the rings at each roller angular location minus the clearance is equal to the sum of the inner and outer raceway maximum contact deformations at the same angular location. Stating this relationship in equation format:

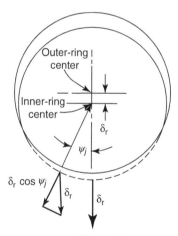

FIGURE 1.14 Displacement of ring centers caused by radial loading showing relative radial movement.

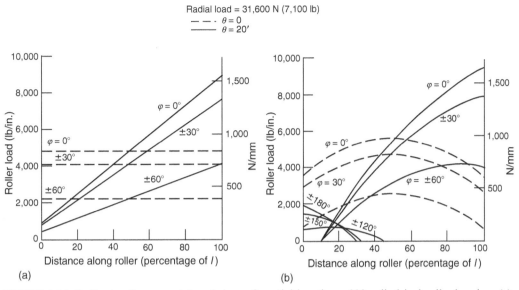

FIGURE 1.15 Roller loading vs. axial and circumferential location—309 cylindrical roller bearing: (a) ideally crowned rollers; (b) fully crowned rollers.

$$\left[\delta_r \pm \frac{1}{2}l\theta\right]\cos\psi_j - \frac{P_d}{2} - 2\left[\Delta_j \pm \frac{1}{2}\theta\left(\lambda - \frac{1}{2}\right)w\cos\psi_j - c_\lambda\right]_{max} = 0 \qquad (1.49)$$

Equation 1.39, Equation 1.42, and Equation 1.49 constitute a set of $Z/2 + 3$ simultaneous nonlinear equations that can be solved for δ_r, θ, and Δ_j using numerical analysis techniques. Thereafter, the variation of roller load per unit length, and subsequently the roller load, may be determined for each roller location using Equation 1.36 and Equation 1.37, respectively.

Using this method of digital computation, Harris [5] analyzed a 309 cylindrical roller bearing having the following dimensions and loading:

Number of rollers	12
Roller effective length	12.6 mm (0.496 in.)
Roller straight lengths	4.78, 7.770, 12.6 mm
Roller crown radius	1,245 mm (49 in.)
Roller diameter	14 mm (0.551 in.)
Bearing pitch diameter	72.39 mm (2.85 in.)
Applied radial load	31,600 N (7,100 lb)

For these conditions, Figure 1.15 shows the loading on various rollers for the bearing with ideally crowned rollers ($l_s = 12.6$ mm [0.496 in.]) and with fully crowned rollers ($l_s = 0$).

Figure 1.16 shows the effect of roller crowning on bearing radial deflection as a function of misalignment.

1.4 THRUST LOADING OF RADIAL CYLINDRICAL ROLLER BEARINGS

When radial cylindrical roller bearings have fixed flanges on both inner and outer rings, they can carry some thrust load in addition to radial load. The greater the amount of radial load applied, the more is the thrust load that can be carried. As shown by Harris [6] and seen in Figure 1.17, the thrust load causes each roller to tilt an amount ζ_j.

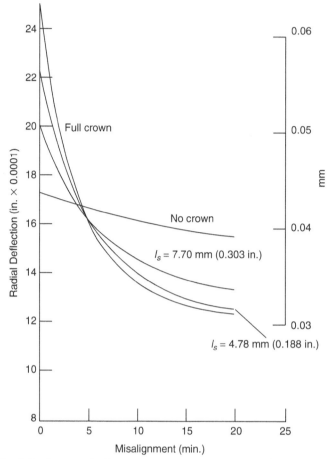

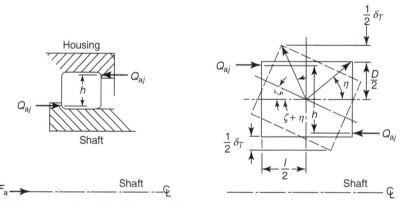

FIGURE 1.16 Roller deflection vs. misalignment and crowning—309 cylindrical roller bearing at 31,600 N (7,100 lb) radial load.

Again, it is assumed that a roller–raceway contact can be subdivided into laminae in planes parallel to the radial plane of the bearing. When a radial cylindrical roller bearing is subjected to applied thrust load, the inner ring shifts axially relative to the outer ring.

FIGURE 1.17 Thrust couple, roller tilting, and interference owing to applied thrust load.

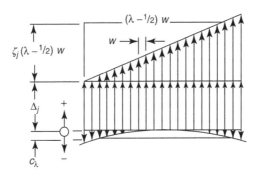

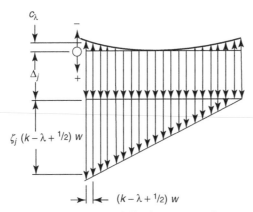

FIGURE 1.18 Components of roller–raceway deflection at opposing raceways due to radial load, thrust load, and crowning.

Assuming deflections owing to roller end–flange contacts are negligible, the interference at any axial location (lamina) is

$$\delta_{\lambda j} = \Delta_j + \zeta_j\left(\lambda - \frac{1}{2}\right)w - c_\lambda, \quad \lambda = 1, k_j \tag{1.50}$$

where c_λ is given by Equation 1.26 through Equation 1.30. Figure 1.18 illustrates the component deflections in Equation 1.50. Substituting Equation 1.50 into Equation 1.35 yields

$$q_{\lambda j} = \frac{\left[\Delta_j + \zeta_j\left(\lambda - \frac{1}{2}\right)w - c_\lambda\right]^{1.11}}{1.24 \times 10^{-5}(k_j w)^{0.11}} \tag{1.51}$$

and at any azimuth ψ_j, the total roller loading is

$$Q_j = \frac{w^{0.89}}{1.24 \times 10^{-5}k_j^{0.11}} \sum_{\lambda=1}^{\lambda=k_j}\left[\Delta_j + \zeta_j\left(\lambda - \frac{1}{2}\right)w - c_\lambda\right]^{1.11} \tag{1.52}$$

1.4.1   Equilibrium Equations

To determine roller loading, it is necessary to satisfy static equilibrium requirements. Hence, for applied radial load

$$\frac{F_r}{2} - \sum_{j=1}^{j=Z/2+1} \tau_j Q_j \cos \psi_j = 0 \qquad \begin{array}{ll} \tau_j = 0.5; & \psi_j = 0, \pi \\ \tau_j = 1; & \psi_j \neq 0, \pi \end{array} \tag{1.53}$$

Substituting Equation 1.52 into Equation 1.53 yields

$$\frac{0.62 \times 10^{-5} F_r}{w^{0.89}} - \sum_{j=1}^{j=Z/2+1} \frac{\tau_j \cos \psi_j}{k_j^{0.11}} \sum_{\lambda=1}^{\lambda=k_j} \left[\Delta_j + \zeta_j \left(\lambda - \frac{1}{2} \right) w - c_\lambda \right]^{1.11} = 0 \tag{1.54}$$

For an applied centric thrust load, the equilibrium condition to be satisfied is

$$\frac{F_a}{2} - \sum_{j=1}^{j=Z/2+1} \tau_j Q_{aj} = 0 \tag{1.55}$$

At each roller location, the thrust couple is balanced by a radial load couple caused by the skewed axial load distribution. Thus, $hQ_{aj} = 2Q_j e_j$ and

$$\frac{F_a}{2} - \frac{2}{h} \sum_{j=1}^{j=Z/2+1} \tau_j Q_j e_j = 0 \qquad \begin{array}{ll} \tau_j = 0.5; & \psi_j = 0, \ \pi \\ \tau_j = 1; & \psi_j \neq 0, \ \pi \end{array} \tag{1.56}$$

where e_j is the eccentricity of loading indicated in Figure 1.12 and defined by

$$e_j = \frac{\displaystyle\sum_{\lambda=1}^{\lambda=k_j} q_{\lambda j} \left(\lambda - \frac{1}{2} \right) w}{\displaystyle\sum_{\lambda=1}^{\lambda=k_j} q_{\lambda j}} - \frac{l}{2} \tag{1.57}$$

Substitution of Equation 1.52 and Equation 1.57 into Equation 1.56 yields

$$\frac{0.31 \times 10^{-5} F_a}{w^{0.89}} - \sum_{j=1}^{j=Z/2+1} \frac{\tau_j}{k_j^{0.11}}$$

$$\times \left\{ \sum_{\lambda=1}^{\lambda=k_j} \left[\Delta_j \pm \zeta_j \left(\lambda - \frac{1}{2} \right) w - c_\lambda \right]^{1.11} \left(\lambda - \frac{1}{2} \right) w - \frac{l}{2} \right. \tag{1.58}$$

$$\left. \times \sum_{\lambda=1}^{\lambda=k_j} \left[\Delta_j \pm \zeta_j \left(\lambda - \frac{1}{2} \right) w - c_\lambda \right]^{1.11} \right\} = 0 \qquad \begin{array}{ll} \tau_j = \frac{1}{2}; & \psi_j = 0, \pi \\ \tau_j = 1; & \psi_j \neq 0, \pi \end{array}$$

1.4.2 DEFLECTION EQUATIONS

Radial deflection relationships remain to be established. It is necessary to determine the relative radial movement of the bearing rings caused by the thrust loading as well as that due to radial loading. To assist in this derivation, Figure 1.17 shows schematically a thrust-loaded roller–ring assembly. From this sketch, a roller angle is described by

$$\tan \eta = \frac{D}{l} \tag{1.59}$$

The maximum radial interference between a roller and both rings is given by

$$\delta_j = D \left[\frac{\sin(\zeta_j + \eta)}{\sin \eta} - 1 \right] \tag{1.60}$$

In developing the above equation, the effect of crown drop was found to be negligible. Expanding Equation 1.60 in terms of the trigonometric identity and recognizing that ζ_j is small and $l = D \operatorname{ctn} \eta$ yields

$$\delta_{tj} = l\zeta_j \tag{1.61}$$

Although δ_{tj} is the radial deflection due to roller tilting, it can be similarly shown that axial deflection owing to roller tilting is

$$\delta_{aj} = D\zeta_j \tag{1.62}$$

Therefore, the radial interference caused by axial deflection is

$$\delta_{ra} = \delta_a \frac{l}{D} \tag{1.63}$$

The sum of the relative radial movements of the inner and outer rings at each roller azimuth minus the radial clearance is equal to the sum of the inner and outer raceway maximum contact deformations at the same azimuth, or

$$\delta_a \frac{l}{D} + \delta_r \cos \psi_j - \frac{P_d}{2} - 2 \left[\Delta_j + \zeta_j \left(\lambda - \frac{1}{2} \right) w - c_\lambda \right]_{max} = 0 \tag{1.64}$$

The set of simultaneous equations, Equation 1.54, Equation 1.58, and Equation 1.64, can be solved for ζ_j, Δ_j, δ_r, and δ_a. Thereafter, the variation of the roller load per unit length q and roller load Q may be determined for each roller azimuth using Equation 1.51 and Equation 1.52, respectively. The axial load on each roller may be determined from

$$Q_{aj} = \frac{w^{0.89}}{3.84 \times 10^{-5} k_j^{0.11} h}$$

$$\times \left\{ \sum_{\lambda=1}^{\lambda=k_j} \left[\Delta_j + \zeta_j \left(\lambda - \frac{1}{2} \right) w - c_\lambda \right]^{1.11} \left(\lambda - \frac{1}{2} \right) w - \frac{l}{2} \sum_{\lambda=1}^{\lambda=k_j} \left[\Delta_j + \zeta_j \left(\lambda - \frac{1}{2} \right) w - c_\lambda \right]^{1.11} \right\}$$

$$\tag{1.65}$$

1.4.3 Roller–Raceway Deformations Due to Skewing

When rollers are subjected to axial loading as shown in Figure 1.17, due to sliding motions between the roller ends and ring flanges, friction forces occur. For example, $F_{aj} = \mu Q_{aj}$, in which μ is the coefficient of friction. In a misaligned bearing, each roller that carries load is squeezed at one end and forced against the opposing flange with a load Q_{aj}, creating friction force F_{aj} at the roller end. Because of F_{aj}, a moment occurs creating, in addition to the predominant rolling motion about the roller axis, a yawing or skewing motion and secondary roller tilting. The tilting and skewing motions occur in orthogonal planes that contain the roller axis. Roller skewing is resisted by the concave curvature of the outer raceway. The resisting forces and accompanying deformations alter the distribution of load along both the outer and inner raceway–roller contacts. Figure 1.19 illustrates the forces that occur on a roller subjected to radial and thrust loadings. Frictional stresses $\sigma_{s1j\lambda}$ and $\sigma_{s2j\lambda}$ in Figure 1.19 tend not to significantly influence the roller–raceway normal loadings per unit length $q_{1j\lambda}$ and $q_{2j\lambda}$ on the outer and inner raceways, respectively.

Figure 1.20 shows the roller skewing angle ξ_j and the roller–outer raceway loading that result.

The roller–raceway contact deformations that result from skewing as demonstrated by Harris et al. [7] may be described by

$$\delta_{mj\lambda} = \Delta_{mj} + w\left(\lambda - \frac{1}{2}\right)\zeta_j + \phi_{mj\lambda} - c_\lambda \tag{1.66}$$

In the above equation, subscript m refers to the outer and inner raceway contacts; $m = 1$ and 2, respectively, and deformations due to skewing $\varphi_{mj\lambda}$ are given by

$$\phi_{1j\lambda} = \frac{\left[\left(\frac{k}{2} - \left|\lambda - \frac{1}{2}(k+1)\right|\right)w\right]}{d_m + D}\xi_j^2 \tag{1.67}$$

$$\phi_{2j\lambda} = \frac{\left[\left|\lambda - \frac{1}{2}(k+1)\right|w\right]^2}{d_m + D}\xi_j^2 \tag{1.68}$$

It can be further seen that Equation 1.64 must become

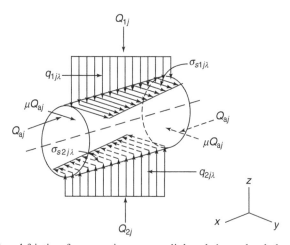

FIGURE 1.19 Normal and friction forces acting on a radial and thrust-loaded roller.

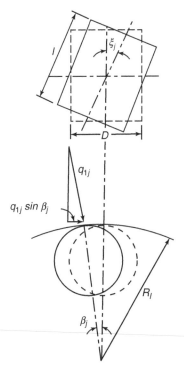

FIGURE 1.20 Roller–outer raceway contact showing roller skewing angle ξ_j and restoring forces.

$$\delta_a \frac{l}{D} + \delta_r \cos \psi_j - \frac{P_d}{2} - (\delta_{1mj,\,\max} + \delta_{2mj,\,\max}) = 0 \tag{1.69}$$

Owing to the unknown variables ξ_j and Δ_{mj}, the latter replacing Δ_j, additional equilibrium equations must be established. For equilibrium of roller loading in the radial direction,

$$\sum_{m=1}^{m=2} Q_{mj} = w \sum_{m=1}^{m=2} \sum_{\lambda=1}^{\lambda=k} q_{mj\lambda} = 0 \tag{1.70}$$

Referring to Figure 1.20 and considering equilibrium of moments in the plane of roller skewing,

$$
\begin{aligned}
lF_{aj} &+ \sum_{m=1}^{m=2} \sum_{\lambda=1}^{\lambda=k} w^2 \left[\lambda - \frac{1}{2}(k+1) \right] \sigma_{mj\lambda} \\
&- \sum_{m=1}^{m=2} \sum_{\lambda=1}^{\lambda=k} w^2 \left[\lambda - \frac{1}{2}(k+1) \right] q_{mj\lambda} \sin \beta_j = 0
\end{aligned}
\tag{1.71}
$$

As the angle $\beta_j \rightarrow 0$, $\sin \beta_j \rightarrow \beta_j$,

$$\sin \beta_j = \frac{2w}{d_m + D} \left[\lambda - \frac{1}{2}(k+1) \right] \xi_j \tag{1.72}$$

As indicated above, the frictional stresses, $\sigma_{s1j\lambda}$ and $\sigma_{s2j\lambda}$, tend not to influence the roller–raceway normal loading significantly, meaning that the frictional moment loading is rather small compared with those caused by the restoring forces $q_{1j} \sin \beta_j$ shown in Figure 1.20 and roller end–flange friction forces. Therefore, substituting Equation 1.72 into Equation 1.71 yields

$$lF_{aj} - \frac{2w^3 \xi_j}{d_m + D} \sum_{m=1}^{m=2} \sum_{\lambda=1}^{\lambda=k} \left[\lambda - \frac{1}{2}(k+1)\right]^2 q_{mj\lambda} = 0 \qquad (1.73)$$

Considering that the contact deformations due to roller radial loading are different for each roller–raceway contact, bearing load equilibrium equations, Equation 1.54 and Equation 1.58, must be changed accordingly; hence,

$$\frac{0.62 \times 10^{-5} F_r}{w^{0.89}} - \sum_{j=1}^{j=Z/2+1} \frac{\tau_j \cos \psi_j}{k_{2j}^{0.11}} \sum_{\lambda=1}^{\lambda=k} \left[\Delta_{2j} + w\left(\lambda - \frac{1}{2}\right)\zeta_j + \phi_{2j\lambda} - c_\lambda\right]^{1.11} = 0 \quad (1.74)$$

and

$$\frac{0.31 \times 10^{-5} F_a}{w^{0.89}} - \sum_{j=1}^{j=Z/2+1} \frac{\tau_j}{k_{2j}^{0.11}} \times \left\{ \sum_{\lambda=1}^{\lambda=k} \left[\Delta_{2j} + w\left(\lambda - \frac{1}{2}\right)\zeta_j + \phi_{2j\lambda} - c_\lambda\right]^{1.11} w\left(\lambda - \frac{1}{2}\right) \right.$$
$$\left. - \frac{l}{2} \sum_{\lambda=1}^{\lambda=k} \left[\Delta_{2j} + w\left(\lambda - \frac{1}{2}\right)\zeta_j + \phi_{2j\lambda} - c_\lambda\right]^{1.11} \right\} = 0 \qquad (1.75)$$

Equation 1.56, Equation 1.69, Equation 1.70, and Equation 1.73 through Equation 1.75 constitute a set of simultaneous, nonlinear equations that may be solved for Δ_{mj}, ζ_j, ξ_j, δ_r, and δ_a. Subsequently, the roller–raceway loads Q_j and roller end–flange loads Q_{aj} may be determined.

The skewing angles determined using the earlier equations strictly pertain to full complement bearings and bearings having no roller guide flanges. For a bearing with a substantially robust and rigid cage, the skewing angle may be limited by the clearances between the rollers and the cage pockets. For a bearing with guide flanges, the skewing may be limited by the endplay between the roller ends and guide flanges. In general, the latter situation is obtained; however, to the extent that skewing is permitted, the earlier analysis is applicable.

1.5 RADIAL, THRUST, AND MOMENT LOADINGS OF RADIAL ROLLER BEARINGS

1.5.1 CYLINDRICAL ROLLER BEARINGS

For radial cylindrical roller bearings, it is possible to apply general combined loading. The equations for load equilibrium defined earlier apply; however, the interference at any lamina in the roller–raceway contact is given by

$$\delta_{mj\lambda} = \Delta_{mj} + w\left(\lambda - \frac{1}{2}\right)\left(v_m \zeta_j \pm \frac{1}{2}\theta \cos \psi_j\right) + \phi_{mj\lambda} - c_\lambda \qquad (1.76)$$

where subscript $m = 1$ refers to the outer raceway and $m = 2$ refers to the inner raceway. Coefficient $v_1 = -1$ and $v_2 = +1$. The contact load per unit length is given by

$$q_{mj\lambda} = \frac{\left[\Delta_{mj} + w\left(\lambda - \frac{1}{2}\right)\left(\nu_m \zeta_j \pm \frac{1}{2}\theta \cos \psi_j\right) + \phi_{mj\lambda} - c_\lambda\right]^{1.11}}{1.24 \times 10^{-5}\left(k_j w\right)^{0.11}} \tag{1.77}$$

1.5.2 Tapered Roller Bearings

Similar equations may be developed for tapered roller bearings. As shown in Chapter 5 of the first volume of this book, roller end–flange loading occurs during all conditions of applied loading, and bearing equilibrium equations must be altered accordingly. Figure 1.21 illustrates the geometry and loading of a tapered roller in a bearing.

Considering Figure 1.21 and establishing the following dimensions:

r_2 Radius in a radial plane from the inner-ring axis of rotation to the center of the inner raceway contact

r_{fz} Radius in a radial plane from the inner-ring axis of rotation to the center of the roller end–inner ring flange contact

r_{fx} x Direction distance in an axial plane from the center of the inner raceway contact to the center of the roller end–inner ring flange contact

The roller load equilibrium equations are

$$w \sum_{m=1}^{m=2} c_m \cos \alpha_m \sum_{\lambda=1}^{\lambda=k} q_{mj\lambda} - Q_{fj} \cos \alpha_f = 0 \tag{1.78}$$

$$w \sum_{m=1}^{m=2} c_m \sin \alpha_m \sum_{\lambda=1}^{\lambda=k} q_{mj\lambda} + Q_{fj} \sin \alpha_f = 0 \tag{1.79}$$

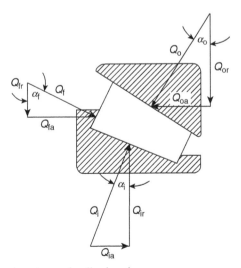

FIGURE 1.21 Roller loading in a tapered roller bearing.

In Equation 1.78 and Equation 1.79, coefficient $c_1 = -1$ and $c_2 = +1$. The equation for radial plane moment equilibrium of the roller is

$$w^2 \sum_{m=1}^{m=2} \sum_{\lambda=1}^{\lambda=k} q_{mj\lambda}\left[\lambda - \frac{1}{2}(k+1)\right] - R_f Q_{fj} = 0 \tag{1.80}$$

where R_f is the radius from the roller axis of rotation to the center of the roller end–flange contact. Equilibrium of actuating and resisting moments pertaining to roller skewing is given by

$$\frac{1}{2}l\mu Q_{fj} - \frac{w^3 \xi_j}{(d_m + D)} \sum_{m=1}^{m=2} \sum_{\lambda=1}^{\lambda=k}\left[\lambda - \frac{1}{2}(k+1)\right]^2 q_{mj\lambda} = 0 \tag{1.81}$$

The force and moment equilibrium equations with respect to the bearing inner ring are as follows:

$$F_r - w\sum_{j=1}^{j=Z}\cos\psi_j\left[\sum_{\lambda=1}^{\lambda=k} q_{2j\lambda}\cos\alpha_2 - Q_{fj}\cos\alpha_f\right] = 0 \tag{1.82}$$

$$F_a - w\sum_{j=1}^{j=Z}\left[\sum_{\lambda=1}^{\lambda=k} q_{2j\lambda}\sin\alpha_2 + Q_{fj}\sin\alpha_f\right] = 0 \tag{1.83}$$

$$M - w\sum_{j=1}^{j=Z}\cos\psi_j\left[\sum_{\lambda=1}^{\lambda=k} q_{2j\lambda}r_2\cos\alpha_2 - Q_{fj}(r_{fz}\sin\alpha_f - r_{fx}\cos\alpha_f)\right] = 0 \tag{1.84}$$

In these equations, the subscript 2 refers to the inner raceway.

1.5.3 SPHERICAL ROLLER BEARINGS

Spherical roller bearings are internally self-aligning and therefore cannot carry moment loading. Moreover, for slow- or moderate-speed applications causing insignificant roller centrifugal forces, gyroscopic moments, and friction (see Chapter 2 and Chapter 3), rollers in spherical roller bearings will not exhibit a tendency to tilt. Therefore, the simpler analytical methods provided in Chapter 7 of the first volume of this book will yield accurate results. For spherical roller bearings that have asymmetrical contour rollers (for example, spherical roller thrust bearings) roller tilting and hence skewing are not eliminated. In this case for the purpose of analysis, the bearing may be considered a special type of tapered roller bearing with fully crowned rollers. Then, the methods of analysis discussed in Section 1.5.2 may be applied.

1.6 STRESSES IN ROLLER–RACEWAY NONIDEAL LINE CONTACTS

In practice, the contact between rollers and raceways is rarely an ideal line contact nor is it truly a series of independent laminae without interactions. The laminae approach used earlier is sufficient for determining the distribution of load within the contacts as the stresses due to truncation at the roller ends and other transitions with profile design cover very small areas. However, as bearing fatigue life is a function of the subsurface and hence surface contact stresses, the laminae approach is not always sufficient to estimate the contact stress distribution. Therefore, more sophisticated methods for the analysis of contact stresses are typically performed after the load distributions of the bearing have been estimated.

Starting with Thomas and Hoersch [8], several researchers have advanced the contact solution of Hertz for the nonideal situations. Using stress functions with Equation 6.7, Equation 6.9, Equation 6.10, Equation 6.13, and Equation 6.14 in the first volume of this book, Hartnett [9] defined the following relationship between the normal contact pressure at a location (x', y') and the surface deflection at a distant point (x, y) on an elastic half space as

$$w(x,y) = \left(\frac{1-\xi^2}{\pi E}\right) \frac{P(x',y')}{\sqrt{(x-x')^2+(y-y')^2}} \tag{1.85}$$

By breaking the contact surface into several small, rectangular patches of dimensions $2g$ along the y-axis and $2c$ along the x-axis directions with a node at the center of each patch, and assuming constant pressure over the area, Equation 1.85 can be integrated to determine the effect of contact pressure at a given node, i, on the deflection at another node, j. This is done by the use of influence coefficients, D_{ij}:

$$
\begin{aligned}
D_{ij} = &\left(|x_i - x_j| + c\right) \ln \left[\frac{(|y_i - y_j| + g) + \sqrt{(|y_i - y_j| + g)^2 + (|x_i - x_j| + c)^2}}{(|y_i - y_j| - g) + \sqrt{(|y_i - y_j| - g)^2 + (|x_i - x_j| + c)^2}}\right] \\
&+ \left(|y_i - y_j| + g\right) \ln \left[\frac{(|x_i - x_j| + c) + \sqrt{(|y_i - y_j| + g)^2 + (|x_i - x_j| + c)^2}}{(|x_i - x_j| - c) + \sqrt{(|y_i - y_j| + g)^2 + (|x_i - x_j| - c)^2}}\right] \\
&+ \left(|x_i - x_j| - c\right) \ln \left[\frac{(|y_i - y_j| - g) + \sqrt{(|y_i - y_j| - g)^2 + (|x_i - x_j| - c)^2}}{(|y_i - y_j| + g) + \sqrt{(|y_i - y_j| + g)^2 + (|x_i - x_j| - c)^2}}\right] \\
&+ \left(|y_i - y_j| - g\right) \ln \left[\frac{(|x_i - x_j| - c) + \sqrt{(|y_i - y_j| - g)^2 + (|x_i - x_j| - c)^2}}{(|x_i - x_j| + c) + \sqrt{(|y_i - y_j| - g)^2 + (|x_i - x_j| + c)^2}}\right]
\end{aligned}
\tag{1.86}
$$

Using the influence coefficients, the interference of two bodies in contact with a given approach δ is given by

$$\left\langle \delta - z_j - \frac{y^2}{2}\rho_y \right\rangle - \left(\frac{1-\xi_1^2}{\pi E_1} + \frac{1-\xi_2^2}{\pi E_2}\right) \sum_{i=1}^{i=n} D_{ij}\sigma_j = 0 \tag{1.87}$$

where z_j is the drop at location j from the highest point on the body due to profiling, and the term $\langle \delta - z_j - (y^2 / 2)\rho_y \rangle = 0$ when the computed value is less than zero. Finally, the equilibrium of applied contact force and the integral of the pressure over the contact yields

$$Q - 4gc \sum_{j=1}^{j=n} \sigma_j = 0 \tag{1.88}$$

Equation 1.86 and Equation 1.87 allow for the nonideal contact pressure to be estimated for any given contact geometry by varying δ until Equation 1.88 is satisfied within acceptable error limits.

1.7 FLEXIBLY SUPPORTED ROLLING BEARINGS

1.7.1 RING DEFLECTIONS

The preceding discussion of distribution of load among the bearing rolling elements pertains to bearings that have rigidly supported rings. Such bearings are assumed to be supported in infinitely stiff (rigid) housings and on solid shafts of rigid material. The deflections considered in the determination of load distribution were contact deformations. This assumption is an excellent approximation for most bearing applications.

In some radial bearing applications, however, the outer ring of the bearing may be supported at one or two azimuth positions only, and the shaft on which the inner ring is positioned may be hollow. The condition of two-point outer-ring support, as shown in Figure 1.22 and Figure 1.23, occurs in the planet gear bearings of a planetary gear power transmission system, and was analyzed by Jones and Harris [10]. In certain rolling mill applications, the backup roll bearings may be supported at only one point on the outer ring or possibly at two points as shown in Figure 1.24. These conditions were analyzed by Harris [11]. In certain high-speed radial bearings, to prevent skidding it is desirable to preload the rolling elements by using an elliptical raceway, thus achieving essentially two-point ring loading under conditions of light applied load. The case of a flexible outer ring and an elliptical inner ring was investigated by Harris and Broschard [12]. In each of these applications, the outer ring must be considered flexible to achieve a correct analysis of rolling element loading.

In many aircraft applications, to conserve weight the power transmission shafting is made hollow. In these cases, the inner-ring deflections will alter the load distribution from that considering only contact deformation.

To determine the load distribution among the rolling elements when one or both of the bearing rings is flexible, it is necessary to determine the deflections of a ring loaded at various

FIGURE 1.22 Planet gear bearing.

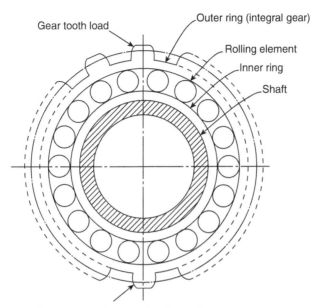

FIGURE 1.23 Planet gear bearing showing gear tooth loading.

points around its periphery. This analysis may be achieved by the application of classical energy methods for the bending of thin rings.

As an example of the method of analysis, consider a thin ring subjected to loads of equal magnitude equally spaced at angles $\Delta\psi$ (see Figure 1.25). According to Timoshenko [13], the

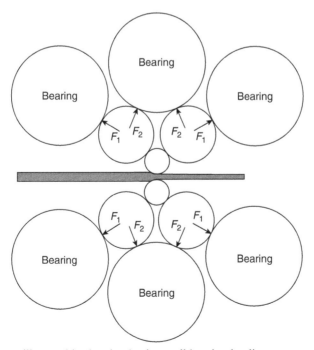

FIGURE 1.24 Cluster mill assembly showing backup roll bearing loading.

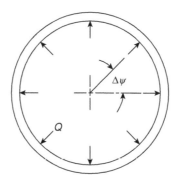

FIGURE 1.25 Thin ring loaded by equally spaced loads of equal magnitude.

differential equation describing radial deflection u for bending of a thin bar with a circular center line is

$$\frac{\mathrm{d}^2 u}{\mathrm{d}\phi^2} + u = -\frac{M \Re^2}{EI} \tag{1.89}$$

where I is the section moment of inertia in bending and E is the modulus of elasticity. It can be shown that the complete solution of Equation 1.89 consists of a complementary solution and a particular solution. The complementary solution is

$$u_{\mathrm{c}} = C_1 \sin \phi + C_2 \cos \phi \tag{1.90}$$

where C_1 and C_2 are arbitrary constants.

Consider that the ring is cut at two positions: at the position of loading, $\phi = \frac{1}{2}\Delta\psi$, and at the position $\phi = 0$, midway between the loads. The loads required to maintain equilibrium over the section are shown in Figure 1.26. From Figure 1.26 it can be seen that since horizontal forces are balanced,

$$Q = 2F_0 \sin \phi \tag{1.91}$$

or

$$F_0 = \frac{Q}{2 \sin \phi} \tag{1.92}$$

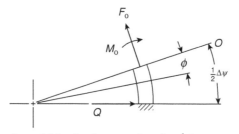

FIGURE 1.26 Loading of section of thin ring between $0 \leq \phi \leq \frac{1}{2}\Delta\psi$.

The moment at any angle ϕ between 0 and $\frac{1}{2}\Delta\psi$ is apparently

$$M = M_0 - F_0\Re(1 - \cos\phi) \tag{1.93}$$

or

$$M = M_0 - \frac{Q\Re}{2\sin\phi}(1 - \cos\phi) \tag{1.94}$$

Since the section at $\phi = 0$ is midway between loads, it cannot rotate. According to Castigliano's theorem [13] the angular rotation at any section is

$$\theta = \frac{\partial U}{\partial M} \tag{1.95}$$

where U is the strain energy in the beam at the position of loading. Timoshenko [13] shows that for a curved beam

$$U = \int_0^\phi \frac{M^2\Re}{2EI}\,\mathrm{d}\phi \tag{1.96}$$

At $\phi = 0$, $M = M_0$ and since the section is constrained from rotation,

$$\frac{\partial U}{\partial M_0} = 0 = \frac{\Re}{EI}\int_0^{1/2\Delta\psi} M\frac{\partial M}{\partial M_0}\,\mathrm{d}\phi \tag{1.97}$$

Substituting Equation 1.94 into Equation 1.97 and integrating yields

$$M_0 = \frac{Q\Re}{2}\left[\frac{1}{\sin\left(\frac{1}{2}\Delta\psi\right)} - \frac{2}{\Delta\psi}\right] \tag{1.98}$$

Hence,

$$M = \frac{Q\Re}{2}\left[\frac{\cos\phi}{\sin\left(\frac{1}{2}\Delta\psi\right)} - \frac{2}{\Delta\psi}\right] \tag{1.99}$$

Equation 1.99 may be substituted for M in Equation 1.89 such that the particular solution is

$$u_\mathrm{p} = \frac{Q\Re^3}{2EI}\left[\frac{\phi\sin\phi}{2\sin\left(\frac{1}{2}\Delta\psi\right)} - \frac{1}{\Delta\psi}\right] \tag{1.100}$$

The complete solution is

$$u = u_\mathrm{c} + u_\mathrm{p} = C_1\sin\phi + C_2\cos\phi - \frac{Q\Re^3}{2EI}\left[\frac{\phi\sin\phi}{2\sin\left(\frac{1}{2}\Delta\psi\right)} - \frac{1}{\Delta\psi}\right] \tag{1.101}$$

Because the sections at $\phi = 0$ and $\phi = \frac{1}{2}\Delta\psi$ do not rotate,

$$\frac{du}{d\phi}\bigg|_{\phi=0} = 0; \quad C_1 = 0$$

$$\frac{du}{d\phi}\bigg|_{\phi=\Delta\psi/2} = 0; \quad C_2 = -\frac{Q\Re^3}{4EI\sin(\frac{1}{2}\Delta\psi)}\left[\frac{1}{2}\Delta\psi\,\text{ctn}\left(\frac{1}{2}\Delta\psi\right)+1\right]$$

Hence, the radial deflection at any angle ϕ between $\phi = 0$ and $\phi = \frac{1}{2}\Delta\psi$ is

$$u = \frac{Q\Re^3}{2EI}\left\{\frac{2}{\Delta\psi} - \left[\frac{\Delta\psi\cos(\frac{1}{2}\Delta\psi)}{4\sin^2\left(\frac{1}{2}\Delta\psi\right)} + \frac{1}{2\sin\left(\frac{1}{2}\Delta\psi\right)}\right]\cos\phi - \frac{\phi\sin\phi}{2\sin(\frac{1}{2}\Delta\psi)}\right\} \tag{1.102}$$

Equation 1.102 may be expressed in another format as follows:

$$u = C_\phi Q \tag{1.103}$$

where C_ϕ are influence coefficients dependent on angular position and ring dimensions.

$$C_\phi = \frac{\Re^3}{2EI}\left\{\frac{2}{\Delta\psi} - \left[\frac{\Delta\psi\cos\left(\frac{1}{2}\Delta\psi\right)}{4\sin^2\left(\frac{1}{2}\Delta\psi\right)} + \frac{1}{2\sin(\frac{1}{2}\Delta\psi)}\right] \times \cos\phi - \frac{\phi\sin\phi}{2\sin(\frac{1}{2}\Delta\psi)}\right\} \tag{1.104}$$

Lutz [14], using procedures similar to those described earlier, developed influence coefficients for various conditions of point loading of a thin ring. These coefficients have been expressed in infinite series format for the sake of simplicity of use.

For a thin ring loaded by forces of equal magnitude symmetrically located about a diameter as shown in Figure 1.27, the following equation yields radial deflections:

$$\varrho u_i = \varrho C_{ij} Q \tag{1.105}$$

where

$$\varrho C_{ij} = \mp \frac{2\Re^3}{\pi EI}\sum_{m=2}^{m=\infty}\frac{\cos m\psi_j\,\cos m\psi_i}{\left(m^2-1\right)^2} \tag{1.106}$$

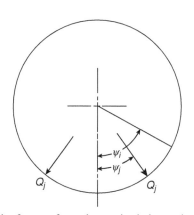

FIGURE 1.27 Thin ring loaded by forces of equal magnitude located asymmetrically about a diameter.

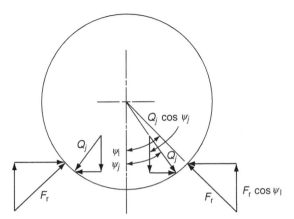

FIGURE 1.28 Thin ring showing equilibrium of forces.

The negative sign in Equation 1.106 is used for internal loads and the positive sign is used for external loads. Equation 1.105 defines radial deflection at angle ψ_i caused by Q_j at position angle ψ_j. When rolling element loads Q_j are such that a rigid body translation δ_1 of the ring occurs in the direction of an applied load, Equation 1.105 is not self-sufficient in establishing a solution; however, a directional equilibrium equation may be used in conjunction with Equation 1.105 to determine the translatory movement. Referring to Figure 1.28, the appropriate equilibrium equation is as follows:

$$F_r \cos \psi_1 - Q_j \cos \psi_j = 0 \qquad (1.107)$$

In the planet gear bearing application demonstrated in Figure 1.23, the gear tooth loads may be resolved into tangential forces, radial forces, and moment loads at $\psi = 90°$ (see Figure 1.29). The ring radial deflections at angle ψ_i due to tangential forces F_t are given by

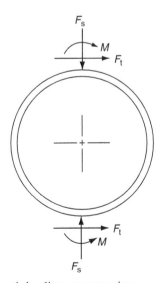

FIGURE 1.29 Resolution of gear tooth loading on outer ring.

$$_t u_i = {_t C_i F_t} \tag{1.108}$$

where

$$_t C_i = \frac{2\mathfrak{R}^3}{\pi EI} \sum_{m=2}^{m=\infty} \frac{\sin \frac{m\pi}{2} \cos m\psi_i}{m(m^2 - 1)^2} \tag{1.109}$$

Equation 1.108 is not self-sufficient and an appropriate equilibrium equation must be used to define a rigid ring translation.

The separating forces F_s are self-equilibrating and thus do not cause a rigid ring translation. The radial deflections at angles ψ_i are given by

$$_s u_i = {_s C_i F_s} \tag{1.110}$$

where

$$_s C_i = \frac{2\mathfrak{R}^3}{\pi EI} \sum_{m=2}^{m=\infty} \frac{\cos \frac{m\pi}{2} \cos m\psi_i}{(m^2 - 1)^2} \tag{1.111}$$

Note that Equation 1.111 is a special case of Equation 1.106 where position angle ψ_j is 90° and loads Q_j are external.

Similarly, the moment loads applied at $\psi = 90°$ are self-equilibrating. The radial deflections are given by

$$_M u_i = {_M C_i M} \tag{1.112}$$

where

$$_M C_i = \frac{2\mathfrak{R}^2}{\pi EI} \sum_{m=2}^{m=\infty} \frac{\sin \frac{m\pi}{2} \cos m\psi_i}{m(m^2 - 1)^2} \tag{1.113}$$

To find the ring radial deflections at any regular position due to the combination of applied and resisting loads, the principle of superposition is used. Hence for the planet gear bearing, the radial deflection at any angular position ψ_i is the sum of the radial deflections due to each individual load, that is,

$$u_i = {_s u_i} + {_M u_i} + {_t u_i} + {_{Qj} u_i} \tag{1.114}$$

or

$$u_i = {_s C_i F_s} + {_M C_i M} + {_t C_i F_t} + \sum {_Q C_{ij} Q_j} \tag{1.115}$$

1.7.2 Relative Radial Approach of Rolling Elements to the Ring

A load may not be transmitted through a rolling element unless the outer ring deflects sufficiently to consume the radial clearance at the angular position occupied by the rolling element. Furthermore, because a contact deformation is caused by loading of the rolling element, the ring deflections cannot be determined without considering these contact deformations. Therefore, the loading of a rolling element at angular position ψ_j depends on the relative radial clearance. The relative radial approach of the rings includes the translatory movement of the center of the outer ring relative to the initial center of that ring, whose position is fixed in space. Hence, for the planet gear bearing, the relative radial approach at angular position ψ_j is

$$\delta_i = \delta_1 \cos \psi_i + u_i \tag{1.116}$$

From Equation 1.12, the relative radial approach is related to the rolling element load as follows:

$$Q_j = \left\{ \begin{array}{cc} K(\delta_j - r_j)^n & \delta_j > r_j \\ 0 & \delta_j \leq r_j \end{array} \right\} \tag{1.117}$$

where r_j is the radial clearance at angular position ψ_j. Here, r_j is the sum of $P_d/2$ and the condition of ring ellipticity.

1.7.3 DETERMINATION OF ROLLING ELEMENT LOADS

Using the example of the planet gear bearing, the complete loading of the outer ring is shown in Figure 1.30, which also illustrates the rigid ring translation δ_1. Combining Equation 1.115 through Equation 1.117 yields

$$\delta_i - \delta_1 \cos \psi_i - {}_sC_iF_s - {}_MC_iM - {}_tC_iF_t - iK \sum_{j=2}^{j=Z/2+2} {}_\varrho C_{ij}(\delta_j - r_j)^n = 0 \tag{1.118}$$

The required equilibrium equation is

$$F_t - iK \sum_{j=2}^{j=Z/2+2} \tau_j(\delta_j - r_j)^n \cos \psi_j = 0 \tag{1.119}$$

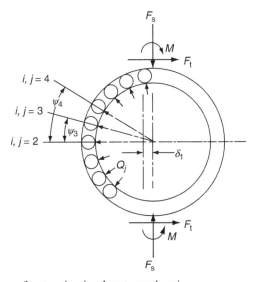

FIGURE 1.30 Total loading of outer ring in planet gear bearing.

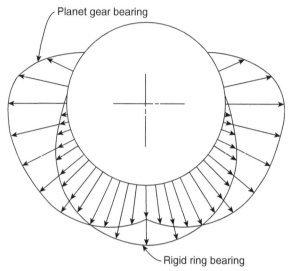

FIGURE 1.31 Comparison of load distribution for a rigid ring bearing and planet gear bearing.

considering the symmetry about the diameter parallel to the load. In Equation 1.119, $\tau_j = 0.5$ if the rolling element is located at $\psi_j = 0°$ or at $\psi_j = 180°$; otherwise $\tau_j = 1$.

Equation 1.118 and Equation 1.119 constitute a set of simultaneous nonlinear equations that may be solved by numerical analysis. The Newton–Raphson method is recommended.

Using these methods, the unknowns δ_j and hence Q_j can be determined at each rolling element location. Figure 1.31 shows a typical distribution of load among the rollers in a planet gear bearing compared with that of a rigid ring bearing subjected to a radial load of $2F_t$. For the backup roll bearings of Figure 1.24 supporting individual line loads F_1, Figure 1.32 compares the load distribution to that of a bearing that has rigid rings. Figure 1.33 shows

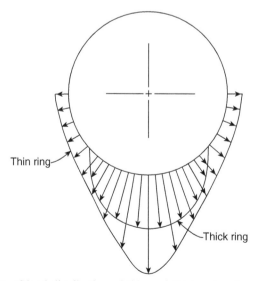

FIGURE 1.32 Comparison of load distribution of thin and thick outer rings, point-loaded backup roll bearing.

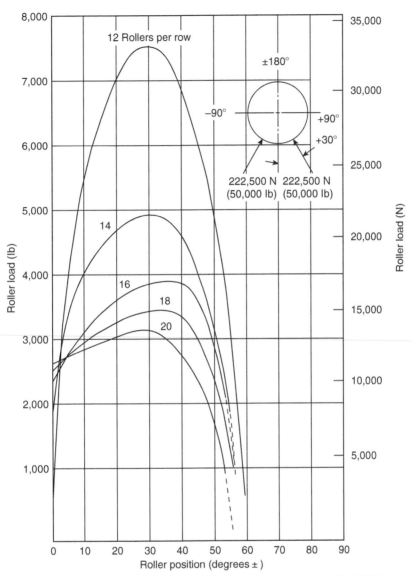

FIGURE 1.33 Roller load vs. number of rollers and position. 222,500 N (50,000 lb) at ±30°, inner dimensions constant. Outer-ring section thickness increases as the number of rollers is increased and roller diameter is subsequently decreased.

typical load distributions for the backup roll bearing of Figure 1.24, which supports paired line loads F_2. Figure 1.34 from Ref. [15], which is a photoelastic study of a similarly loaded bearing, verifies the data in Figure 1.33.

1.7.4 FINITE ELEMENT METHODS

To specify ring deflections, closed form integral analytical methods as well as influence coefficients calculated using infinite series techniques have been indicated for ring shapes, which are assumed simple both in circumference and cross-section in the previous discussions. For more complex structures, the finite element methods of calculations can be used to obtain

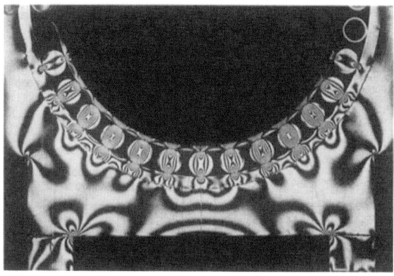

FIGURE 1.34 Photoelastic study of a roller bearing supporting loads aligned at approximately $\pm 30°$ to the bearing axis. (From Eimer, H., Aus dem Gebiet der Wälzlagertechnik, Semesterentwurf, Technische Hochschule, München, June 1964.)

a solution whose accuracy depends only on the fineness of the grid selected to represent the structure.

In finite element methods a function, customarily a polynomial, is chosen to define uniquely the displacement in each element (in terms of nodal displacements). The element

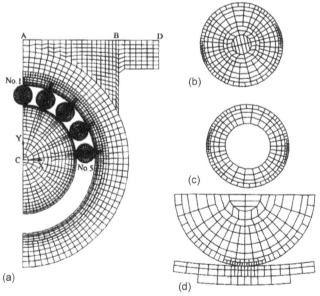

FIGURE 1.35 Finite element meshes for analyzing (a) cylindrical roller bearing rings, (b) solid rollers, (c) hollow rollers, and (d) contact zone. (From Zhao, H., *ASME Trans. J. Tribol.*, 120, 134–139, January 1998. With permission.)

stiffness matrix is obtained from equilibrium. The stiffness matrix of the complete structure is assembled, the boundary conditions are introduced, and solution of the resulting matrix equation produces the nodal displacements. A digital computer is required to solve the displacements and load distribution accurately in a rolling bearing mounted on a flexible support. Figure 1.35 from Zhao [16] shows the grids used to analyze a flexibly mounted cylindrical roller bearing assuming both solid and hollow rollers. The load distribution would be similar to that indicated in Figure 1.34.

Bourdon et al. [17,18] provided a method to define stiffness matrices for use in standard finite element models to analyze rolling element bearing loads and deflections, and the loading and deflections of the mechanisms in which they are used. For flexible mechanisms and bearing support systems, they demonstrated the importance of considering the overall mechanical system rather than only the local system in the vicinity of the bearings.

1.8 CLOSURE

The methods developed in this chapter enable the calculation of the internal load distribution of bearings in applications beyond those considered in bearing manufacturers' catalogs as supported by the load rating standards. It must be remembered, however, that these methods still pertain to bearing applications involving slow to moderate rotational speeds. At high speeds of rotation, ball and roller inertial loading (for example, centrifugal forces and gyroscopic moments) influence the internal load distribution, also affecting bearing deflections, friction forces, and moments. In this chapter, the discussion of the effect of speed on bearing performance has been limited to the determination of fatigue life in time units. Commencing with Chapter 3, the detailed effects of speed on overall bearing performance will be investigated.

REFERENCES

1. Jones, A., *Analysis of Stresses and Deflections*, New Departure Engineering Data, Bristol, CT, 1946.
2. Lundberg, G. and Sjövall, H., *Stress and Deformation in Elastic Contacts*, Pub. 4, Institute of Elasticity and Strength of Materials, Chalmers Inst. Tech., Gothenburg, Sweden, 1958.
3. Reussner, H., Druckflächenbelastnung und Overflächenverschiebung in Wälzkontakt von Rotätionkörpern, Dissertation Schweinfurt, Germany, 1977.
4. Palmgren, A., *Ball and Roller Bearing Engineering*, 3rd ed., Burbank, Philadelphia, 1959.
5. Harris, T., The effect of misalignment on the fatigue life of cylindrical roller bearings having crowned rolling members, *ASME Trans. J. Lub. Technol.*, 294–300, April 1969.
6. Harris, T., The endurance of a thrust-loaded, double row, radial cylindrical bearing, *Wear*, 18, 429–438, 1971.
7. Harris, T., Kotzalas, M., and Yu, W.-K., On the causes and effects of roller skewing in cylindrical roller bearings, *Trib. Trans.*, 41(4), 572–578, 1998.
8. Thomas, H. and Hoersch, V., Stresses due to the pressure of one elastic solid upon another, *Univ. Illinois Bull.*, 212, July 15, 1930.
9. Hartnett, M., The analysis of contact stress in rolling element bearings, *ASME Trans. J. Lub. Technol.*, 101, 105–109, January 1979.
10. Jones, A. and Harris, T., Analysis of a rolling element idler gear bearing having a deformable outer race structure, *ASME Trans. J. Basic Eng.*, 273–278, June 1963.
11. Harris, T., Optimizing the design of cluster mill rolling bearings, *ASLE Trans.*, 7, April 1964.
12. Harris, T. and Broschard, J., Analysis of an improved planetary gear transmission bearing, *ASME Trans. J. Basic Eng.*, 457–462, September 1964.
13. Timoshenko, S., *Strength of Materials, Part I*, 3rd ed., Van Nostrand, New York, 1955.

14. Lutz, W., Discussion of Ref. 7, presented at ASME Spring Lubrication Symposium, Miami Beach, FL, June 5, 1962.
15. Eimer, H., *Aus dem Gebiet der Wälzlagertechnik*, Semesterentwurf, Technische Hochschule, München, June 1964.
16. Zhao, H., Analysis of load distributions within solid and hollow roller bearings, *ASME Trans. J. Tribol.*, 120, 134–139, January 1998.
17. Bourdon, A., Rigal, J., and Play, D., Static rolling bearing models in a C.A.D. environment for the study of complex mechanisms: Part I—rolling bearing model, *ASME Trans. J. Tribol.*, 121, 205–214, April 1999.
18. Bourdon, A., Rigal, J., and Play, D., Static rolling bearing models in a C.A.D. environment for the study of complex mechanisms: Part II—complete assembly model, *ASME Trans. J. Tribol.*, 121, 215–223, April 1999.

2 Bearing Component Motions and Speeds

LIST OF SYMBOLS

Symbol	Description	Units
a	Semimajor axis of projected contact ellipse	mm (in.)
b	Semiminor axis of projected contact ellipse	mm (in.)
d_{m}	Pitch diameter	mm (in.)
D	Ball or roller diameter	mm (in.)
f	r/D	
h	Center of sliding	mm (in.)
n	Rotational speed	rpm
n_{m}	Ball or roller orbital speed, cage speed	rpm
n_{R}	Ball or roller speed about its own axis	rpm
Q	Rolling element–raceway contact normal load	N (lb)
r	Raceway groove curvature radius	mm (in.)
r'	Rolling radius	mm (in.)
R	Radius of curvature of deformed surface	mm (in.)
v	Surface velocity	mm/sec (in./sec)
x	Distance in direction of major axis of contact	mm (in.)
y	Distance in direction of minor axis of contact	mm (in.)
α	Contact angle	°, rad
β	Ball pitch angle	°, rad
β'	Ball yaw angle	°, rad
γ	$D \cos \alpha / d_{\mathrm{m}}$	
γ'	D/d_{m}	
θ_{f}	Flange angle	°, rad
ω	Rotational speed	rad/sec
ω_{m}	Orbital speed of ball or roller	rad/sec
ω_{R}	Speed of ball or roller about its own axis	rad/sec

Subscripts

f	Roller guide flange	
i	Inner raceway	
m	Orbital motion	
o	Outer raceway	

r   Radial direction
roll   Rolling motion
R   Rolling element
RE   Roller end
s   Spinning motion
sl   Sliding motion between flange and roller end
x x Direction
z z Direction

2.1 GENERAL

In Chapter 10 of the first volume of this handbook, equations were developed to calculate rolling element orbital speed and speed of the rolling element about its own axis. These equations were constructed using kinematical relationships based on simple rolling motion. Also, as discussed in Chapter 6 of the first volume of this handbook, when a load occurs between a rolling element and raceway, a contact surface is formed. When the rolling element rotates relative to the deformed surface, the simple rolling motion does not occur; rather, a combination of rolling and sliding motions occur. Hence, a system of complex equations needs to be developed to calculate the rolling element speeds.

Also, for angular-contact bearings, if the rolling motion does not occur on a line exactly parallel to the raceway, a parasitic motion called spinning occurs. Such a motion is pure sliding contributing significantly to bearing friction power loss. Finally, motions between roller ends and ring flanges in roller bearings are also pure sliding and can result in substantial power loss. In this chapter, these rolling/sliding relationships will be discussed together with the associated speeds.

2.2 ROLLING AND SLIDING

2.2.1 GEOMETRICAL CONSIDERATIONS

The only conditions that can sustain pure rolling between two contacting surfaces are:

1. Mathematical line contact under zero load
2. Line contact in which the contacting bodies are identical in length
3. Mathematical point contact under zero load

Even when these conditions are achieved, it is possible to have sliding. Sliding is then a condition of overall relative movement of the rolling body over the contact area.

The motion of a rolling element with respect to the raceway consists of a rotation about the generatrix of motion. If the contact surface is a straight line in one of the principal directions, the generatrix of motion may intersect the contact surface at one point only, as in Figure 2.1. The of angular velocity ω, which acts in the plane of the contact surface, produces rolling motion. As indicated in Figure 2.2, the component ω_s of angular velocity ω that acts normal to the surface

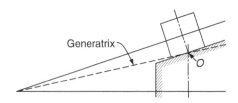

FIGURE 2.1 Roller–raceway contact; generatrix of motion pierces contact surface.

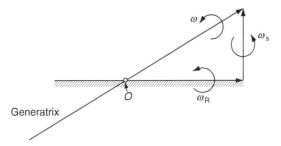

FIGURE 2.2 Resolution of angular velocities into rolling and spinning motions.

causes a spinning motion about a point of pure rolling O. The instantaneous direction of sliding in the contact zone is shown in Figure 2.3.

In ball bearings with nonzero contact angles between balls and raceways, during operation at any shaft or outer-ring speed, a gyroscopic moment occurs on each loaded ball, tending to cause a sliding motion. In most applications, because of relatively slow input speeds or heavy loading, such gyroscopic moments and hence motions can be neglected. In high-speed applications with oil-film lubrication between balls and raceways, such motions will occur.

The sliding velocity due to gyroscopic motion is given by (see Figure 2.4)

$$v_g = \tfrac{1}{2}\omega_g D \tag{2.1}$$

The sliding velocities caused by gyroscopic motion and spinning of the balls are vectorially additive such that at some distance h and O they cancel each other. Thus,

$$v_g = \omega_s h \tag{2.2}$$

and

$$h = \frac{D}{2} \times \frac{\omega_g}{\omega_s} \tag{2.3}$$

The distance h defines the center of sliding about which a rotation of angular velocity ω_s occurs. This center of sliding (spinning) may occur within or outside of the contact surface. Figure 2.5 shows the pattern of sliding lines in the contact area for simultaneous rolling, spinning, and gyroscopic motion in a ball bearing operating under a heavy load and at moderate speed. Figure 2.6, which corresponds to low-load and high-speed conditions (however, not considering skidding*), indicates that the center of sliding is outside of the

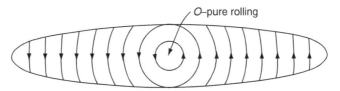

FIGURE 2.3 Contact ellipse showing sliding lines and point of pure rolling.

*Skidding is a very gross sliding condition occurring generally in oil-film lubricated ball and roller bearings operating under relatively light load at very high speed or rapid accelerations and decelerations. When skidding occurs, cage speed will be less than predicted by Equation 8.9 for bearings with inner ring rotation.

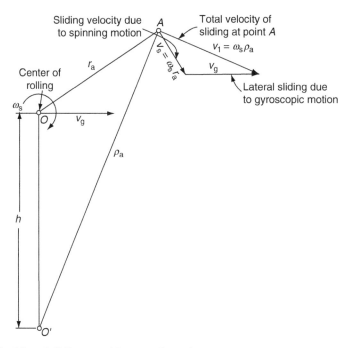

FIGURE 2.4 Velocities of sliding at arbitrary point A in contact area.

contact surface and sliding occurs over the entire contact surface. The distance h between the centers of contact and sliding is a function of the magnitude of the gyroscopic moment that can be compensated by contact surface friction forces.

2.2.2 SLIDING AND DEFORMATION

Even when the generatrix of motion apparently lies in the plane of the contact surface, as for radial cylindrical roller bearings, sliding on the contact surface can occur when a roller is under load. In accordance with the Hertzian radius of the contact surface in the direction transverse to motion, the contact surface has a harmonic mean profile radius, which means that the contact surface is not plane, but generally curved as shown in Figure 2.7 for a radial bearing.* The generatrix of motion, parallel to the tangent plane of the center of the contact

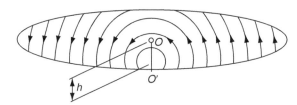

FIGURE 2.5 Sliding lines in contact area for simultaneous rolling, spinning, and gyroscopic motions—low-speed operation of a ball bearing.

---

*The illustration pertains to a spherical roller under relatively light load, that is, the contact ellipse major axis does not exceed the roller length.

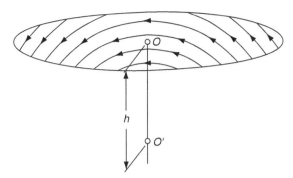

FIGURE 2.6 Sliding lines in contact area for simultaneous rolling, spinning, and gyroscopic motions—high-speed operation of a ball bearing (not considering skidding).

surface, therefore, pierces the contact surface at two points at which rolling occurs. Because the rigid rolling element rotates with a singular angular velocity about its axis, surface points at different radii from the axis have different surface velocities; only two of them that are symmetrically disposed about the roller geometrical center can exhibit pure rolling motion. In Figure 2.7 points within area A–A slide backward with regard to the direction of rolling and points outside of A–A slide forward with respect to the direction of rolling. Figure 2.8 shows the pattern of sliding lines in the elliptical contact area.

If the generatrix of motion is angled with respect to the tangent plane at the center of the contact surface, the center of rolling is positioned asymmetrically in the contact ellipse and, depending on the angle of the generatrix to the contact surface, one point or two points of intersection may occur at which rolling obtains. Figure 2.9 shows the sliding lines for this condition.

For a ball bearing in which rolling, spinning, and gyroscopic motions occur simultaneously, the pattern of sliding lines in the elliptical contact area is as shown in Figure 2.10 and Figure 2.11. More detailed information on sliding in the elliptical contact area may be found in the work by Lundberg [1].

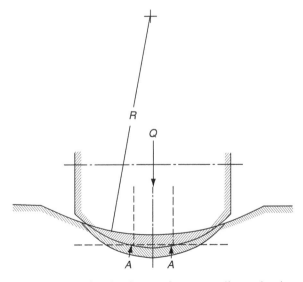

FIGURE 2.7 Roller–raceway contact showing harmonic mean radius and points of rolling A–A.

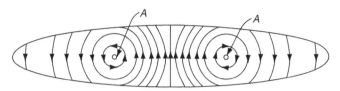

FIGURE 2.8 Sliding lines in contact area of Figure 2.7.

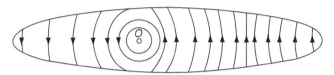

FIGURE 2.9 Sliding lines for rolling element–raceway contact area when load is applied; generatrix of motion pierces contact area.

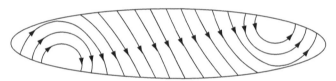

FIGURE 2.10 Sliding lines for ball–raceway contact area for simultaneous rolling, spinning, and gyroscopic motions—high-load and low-speed operation of an angular-contact ball bearing.

2.3 ORBITAL, PIVOTAL, AND SPINNING MOTIONS IN BALL BEARINGS

2.3.1 General Motions

Figure 2.12 illustrates the speed vector for a single ball in a bearing. The bearing is associated with the coordinate system x, y, z with the bearing axis collinear with the x axis. In Figure 2.12, the ball center O' is displaced angular distance ψ from the xz plane, and the x' axis passing through O' is distance $\frac{1}{2} d_m$ from, and parallel to, the x axis. The bearing is seen to rotate at speed ω about the x axis while the ball rotates at speed ω_R about an axis displaced at pitch and yaw angles β and β', respectively, from the x' axis. Hence, the ball orbits the bearing axis at speed ω_m. If the balls are completely constrained by a cage, then ω_m is the cage speed.

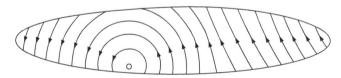

FIGURE 2.11 Sliding lines for ball–raceway contact area for simultaneous rolling, spinning, and gyroscopic motions—low-load and high-speed operation of an angular-contact ball bearing (not considering skidding).

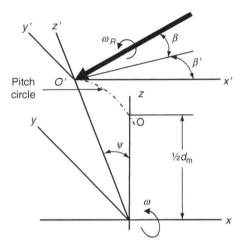

FIGURE 2.12 Ball speed vector in a nonzero ball–raceway contact.

In the same bearing, Figure 2.13 shows a ball contacting the outer raceway such that the normal force Q between the ball and the raceway is distributed over an elliptical surface defined by the projected major and minor axes $2a_o$ and $2b_o$, respectively. The radius of curvature of the deformed pressure surface as defined by Hertz is

$$R_o = \frac{2r_oD}{2r_o + D} \tag{2.4}$$

where r_o is the outer raceway groove curvature radius. In terms of curvature f_o,

$$R_o = \frac{2f_oD}{2f_o + 1} \tag{2.5}$$

Assume for the present purpose that the ball center is fixed in space and that the outer raceway rotates with angular speed ω_o. (The vector of ω_o is perpendicular to the plane of rotation and therefore collinear with the x axis.) Moreover, it can be seen from Figure 2.12 that ball rotational speed ω_R has components $\omega_{x'}$ and $\omega_{z'}$ lying in the plane of the paper when $\psi = 0$.

Because of the deformation at the pressure surface defined by a_o and b_o, the radius from the ball center to the raceway contact point varies in length as the contact ellipse is traversed from $+a_o$ to $-a_o$. Therefore, because of symmetry about the minor axis of the contact ellipse, pure rolling motion of the ball over the raceway occurs at most at two points. The radius at which pure rolling occurs is defined as r'_o and must be determined by methods of contact deformation analysis.

It can be seen from Figure 2.13 that the outer raceway has a component $\omega_o \cos \alpha_o$ of the angular velocity vector in a direction parallel to the major axis of the contact ellipse. Therefore, a point (x_o, y_o) on the outer raceway has a linear velocity v_{1o} in the direction of rolling as defined below:

$$v_{1o} = -\frac{d_m\omega_o}{2} - \left\{ (R_o^2 - x_o^2)^{1/2} - (R_o^2 - a_o^2)^{1/2} + \left[\left(\frac{D}{2}\right)^2 - a_o^2 \right]^{1/2} \right\} \omega_o \cos \alpha_o \tag{2.6}$$

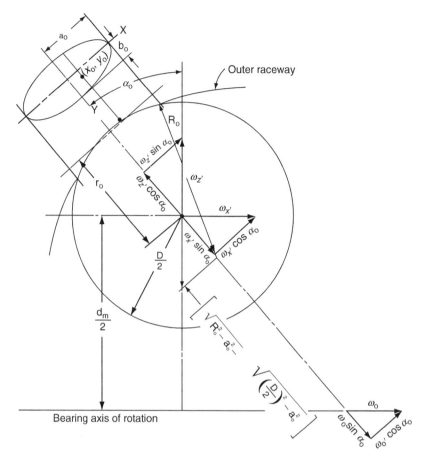

FIGURE 2.13 Outer raceway contact.

Similarly, the ball has angular velocity components, $\omega_{x'} \cos \alpha_o$ and $\omega_{z'} \sin \alpha_o$, of the angular velocity vector ω_R lying in the plane of the paper and parallel to the major axis of the contact ellipse. Thus, a point (x_o, y_o) on the ball has a linear velocity v_{2o} in the direction of rolling defined as follows:

$$v_{2o} = -(\omega_{x'} \cos \alpha_o + \omega_{z'} \sin \alpha_o) \times \left\{ (R_o^2 - x_o^2)^{1/2} - (R_o^2 - a_o^2)^{1/2} + \left[\left(\frac{D}{2} \right)^2 - a_o^2 \right]^{1/2} \right\} \quad (2.7)$$

Slip or sliding of the outer raceway over the ball in the direction of rolling is determined by the difference between the linear velocities of raceway and ball. Hence,

$$v_{yo} = v_{1o} - v_{2o} \quad (2.8)$$

or

$$v_{yo} = o - \frac{d_m \omega_o}{2} + (\omega_{x'} \cos \alpha_o + \omega_{z'} \sin \alpha_o - \omega_o \cos \alpha_o)$$

$$\times \left\{ (R_o^2 - x_o^2)^{1/2} - (R_o^2 - a_o^2)^{1/2} + \left[\left(\frac{D}{2} \right)^2 - a_o^2 \right]^{1/2} \right\} \quad (2.9)$$

Additionally, the ball angular velocity vector ω_R has a component $\omega_{y'}$ in a direction perpendicular to the plane of the paper. This component causes a slip v_{xo} in the direction transverse to the rolling, that is, in the direction of the major axis of the contact ellipse. This slip velocity is given by

$$v_{xo} = -\omega_{y'} \left\{ (R_o^2 - x_o^2)^{1/2} - (R_o^2 - a_o^2)^{1/2} + \left[\left(\frac{D}{2} \right)^2 - a_o^2 \right]^{1/2} \right\} \tag{2.10}$$

From Figure 2.13, it can be observed that both the ball angular velocity vectors $\omega_{x'}$ and $\omega_{z'}$, and the raceway angular velocity vector ω_o have components normal to the contact area. Hence, there is a rotation about a normal to the contact area; in other words, a spinning of the outer raceway relative to the ball, the net magnitude of which is given by

$$\omega_{so} = -\omega_o \sin \alpha_o + \omega_{x'} \sin \alpha_o - \omega_{z'} \cos \alpha_o \tag{2.11}$$

From Figure 2.12, it can be determined that

$$\omega_{x'} = \omega_R \cos \beta \cos \beta' \tag{2.12}$$

$$\omega_{y'} = \omega_R \cos \beta \sin \beta' \tag{2.13}$$

$$\omega_{z'} = \omega_R \sin \beta \tag{2.14}$$

Substitution of Equation 2.12 and Equation 2.14 into Equation 2.9 through Equation 2.11 yields

$$
\begin{aligned}
v_{yo} = &-\frac{d_m \omega_o}{2} + \left\{ (R_o^2 - x_o^2)^{1/2} - (R_o^2 - a_o^2)^{1/2} + \left[\left(\frac{D}{2} \right)^2 - a_o^2 \right]^{1/2} \right\} \\
&\times \left(\frac{\omega_R}{\omega_o} \cos \beta \cos \beta' \cos \alpha_o + \frac{\omega_R}{\omega_o} \sin \beta \sin \alpha_o - \cos \alpha_o \right) \omega_o
\end{aligned}
\tag{2.15}
$$

$$v_{xo} = -\left\{ (R_o^2 - x_o^2)^{1/2} - (R_o^2 - a_o^2)^{1/2} + \left[\left(\frac{D}{2} \right)^2 - a_o^2 \right]^{1/2} \right\} \omega_o \left(\frac{\omega_R}{\omega_o} \right) \cos \beta \sin \beta' \tag{2.16}$$

$$\omega_{so} = \left(\frac{\omega_R}{\omega_o} \cos \beta \cos \beta' \sin \alpha_o - \frac{\omega_R}{\omega_o} \sin \beta \cos \alpha_o - \sin \alpha_o \right) \omega_o \tag{2.17}$$

Note that at the radius of rolling r'_o on the ball, the translation velocity of the ball is identical to that of the outer raceway. From Figure 2.13, therefore,

$$\left(\frac{d_m}{2 \cos \alpha_o} + r'_o \right) \omega_o \cos \alpha_o = r'_o (\omega_{x'} \cos \alpha_o + \omega_{z'} \sin \alpha_o) \tag{2.18}$$

Substituting Equation 2.12 and Equation 2.13 into Equation 2.18, and rearranging the terms yields

$$\frac{\omega_R}{\omega_o} = \frac{(d_m/2) + r'_o \cos \alpha_o}{r'_o (\cos \beta \cos \beta' \cos \alpha_o + \sin \beta \sin \alpha_o)} \tag{2.19}$$

A similar analysis may be applied to the inner raceway contact as illustrated in Figure 2.14. The following equations can be determined:

$$
v_{yi} = -\frac{d_m \omega_i}{2} - \left\{ (R_i^2 - x_i^2)^{1/2} - (R_i^2 - a_i^2)^{1/2} + \left[\left(\frac{D}{2}\right)^2 - a_i^2 \right]^{1/2} \right\}
$$
$$
\times \left(\frac{\omega_R}{\omega_i} \cos\beta \cos\beta' \cos\alpha_i + \frac{\omega_R}{\omega_i} \sin\beta \sin\alpha_i - \cos\alpha_i \right) \omega_i \tag{2.20}
$$

$$
v_{xi} = -\left\{ (R_i^2 - x_i^2)^{1/2} - (R_i^2 - a_i^2)^{1/2} + \left[\left(\frac{D}{2}\right)^2 - a_i^2 \right]^{1/2} \right\} \omega_i \left(\frac{\omega_R}{\omega_i} \right) \cos\beta \sin\beta' \tag{2.21}
$$

$$
\omega_{si} = \left(-\frac{\omega_R}{\omega_i} \cos\beta \cos\beta' \sin\alpha_i + \frac{\omega_R}{\omega_i} \sin\beta \cos\alpha_i + \sin\alpha_i \right) \omega_i \tag{2.22}
$$

$$
\frac{\omega_R}{\omega_i} = \frac{-(d_m/2) + r_i' \cos\alpha_i}{r_i'(\cos\beta \cos\beta' \cos\alpha_i + \sin\beta \sin\alpha_i)} \tag{2.23}
$$

If instead of the ball center fixed in space, the outer raceway is fixed, then the ball center must orbit about the center 0 of the fixed coordinate system with an angular speed $\omega_m = -\omega_o$.

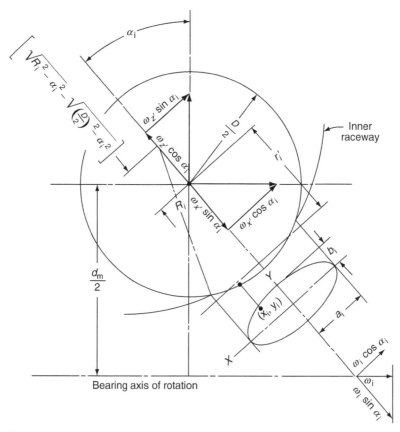

FIGURE 2.14 Inner raceway contact.

Therefore, the inner raceway must rotate with absolute angular speed $\omega = \omega_i + \omega_m$. By using these relationships, the relative angular speeds ω_i and ω_o can be described in terms of the absolute angular speed of the inner raceway as follows:

$$\omega_i = \frac{\omega}{1 + \dfrac{r'_o[(d_m/2) - r'_i \cos \alpha_i](\cos \beta \cos \beta' \cos \alpha_o + \sin \beta \sin \alpha_o)}{r'_i[(d_m/2) + r'_o \cos \alpha_o](\cos \beta \cos \beta' \cos \alpha_i + \sin \beta \sin \alpha_i)}} \tag{2.24}$$

$$\omega_o = \frac{-\omega}{1 + \dfrac{r'_i[(d_m/2) + r'_o \cos \alpha_o](\cos \beta \cos \beta' \cos \alpha_i + \sin \beta \sin \alpha_i)}{r'_o[(d_m/2) - r'_i \cos \alpha_i](\cos \beta \cos \beta' \cos \alpha_o + \sin \beta \sin \alpha_o)}} \tag{2.25}$$

Further,

$$\omega_R = \frac{-\omega}{\dfrac{r'_o(\cos \beta \cos \beta' \cos \alpha_o + \sin \beta \sin \alpha_o)}{(d_m/2) + r'_o \cos \alpha_o} + \dfrac{r'_i(\cos \beta \cos \beta' \cos \alpha_i + \sin \beta \sin \alpha_i)}{(d_m/2) - r'_i \cos \alpha_i}} \tag{2.26}$$

Similarly, if the outer raceway rotates with absolute angular speed ω and the inner raceway is stationary, $\omega_m = \omega_i$ and $\omega = \omega_m + \omega_o$. Therefore,

$$\omega_o = \frac{\omega}{1 + \dfrac{r'_i[(d_m/2) + r'_o \cos \alpha_o](\cos \beta \cos \beta' \cos \alpha_i + \sin \beta \sin \alpha_i)}{r'_o[(d_m/2) - r'_i \cos \alpha_i](\cos \beta \cos \beta' \cos \alpha_o + \sin \beta \sin \alpha_o)}} \tag{2.27}$$

$$\omega_i = \frac{-\omega}{1 + \dfrac{r'_o[(d_m/2) - r'_i \cos \alpha_i](\cos \beta \cos \beta' \cos \alpha_o + \sin \beta \sin \alpha_o)}{r'_i[(d_m/2) + r'_o \cos \alpha_o](\cos \beta \cos \beta' \cos \alpha_i + \sin \beta \sin \alpha_i)}} \tag{2.28}$$

$$\omega_R = \frac{\omega}{\dfrac{r'_o(\cos \beta \cos \beta' \cos \alpha_o + \sin \beta \sin \alpha_o)}{(d_m/2) + r'_o \cos \alpha_o} + \dfrac{r'_i(\cos \beta \cos \beta' \cos \alpha_i + \sin \beta \sin \alpha_i)}{(d_m/2) - r'_i \cos \alpha_i}} \tag{2.29}$$

Inspection of the final equations relating to the relative motions of the balls and raceways reveals the following unknown quantities: r'_o, r'_i, β, β', α_i, and α_o. It is apparent that an analysis of the forces and moments acting on each ball will be required to evaluate the unknown quantities. As a practical matter, however, it is sometimes possible to avoid this lengthy procedure requiring digital computation by using the simplifying assumption that a ball will roll on one raceway without spinning and spin and roll simultaneously on the other raceway. The raceway on which only rolling occurs is called the "controlling" raceway. Moreover, it is also possible to assume that gyroscopic pivotal motion is negligible; some criteria for this will be discussed.

2.3.2 No Gyroscopic Pivotal Motion

In the event that gyroscopic rotation is minimal, the angle β' approaches $0°$ (see Figure 2.12). Therefore, the angular rotation $\omega_{y'}$ is zero and further

$$\omega_{x'} = \omega_R \cos \beta \tag{2.30}$$

$$\omega_{z'} = \omega_R \sin \beta \tag{2.31}$$

A second consequence of $\beta' = 0$ is that

$$\frac{\omega_R}{\omega_o} = \frac{(d_m/12) + r'_o \cos \alpha_o}{r'_o (\cos \alpha_o \cos \beta + \sin \beta \sin \alpha_o)} \tag{2.32}$$

and

$$\frac{\omega_R}{\omega_i} = \frac{-(d_m/2) + r'_i \cos \alpha_i}{r'_i (\cos \beta \cos \alpha_i + \sin \beta \sin \alpha_i)} \tag{2.33}$$

2.3.3 SPIN-TO-ROLL RATIO

Assuming for this calculation that r_i, r_o, and $\frac{1}{2} D$ are essentially equal, the ball rolling speed relative to the outer raceway is

$$\omega_{roll} = -\omega_o \frac{d_m}{D} = -\frac{\omega_o}{\gamma'} \tag{2.34}$$

From Equation 2.17 for negligible gyroscopic moment ($\beta' = 0$),

$$\omega_{so} = \omega_R \cos \beta \sin \alpha_o - \omega_R \sin \beta \cos \alpha_o - \omega_o \sin \alpha_o \tag{2.35}$$

or

$$\omega_{so} = \omega_R \sin(\alpha_o - \beta) - \omega_o \sin \alpha_o \tag{2.36}$$

Dividing by ω_{roll} according to Equation 2.34 yields

$$\left(\frac{\omega_s}{\omega_{roll}}\right)_o = -\gamma' \frac{\omega_R}{\omega_o} \sin(\alpha_o - \beta) + \gamma' \sin \alpha_o \tag{2.37}$$

According to Equation 2.32, replacing $2r'_o/d_m$ by γ':

$$\frac{\omega_R}{\omega_o} = \frac{1 + \gamma' \cos \alpha_o}{\gamma' (\cos \beta \cos \alpha_o + \sin \beta \sin \alpha_o)} \tag{2.38}$$

or

$$\frac{\omega_R}{\omega_o} = \frac{1 + \gamma' \cos \alpha_o}{\gamma' \cos(\alpha_o - \beta)} \tag{2.39}$$

Therefore, substitution of Equation 2.39 into Equation 2.37 yields

$$\left(\frac{\omega_s}{\omega_{roll}}\right)_o = -(1 + \gamma' \cos \alpha_o) \tan(\alpha_o - \beta) + \gamma' \sin \alpha_o \tag{2.40}$$

Similarly, for an inner raceway contact,

$$\left(\frac{\omega_s}{\omega_{roll}}\right)_i = (1 - \gamma' \cos \alpha_i) \tan(\alpha_i - \beta) + \gamma' \sin \alpha_i \qquad (2.41)$$

2.3.4 CALCULATION OF ROLLING AND SPINNING SPEEDS

Even assuming that gyroscopic speed ω_y' is zero, the use of Equation 2.40 and Equation 2.41 depends on the knowledge of the ball–raceway contact angles α_i and α_o, and ball speed vector pitch angle β. In Chapter 3, means to calculate β_i and β_o in high-speed, angular-contact ball bearings will be demonstrated. Those equations assume that ball orbital speed ω_m and ball speed about its own axis, ω_R, are known. Unfortunately, unless the ball speed vector pitch angle β is known, the solution of the set of simultaneous equations involving contact deformations, contact angles, and ball speeds cannot be achieved. To determine these parameters in the most elegant manner, ball–raceway friction forces as functions of ball and raceway speeds need to be introduced. This situation will also be investigated later in this text.

In the absence of using a complete set of normal and friction forces and moment balances to solve for speeds, Jones [2] made the simplifying assumption that a ball contacting both inner and outer raceways rolls and spins on one of these raceways and simply rolls on the opposing raceway. He based this assumption on his interpretation of experimental data obtained from gas turbine engine main-shaft ball bearings. The raceway on which pure rolling was assumed to occur was called the controlling raceway; the phenomenon was called raceway control. Assuming the condition that outer raceway control occurs, spinning speed $\omega_{so} = 0$, and substitution of Equation 2.32 into Equation 2.17 yields

$$\tan \beta = \frac{\frac{1}{2}d_m \sin \alpha_o}{\frac{1}{2}d_m \cos \alpha_o + r_o'} \qquad (2.42)$$

As $r_o' \approx \frac{1}{2}D$ and $D/d_m = \gamma'$, Equation 2.42 becomes

$$\tan \beta = \frac{\sin \alpha_o}{\cos \alpha_{o+\gamma'}} \qquad (2.43)$$

Having defined ball speed vector pitch angle β, it is possible to solve the remaining speed equations.

For high-speed operations of very lightly loaded, oil-lubricated, angular-contact ball bearings, Figure 2.15 taken from Ref. [3] indicates that ball–outer raceway spinning speed ω_{so} tends toward zero, approximating the outer raceway control condition. As the applied thrust load increases to normal operating magnitudes, ω_{so} though less than ω_{si} is substantial. This allows one to infer that outer raceway control is a condition that occurs only in a very limited manner for oil-lubricated ball bearings.

Harris [4] also investigated the performance of thrust-loaded, solid-film lubricated, angular-contact ball bearings of the same dimensions assuming a constant coefficient of friction. Figure 2.16 from that analysis demonstrates that outer raceway control does not tend to occur in that application either.

Notwithstanding the above observations, it is of interest to carry the Jones analysis [2] to completion since it has been used for several decades with apparently little negative impact on bearing design.

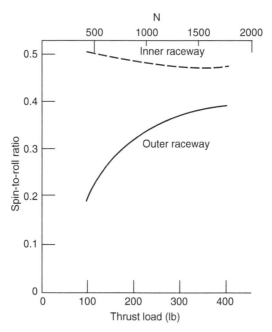

FIGURE 2.15 Spin-to-roll ratio vs. thrust load for an oil-lubricated, angular-contact ball bearing.

From Equation 2.24 and Equation 2.25, setting β' equal to 0 and substituting for Equation 2.43, the ratio between ball and raceway angular velocities is determined:

$$\frac{\omega_R}{\omega} = \frac{\pm 1}{\left(\dfrac{\cos \alpha_o + \tan \beta \sin \alpha_o}{1 + \gamma' \cos \alpha_o} + \dfrac{\cos \alpha_i + \tan \beta \sin \alpha_i}{1 - \gamma' \cos \alpha_i}\right) \gamma' \cos \beta} \tag{2.44}$$

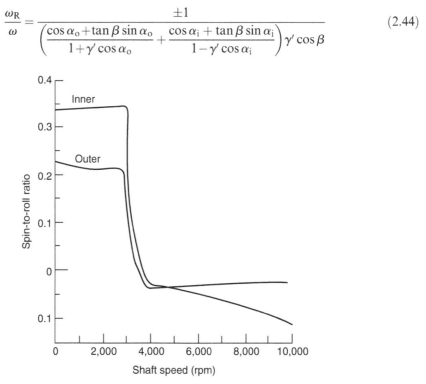

FIGURE 2.16 Spin-to-roll ratio vs. shaft speed for a thrust-loaded, angular-contact ball bearing operating with a solid-film lubricant having a constant coefficient of friction.

The upper sign pertains to outer raceway rotation and the lower sign to inner raceway rotation.

Again, using the condition of outer raceway control as established in Equation 2.43, it is possible to determine the ratio of ball orbital angular velocity to raceway speed. For a rotating inner raceway $\omega_m = -\omega_o$; therefore, from Equation 2.25 for β' equal to 0,

$$\frac{\omega_m}{\omega} = \frac{1}{1 + \dfrac{(1 + \gamma' \cos \alpha_o)(\cos \alpha_i + \tan \beta \sin \alpha_i)}{(1 - \gamma' \cos \alpha_i)(\cos \alpha_o + \tan \beta \sin \alpha_o)}} \tag{2.45}$$

Equation 2.45 is based on the valid assumption that $r_0 \approx r_i \approx D/2$. Similarly, for a rotating outer raceway and by Equation 2.28,

$$\frac{\omega_m}{\omega} = \frac{1}{1 + \dfrac{(1 - \gamma' \cos \alpha_i)(\cos \alpha_o + \tan \beta \sin \alpha_o)}{(1 + \gamma' \cos \alpha_o)(\cos \alpha_i + \tan \beta \sin \alpha_i)}} \tag{2.46}$$

Substitution of Equation 2.43 describing the condition of outer raceway control into Equation 2.45 and Equation 2.46 establishes the equations of the required ratio ω_m/ω. Hence, for a bearing with rotating inner raceway,

$$\frac{\omega_m}{\omega} = \frac{1 - \gamma' \cos \alpha_i}{1 + \cos(\alpha_i - \alpha_o)} \tag{2.47}$$

For a bearing with a rotating outer raceway,

$$\frac{\omega_m}{\omega} = \frac{\cos(\alpha_i - \alpha_o) + \gamma' \cos \alpha_i}{1 + \cos(\alpha_i + \alpha_o)} \tag{2.48}$$

As indicated above, Equation 2.43, Equation 2.44, Equation 2.47, and Equation 2.48 are valid only when ball gyroscopic pivotal motion is negligible, that is, $\beta' = 0$.

2.3.5 GYROSCOPIC MOTION

Palmgren [5] inferred that in an oil-lubricated, angular-contact ball bearing, gyroscopic motions of the balls can be prevented. He stated that the coefficient of sliding friction may be as low as 0.02 and that gyroscopic motion will not occur if the following relationship is satisfied:

$$M_g > 0.02QD \tag{2.49}$$

where Q is the ball–raceway normal load. Jones [2] mentioned that a coefficient of friction from 0.06 to 0.07 suffices for most ball bearing applications to prevent sliding. Both of these statements are inaccurate.

It has been shown that a ball in an angular-contact ball bearing is capable of experiencing both orbital speed ω_m about the bearing axis and speed ω_R about its own axis that is canted at pitch angle β to the x' axis. The latter axis is parallel to the bearing axis (see Figure 2.12). It has been further demonstrated that sliding motion in the direction of rolling motion occurs in the ball–raceway contacts. Additionally, owing to the nonzero contact angle, spinning motion occurs. Given the presence of these sliding motions, it is most probable that motion initiated by a gyroscopic moment will not be prevented. In other words, additional sliding in an orthogonal direction (gyroscopic motion) will occur simultaneously. Subsequent analysis employing complete force and moment balances for each ball shows the speed of ball gyroscopic motion $\omega_{y'}$ to be very small compared with principal ball speed component ω_x' and relatively small compared to $\omega_{z'}$.

2.4 ROLLER END–FLANGE SLIDING IN ROLLER BEARINGS

2.4.1 ROLLER END–FLANGE CONTACT

Roller bearings react with axial roller loads through concentrated contacts between roller ends and flange. Tapered roller bearings and spherical roller bearings (with asymmetrical rollers) require such contact to react with the component of the raceway–roller contact load that acts in the roller axial direction. Some cylindrical roller bearing designs require roller end–flange contacts to react with skewing-induced or externally applied roller axial loads. As these contacts experience sliding motions between roller ends and flange, their contribution to overall bearing frictional heat generation becomes substantial. Furthermore, there are bearing failure modes associated with roller end–flange contact such as wear and smearing of the contacting surfaces. These failure modes are related to the ability of the roller end–flange contact to support the roller axial load under the prevailing speed and lubrication conditions within the contact. Both the frictional characteristics and load-carrying capability of roller end–flange contacts are highly dependent on the geometry of the contacting members.

2.4.2 ROLLER END–FLANGE GEOMETRY

Numerous roller end and flange geometries have been used successfully in roller bearing designs. Typically, performance requirements as well as manufacturing considerations dictate the geometry incorporated into a bearing design. Most designs use either a flat (with corner radii) or sphere end roller contacting an angled flange. The angled flange surface can be described as a portion of a cone at an angle θ_f with respect to a radial plane perpendicular to the ring axis. This angle, known as the flange angle or flange layback angle, can be zero, indicating that the flange surface lies in the radial plane. Examples of cylindrical roller bearing roller end–flange geometries are shown in Figure 2.17. The flat end roller in Figure 2.17a under zero skewing conditions contacts the flange at a single point (in the vicinity of the intersection between the roller end flat and roller corner radius). As the roller skews, the point of contact travels along this intersection on the roller toward the tip of the flange, as shown in Figure 2.18b. If properly designed, a sphere end roller will contact the flange on the roller end sphere surface. For no skewing, the contact will be centrally positioned on the roller, as shown

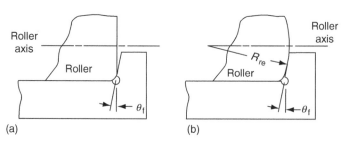

(a)   (b)

FIGURE 2.17 Cylindrical roller bearing, roller end–flange contact geometry. (a) Flat end roller. (b) Sphere end roller.

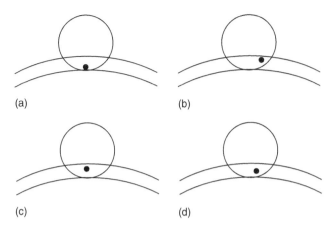

(a) (b)

(c) (d)

FIGURE 2.18 Cylindrical roller bearing, roller end–flange contact location for flat and sphere end rollers. (a) Flat end roller, zero skew angle. (b) Flat end roller, nonzero skew angle. (c) Sphere end roller, zero skew angle. (d) Sphere end roller, nonzero skew angle.

in Figure 2.18c. As the skewing angle is increased, the contact point moves off center and toward the flange tip, as shown in Figure 2.18d for a flanged inner ring. For typical designs, the sphere end roller contact location is less sensitive to skewing than a flat end roller contact.

The location of the roller end–flange contact has been determined analytically [6] for sphere end rollers contacting an angled flange. Consider the cylindrical roller bearing arrangement shown in Figure 2.19. The flanged ring coordinate system X_I, Y_I, Z_I and roller coordinate system X_i, Y_i, Z_i are indicated. The flange contact surface is modeled as a portion of a cone with an apex at point C as shown in Figure 2.20. The equation of this cone, expressed as a function of the x and y ring coordinates is

$$z = [(x - C)^2 \operatorname{ctn}^2 \theta_f - y^2]^{1/2} = f(x, y) \tag{2.50}$$

For a point of flange surface P_x, P_y, P_z, the equation of the surface normal at P can be expressed as

$$\left. \frac{x - P_x}{\frac{\partial f}{\partial x}} \right|_{x = P_x, y = P_y} = \left. \frac{y - P_y}{\frac{\partial f}{\partial y}} \right|_{x = P_x, y = P_y} = -(z - P_z) \tag{2.51}$$

The location of the origin of the roller end sphere radius is defined as point T with coordinates (T_x, T_y, T_z) expressed in the flanged ring coordinate system. As the resultant roller end–flange elastic contact force is normal to the end sphere surface, its line of action must pass through the sphere origin (T_x, T_y, T_z). Evaluating Equation 2.50 and Equation 2.51 at T yields the following three equations:

$$T_x - P_x = \frac{(T_z - P_z)(P_x - C) \operatorname{ctn}^2 \theta_f}{[(P_x - C)^2 \operatorname{ctn}^2 \theta_f - P_y^2]^{1/2}} \tag{2.52}$$

$$T_y - P_y = \frac{(T_z - P_z)P_y}{[(P_x - C)^2 \operatorname{ctn}^2 \theta_f - P_y^2]^{1/2}} \tag{2.53}$$

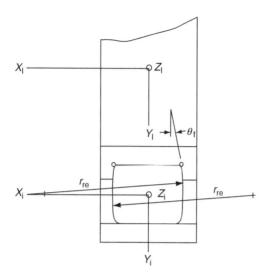

FIGURE 2.19 Cross-section through a cylindrical roller bearing that has a flanged inner ring.

$$P_z = [(P_x - C)^2 \operatorname{ctn}^2 \theta_f - P_y^2]^{1/2} \tag{2.54}$$

Equation 2.52 through Equation 2.54 contain three unknowns (P_x, P_y, P_z) and are sufficient to determine the theoretical point of contact between the roller end and flange. By introducing a fourth equation and unknown, however, namely the length of the line from points (T_x, T_y, T_z) to (P_x, P_y, P_z), the added benefit of a closed-form solution is obtained. The length of a line normal to the flange surface at the point (P_x, P_y, P_z), which joins this point with the sphere origin (T_x, T_y, T_z), is given by

$$\mathcal{D} = [(T_x - P_x)^2 + (T_y - P_y)^2 + (T_z - P_z)^2]^{1/2} \tag{2.55}$$

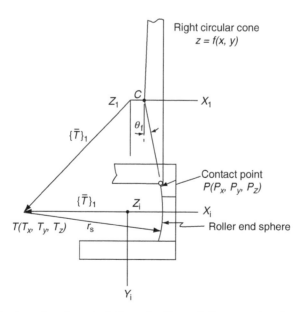

FIGURE 2.20 Coordinate system for calculation of roller end–flange contact location.

After algebraic reduction, $\mathfrak{D}$ is obtained from the positive root of the quadratic equation:

$$\mathfrak{D} = \frac{-\mathfrak{S} \pm (\mathfrak{S}^2 - 4\mathfrak{R}\mathfrak{T})^{1/2}}{2\mathfrak{R}} \tag{2.56}$$

where values for $\mathfrak{S}$, $\mathfrak{R}$, and $\mathfrak{T}$ are

$$\mathfrak{R} = \tan^2 \theta_f - 1$$

$$\mathfrak{S} = \frac{2\sin^2 \theta_f}{\cos \theta_f}[(T_x - C) - \tan \theta_f (T_y^2 + T_z^2)^{1/2}]$$

$$\mathfrak{T} = [(T_x - C) - \tan \theta_f (T_y^2 + T_z^2)^{1/2}]$$

The coordinates $P(P_x, P_y, P_z)$ are given by the following closed-form function of $\mathfrak{D}$:

$$P_x = T_y \tan \theta_f \left[1 + \left(\frac{T_z}{T_y}\right)^2\right]^{1/2} \left[1 - \frac{\mathfrak{D}}{(T_y^2 + T_z^2)^{1/2}}\right] + c \tag{2.57}$$

$$P_y = T_y \left[1 - \frac{\mathfrak{D}\sin \theta_f}{(T_y^2 + T_z^2)^{1/2}}\right] \tag{2.58}$$

$$P_z = T_z \left[1 - \frac{\mathfrak{D}\sin \theta_f}{(T_y^2 + T_z^2)^{1/2}}\right] \tag{2.59}$$

At a point of contact between the roller end and flange, $\mathfrak{D}$ is equal to the roller end sphere radius. Therefore, knowing the roller and flanged ring geometry as well as the coordinate location (with respect to the flanged ring coordinate system) of the roller end sphere origin, it is possible to calculate directly the theoretical roller end–flange contact location.

The analysis, although shown for a cylindrical roller bearing, is general enough to apply to any roller bearing that has sphere end rollers that contact a conical flange. Tapered and spherical roller bearings of this type may be treated if the sphere radius origin is properly defined.

These equations have several notable applications since flange contact location is of interest in bearing design and performance evaluation. It is desirable to maintain contact on the flange below the flange rim (including edge break) and above the undercut at the base of the flange. To do otherwise causes loading on the flange rim (or edge of undercut) and produces higher contact stresses and less than optimum lubrication of the contact. The preceding equations may be used to determine the maximum theoretical skewing angle for a cylindrical roller bearing if the roller axial play (between flanges) is known. Also, by calculating the location of the theoretical contact point, sliding velocities between roller ends and flange can be calculated and used in an estimate of roller end–flange contact friction and heat generation.

2.4.3 SLIDING VELOCITY

The kinematics of a roller end–flange contact causes sliding to occur between the contacting members. The magnitude of the sliding velocity between these surfaces substantially affects friction, heat generation, and load-carrying characteristics of a roller bearing design. The sliding velocity is represented by the difference between the two vectors defining the linear velocities of the flange and the roller end at the point of contact. A graphical representation of

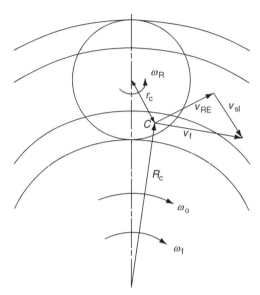

FIGURE 2.21 Roller end–flange contact velocities.

the roller velocity v_{roll} and the flange velocity v_F at their point of contact C is shown in Figure 2.21. The sliding velocity vector v_{sl} is shown as the difference of v_{RE} and v_f. When considering roller skewing motions, v_{sl} will have a component in the flanged ring axial direction, albeit small in comparison with the components in the bearing radial plane, if the roller is not subjected to the components in the bearing radial plane. If the roller is not subjected to skewing, the contact point will lie in the plane containing the roller and flanged ring axes. The roller end–flange sliding velocity may be calculated as

$$v_{sl} = v_f - v_{RE} = \omega_f R_c - (\omega_o R_c + \omega_R r_c) \tag{2.60}$$

where clockwise rotations are considered positive. Varying the position of contact point C over the elastic contact area between roller end and flange allows the distribution of sliding velocity to be determined.

2.5 CLOSURE

In this chapter, methods for calculations of rolling and cage speeds in ball and roller bearings were developed for conditions of rolling and spinning motions. It will be shown in Chapter 3 how the dynamic loading derived from ball and roller speeds can significantly affect ball bearing contact angles, diametral clearance, and subsequently rolling element load distribution. Moreover, spinning motions that occur in ball bearings tend to alter contact area stresses, and hence they affect bearing endurance. Other quantities affected by bearing internal speeds are friction torque and frictional heat generation. It is therefore clear that accurate determinations of bearing internal speeds are necessary for analysis of rolling bearing performance.

It will be demonstrated subsequently that hydrodynamic action of the lubricant in the contact areas can transform what is presumed to be substantially rolling motions into combinations of rolling and translatory motions. In general, this combination of rotation and translation may be tolerated provided the lubricant films resulting from the rolling motions are sufficient to adequately separate the rolling elements and raceways.

REFERENCES

1. Lundberg, G., Motions in loaded rolling element bearings, SKF unpublished report, 1954.
2. Jones, A., Ball motion and sliding friction in ball bearings, *ASME Trans. J. Basic Eng.*, 81, 1959.
3. Harris, T., An analytical method to predict skidding in thrust loaded, angular-contact ball bearings, *ASME Trans. J. Lubrication Technol.*, 17–24, January 1971.
4. Harris, T., Ball motion in thrust-loaded, angular-contact bearings with coulomb friction, *ASME J. Lubrication Technol.*, 93, 17–24, 1971.
5. Palmgren, A., *Ball and Roller Bearing Engineering*, 3rd ed., Burbank, Philadelphia, 1959, pp. 70–72.
6. Kleckner, R. and Pirvics, J., High speed cylindrical roller bearing analysis—SKF Computer Program CYBEAN, Vol. 1: Analysis, SKF Report AL78P022, NASA Contract NAS3-20068, July 1978.

3 High-Speed Operation: Ball and Roller Dynamic Loads and Bearing Internal Load Distribution

LIST OF SYMBOLS

Symbol	Description	Units
B	$f_i + f_o - 1$	
d_m	Pitch diameter	mm (in.)
D	Ball or roller diameter	mm (in.)
f	r/D	
F	Force	N (lb)
F_c	Centrifugal force	N (lb)
F_f	Friction force	N (lb)
g	Gravitational constant	mm/sec^2 (in./sec^2)
H	Roller hollowness ratio	
J	Mass moment of inertia	kg $\cdot$ mm^2 (in. $\cdot$ lb $\cdot$ sec^2)
K	Load–deflection constant	N/mm^x (lb/in.^{x})
l	Roller length	mm (in.)
m	Ball or roller mass	kg (lb $\cdot$ sec^2/in.)
M	Moment	N $\cdot$ mm (lb $\cdot$ in.)
M_g	Gyroscopic moment	N $\cdot$ mm (lb $\cdot$ in.)
M	Applied moment	N $\cdot$ mm (lb $\cdot$ in.)
n	Rotational speed	rpm
n_m	Ball or roller orbital speed, cage speed	rpm
n_R	Ball or roller speed about its own axis	rpm
P_d	Radial or diametral clearance	N (lb.)
q	Roller–raceway load per unit length	N/mm (lb/in.)
Q	Ball or roller normal load	N (lb)
Q_a	Axial direction load on ball or roller	N (lb)
Q_r	Radial direction load on ball or roller	N (lb)
R	Radius to locus of raceway groove curvature centers	mm (in.)
s	Distance between inner and outer groove curvature center loci	mm (in.)
X_1	Axial projection of distance between ball center and outer raceway groove curvature center	mm (in.)

X_2	Radial projection of distance between ball center and outer raceway groove curvature center	mm (in.)
α	Contact angle	°, rad
β	Ball attitude angle	°, rad
γ	$(D\cos\alpha)/d_m$	
δ	Deflection or contact deformation	mm (in.)
θ	Bearing misalignment or angular deflection	°, rad
ρ	Mass density	kg/mm^3 (lb z · sec^2/in.^{4})
ϕ	Angle in WV plane	°, rad
ψ	Angle in yz plane	°, rad
ω	Rotational speed	rad/sec
ω_m	Orbital speed of ball or roller	rad/sec
ω_R	Speed of ball or roller about its own axis	rad/sec
$\Delta\psi$	Angular distance between rolling elements	rad

Subscripts

a	Axial direction
e	Rotation about an eccentric axis
f	Roller guide flange
i	Inner raceway
j	Rolling element at angular location
m	Cage motion and orbital motion
o	Outer raceway
r	Radial direction
R	Rolling element
x	x direction
z	z direction

3.1 GENERAL

Dynamic (inertial) loading occurs between rolling elements and bearing raceways because of rolling element orbital speeds and speeds about their own axes. At slow-to-moderate operating speeds, these dynamic loads are very small compared with the ball or roller loads caused by the loading applied to the bearing. At high operating speeds, however, these rolling element dynamic loads, centrifugal forces, and gyroscopic moments will alter the distribution of the applied loading among the balls or rollers. In roller bearings, the increase in loading on the outer raceway due to roller centrifugal forces causes larger contact deformations in that member; this effect is similar to that of increasing clearance. Increase of clearance, as demonstrated in Chapter 7 of the first volume of this handbook, causes increased maximum roller load due to a decrease in the extent of the load zone. For relatively thin section bearings supported at only a few points on the outer ring, for example, an aircraft gas turbine mainshaft bearing, the centrifugal forces cause bending of the outer ring, also affecting the distribution of loading among the rollers.

In high-speed ball bearings, depending on the contact angles, ball gyroscopic moments and ball centrifugal forces can be of significant magnitude such that inner-ring contact angles increase and outer-ring contact angles decrease. This affects the deflection vs. load characteristics of the bearing and therefore also affects the dynamics of the ball bearing–supported rotor system.

High speed also affects the lubrication and friction characteristics in both ball and roller bearings. This influences bearing internal speeds, and hence rolling element dynamic loads. It is possible, however, to determine the internal load distribution and hence contact stresses in many high-speed rolling bearing applications with sufficient accuracy while not considering the frictional loading on the rolling elements. This will be demonstrated in this chapter. The effects of friction on load distribution will be considered in a later chapter.

3.2 DYNAMIC LOADING OF ROLLING ELEMENTS

3.2.1 Body Forces Due to Rolling Element Rotations

The development of equations in this section is based on the motions occurring in an angular-contact ball bearing because it is the most general form of rolling bearing. Subsequently, the equations developed can be so restricted as to apply to other ball bearings and also to roller bearings.

Figure 3.1 illustrates the instantaneous position of a particle of mass m in a ball of an angular-contact ball bearing operating at a high rotational speed about an axis x. To simplify the analysis, the following coordinate axes systems are introduced:

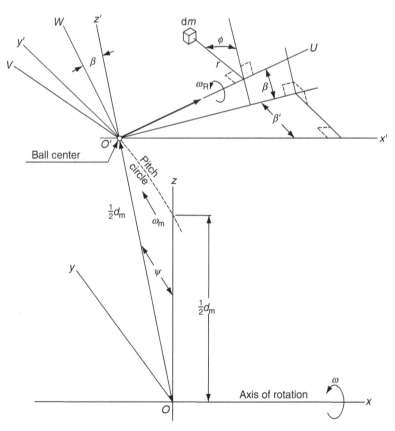

FIGURE 3.1 Instantaneous position of ball mass element dm.

x, y, z	A fixed set of Cartesian coordinates with the x axis coincident with the bearing rotational axis
x', y', z'	A set of Cartesian coordinates with the x' axis parallel to the x axis of the fixed set. This set of coordinates has its origin O' at the ball center and rotates at orbital speed about the fixed x axis at radius $\frac{1}{2}d_m$
U, V, W	A set of Cartesian coordinates with origin at the ball center O' and rotating at orbital speed ω_m. The U axis is collinear with the axis of rotation of the ball about its own center. The W axis is in the plane of the U axis and z' axis; the angle between the W axis and z' axis is β
U, r, ϕ	A set of polar coordinates rotating with the ball

In addition, the following symbols are introduced:

β'	The angle between the projection of the U axis on the $x'y'$ plane and the x' axis
ψ	The angle between the z axis and z' axis, that is, the angular position of the ball on the pitch circle

Consider that an element of mass dm in the ball has the following instantaneous location in the system of rotating coordinates: U, r, ϕ. As

$$
\begin{aligned}
U &= U \\
V &= r \sin \phi \\
W &= r \cos \phi
\end{aligned}
\tag{3.1}
$$

and

$$
\begin{aligned}
x' &= U \cos \beta \cos \beta' - V \sin \beta' - W \sin \beta \cos \beta' \\
y' &= U \cos \beta \sin \beta' + V \cos \beta' - W \sin \beta \sin \beta' \\
z' &= U \sin \beta + W \cos \beta
\end{aligned}
\tag{3.2}
$$

and

$$
\begin{aligned}
x &= x' \\
y &= \tfrac{1}{2}d_m \sin \psi + y' \cos \psi + z' \sin \psi \\
z &= \tfrac{1}{2}d_m \cos \psi - y' \sin \psi + z' \cos \psi
\end{aligned}
\tag{3.3}
$$

by substitution of Equation 3.1 into Equation 3.2 and thence into Equation 3.3, the following expressions relating the instantaneous position of the element of mass dm to the fixed system of Cartesian coordinates can be formulated:

$$
x = U \cos \beta \cos \beta' - r(\sin \beta' \sin \phi + \sin \beta \cos \beta' \cos \phi)
\tag{3.4}
$$

$$
\begin{aligned}
y = \frac{d_m}{2} \sin \psi &+ U(\cos \beta \sin \beta' \cos \psi + \sin \beta \sin \psi) \\
&+ r(\cos \beta \sin \phi \cos \psi + \cos \beta \cos \phi \sin \psi \\
&- \sin \beta \sin \beta' \cos \phi \cos \psi)
\end{aligned}
\tag{3.5}
$$

$$z = \frac{d_m}{2} \cos \psi + U(-\cos \beta \sin \beta' \sin \psi + \sin \beta \cos \psi)$$
$$+ r(-\cos \beta' \sin \phi \sin \psi + \cos \beta \cos \phi \cos \psi \tag{3.6}$$
$$+ \sin \beta \sin \beta' \cos \phi \cos \psi)$$

In accordance with Newton's second law of motion, the following relationships can be determined if the rolling element position angle ψ is arbitrarily set equal to $0°$:

$$dF_x = \ddot{x} \, dm \tag{3.7}$$

$$dF_y = \ddot{y} \, dm \tag{3.8}$$

$$dF_z = \ddot{z} \, dm \tag{3.9}$$

$$dM_z' = \{-\ddot{x}[U \cos \beta \sin \beta' + r(\cos \beta' \sin \phi - \sin \beta \sin \beta' \cos \phi)]$$
$$+ \ddot{y}[U \cos \beta \cos \beta' - r(\sin \beta' \sin \phi + \sin \beta \cos \beta' \cos \phi)]\} \, dm \tag{3.10}$$

$$dM_y' = \{\ddot{x}(U \sin \beta + r \cos \beta \cos \phi)$$
$$- \ddot{z}[U \cos \beta \cos \beta' - r(\sin \beta' \sin \phi + \sin \beta \cos \beta' \cos \phi)]\} \, dm \tag{3.11}$$

The net moment about the x axis must be zero for constant speed motion. At each ball location (ψ, β), ω_R (rotational speed $d\phi/dt$ of the ball about its own axis U) and ω_m (orbital speed $d\psi/dt$ of the ball about the bearing axis x) are constant; therefore, at $\psi = 0$,

$$\ddot{x} = \frac{d^2 x}{dt^2} = r\omega_R^2 (\sin \beta' \sin \phi + \sin \beta \cos \beta' \cos \phi) \tag{3.12}$$

$$\ddot{y} = \frac{d^2 y}{dt^2} = -2\omega_R \omega_m r \cos \beta \sin \phi$$
$$+ \omega_m^2 [-U \cos \beta \sin \beta' + r(-\cos \beta' \sin \phi + \sin \beta \cos \phi \sin \beta')] \tag{3.13}$$
$$+ \omega_R^2 r(-\cos \beta' \cos \phi + \sin \beta \sin \beta' \sin \phi)$$

$$\ddot{z} = \frac{d^2 z}{dt^2} = -2\omega_R \omega_m r(\cos \beta' \cos \phi + \sin \beta \sin \beta' \sin \phi)$$
$$- \omega_m^2 \left(\frac{d_m}{2} + U \sin \beta + r \cos \beta \cos \phi\right) \tag{3.14}$$
$$- \omega_R^2 r \cos \beta \cos \phi$$

Substitution of Equation 3.12 through Equation 3.14 into Equation 3.7 through Equation 3.11 and placing the latter into integral format yields

$$F_{x'} = -\rho \int_{-r_R}^{+r_R} \int_0^{(r_R^2 - U^2)^{1/2}} \int_0^{2\pi} \ddot{x} r \, dr \, dU \, d\phi \tag{3.15}$$

$$F_{y'} = -\rho \int_{-r_R}^{+r_R} \int_0^{(r_R^2 - U^2)^{1/2}} \int_0^{2\pi} \ddot{y} r \, dr \, dU \, d\phi \tag{3.16}$$

$$F_{z'} = -\rho \int_{-r_R}^{+r_R} \int_0^{(r_R^2 - U^2)^{1/2}} \int_0^{2\pi} \ddot{z} r \, dr \, dU \, d\phi \tag{3.17}$$

$$M_{z'} = -\rho \int_{-r_R}^{+r_R} \int_0^{(r_R^2-U^2)^{1/2}} \int_0^{2\pi} \{-\ddot{x}[U\cos\beta\sin\beta'$$
$$+ r(\cos\beta'\sin\phi - \sin\beta\sin\beta'\cos\phi)]$$
$$+ \ddot{y}[U\cos\beta\cos\beta' - r(\sin\beta'\sin\phi \qquad (3.18)$$
$$+ \sin\beta\cos\beta'\cos\phi)]\}r\,dr\,dU\,d\phi$$

$$M_{y'} = -\rho \int_{-r_R}^{+r_R} \int_0^{(r_R^2-U^2)^{1/2}} \int_0^{2\pi} \{\ddot{x}(U\sin\beta + r\cos\beta\cos\phi)$$
$$- \ddot{z}[U\cos\beta\cos\beta' - r(\sin\beta'\sin\phi \qquad (3.19)$$
$$+ \sin\beta\cos\beta'\cos\phi)]\}\,r\,dr\,dU\,d\phi$$

In Equation 3.15 through Equation 3.19, ρ is the mass density of the ball material and r_R is the ball radius.

Performing the integrations indicated by Equation 3.15 through Equation 3.19 establishes that the net forces in the x' and y' directions are zero and that

$$F_{z'} = \tfrac{1}{2}md_m\omega_m^2 \qquad (3.20)$$

$$M_{y'} = J\omega_R\omega_m\sin\beta \qquad (3.21)$$

$$M_{z'} = -J\omega_R\omega_m\cos\beta\sin\beta' \qquad (3.22)$$

where m is the mass of the ball and J is the mass moment of inertia, and are defined as follows:

$$m = \tfrac{1}{6}\rho\pi D^3 \qquad (3.23)$$

$$J = \tfrac{1}{60}\rho\pi D^5 \qquad (3.24)$$

3.2.2 Centrifugal Force

3.2.2.1 Rotation about the Bearing Axis

Substituting Equation 3.23 into Equation 3.20 and recognizing that

$$\omega_m = \frac{2\pi n_m}{60} \qquad (3.25)$$

Equation 3.26 yielding the ball centrifugal force is obtained:

$$F_c = \frac{\pi^3\rho}{10800g} D^3 n_m^2 d_m \qquad (3.26)$$

For steel balls,

$$F_c = 2.26 \times 10^{-11} D^3 n_m^2 d_m \qquad (3.27)$$

For an applied thrust load per ball Q_{ia} and a ball centrifugal load F_c directed radially outward, the ball loading is as shown in Figure 3.2. For conditions of equilibrium, assuming the bearing rings are not flexible,

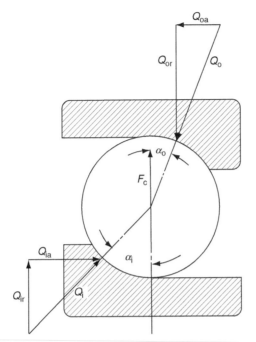

FIGURE 3.2 Ball under thrust load and centrifugal load.

$$Q_{ia} - Q_{oa} = 0 \tag{3.28}$$

$$Q_{ir} + F_c - Q_{or} = 0 \tag{3.29}$$

or

$$Q_{ia} - Q_o \sin \alpha_o = 0 \tag{3.30}$$

$$Q_{ia} \cot \alpha_i + F_c - Q_o \cos \alpha_o = 0 \tag{3.31}$$

Equation 3.30 and Equation 3.31 must be solved simultaneously for unknowns Q_o and α_o. Thus,

$$\alpha_o = \cot^{-1} \left(\cot \alpha_i + \frac{F_c}{Q_{ia}} \right) \tag{3.32}$$

$$Q_o = \left[1 + \left(\cot \alpha_i + \frac{F_c}{Q_{ia}} \right)^2 \right]^{1/2} Q_{ia} \tag{3.33}$$

Further,

$$Q_i = \frac{Q_{ia}}{\sin \alpha_i} \tag{3.34}$$

From Equation 3.32, because of centrifugal force F_c, it is apparent that $\alpha_o < \alpha_i$. α_i is the contact angle under thrust load, and $\alpha_i > \alpha^o$ the free contact angle. This condition is discussed in detail in Chapter 7 in the first volume of this handbook.

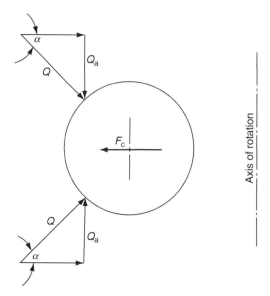

FIGURE 3.3 Ball loading in a 90° contact-angle, thrust ball bearing.

See Example 3.1.

Thrust ball bearings with nominal contact angle $\alpha = 90°$ operating at high speeds and light loads tend to permit the balls to override the land on both rings (washers). The contact angle thus deviates from 90° in the same direction on both raceways. From Figure 3.3, which depicts this condition,

$$Q = \frac{F_c}{2\cos\alpha} \tag{3.35}$$

and

$$\alpha = \tan^{-1}\left(\frac{2Q_a}{F_c}\right) \tag{3.36}$$

See Example 3.2.

Equation 3.20 is not restrictive as to geometry and since the mass of a cylindrical (or nearly cylindrical) roller is given by

$$m = \tfrac{1}{4}\rho\pi D^2 l_t \tag{3.37}$$

the centrifugal force for a steel roller orbiting at speed n_m about a bearing axis is given by

$$F_c = 3.39 \times 10^{-11} D^2 l_t d_m n_m^2 \tag{3.38}$$

For a tapered roller bearing, however, roller centrifugal force alters the distribution of load between the outer raceway and inner-ring guide flange. Figure 3.4 demonstrates this condition for an applied thrust load Q_{ia}.

For equilibrium to exist,

$$Q_{ia} + Q_{fa} - Q_{oa} = 0 \tag{3.39}$$

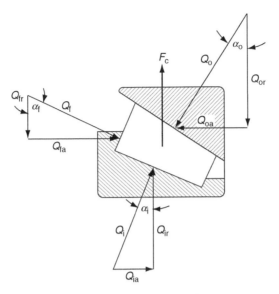

FIGURE 3.4 Tapered roller under thrust load and centrifugal force.

$$Q_{ir} - Q_{fr} + F_c - Q_{or} = 0 \qquad (3.40)$$

or

$$Q_{ia} + Q_f \sin \alpha_f - Q_o \sin \alpha_o = 0 \qquad (3.41)$$

$$Q_{ia} \cot \alpha_i - Q_f \cos \alpha_f + F_c - Q_o \cos \alpha_o = 0 \qquad (3.42)$$

Solving Equation 3.41 and Equation 3.42 simultaneously yields

$$Q_o = \frac{Q_{ia}(\cot \alpha_i \sin \alpha_f + \cos \alpha_f) + F_c \sin \alpha_f}{\sin(\alpha_o + \alpha_f)} \qquad (3.43)$$

$$Q_o = \frac{Q_{ia}(\cot \alpha_i \sin \alpha_o - \cos \alpha_o) + F_c \sin \alpha_o}{\sin(\alpha_o + \alpha_f)} \qquad (3.44)$$

See Example 3.3.

Care must be exercised in operating a tapered roller bearing at a very high speed. At some critical speed related to the magnitude of the applied load, the force at the inner-ring raceway contact approaches zero, and the entire axial load is carried at the roller end–inner-ring flange contact. Because this contact has only sliding motion, very high friction results with attendant high heat generation.

Most modern radial spherical roller bearings have complements of symmetrical contour (barrel-shaped) rollers and relatively small contact angles; for example, $\alpha \leq 15°$. When the bearings are operated at a high speed, roller loading is as illustrated in Figure 3.5.

Equilibrium of forces in the radial and axial directions gives

$$Q_o \cos \alpha_o - Q_i \cos \alpha_i - F_c = 0 \qquad (3.45)$$

$$Q_o \sin \alpha_o - Q_i \sin \alpha_i = 0 \qquad (3.46)$$

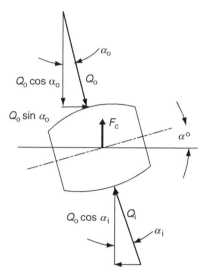

FIGURE 3.5 Loading of a barrel-shaped roller subjected to applied and centrifugal forces.

Solving these equations simultaneously gives

$$Q_o = \frac{F_c \sin \alpha_i}{\sin(\alpha_i - \alpha_o)} \tag{3.47}$$

$$Q_i = \frac{F_c \sin \alpha_o}{\sin(\alpha_i - \alpha_o)} \tag{3.48}$$

Therefore, it appears that roller–raceway loading is uniquely determined by roller centrifugal loading. Clearly, in this instance the inner and outer raceway contact angles are functions of the applied radial and thrust loadings of the bearing, and these must be determined from the equilibrium of loading on the bearing. Doing this requires the determination of the bearing contact deformations.

Another way to view the operation of a loaded spherical roller operating at a high speed is to resolve the centrifugal force into components collinear with, and normal to, the roller axis of rotation. Hence,

$$F_{ca} = F_c \sin \alpha^o \tag{3.49}$$

$$F_{cr} = F_c \cos \alpha^o \tag{3.50}$$

where α^o is the nominal contact angle. Equilibrium of forces acting in the radial plane of the roller gives

$$Q_o = Q_i + F_c \cos \alpha^o \tag{3.51}$$

and the component $F_c \sin \alpha^o$ causes the inner and outer raceway contact angles to shift slightly from α^o to accommodate the roller axial loading. In general, spherical roller bearings do not operate at speeds that will cause significant change in the nominal contact angle. Also, consider a double-row spherical roller bearing with barrel-shaped rollers subjected to a radial load while rotating at high speed. The speed-induced roller axial loads are self-equilibrated within the bearing; however, the outer raceways carry larger thrust components than do the inner raceways.

3.2.2.2 Rotation about an Eccentric Axis

Section 3.2.2.1 dealt with rolling element centrifugal loading when the bearing rotates about its own axis; this is the usual case. In planetary gear transmissions, however, the planet gear bearings rotate about the input and output shaft axes as well as about their own axes. Hence, an additional inertial or centrifugal force is induced in the rolling element. Figure 3.6 shows a schematic diagram of such a system. From Figure 3.6, it can be seen that the instantaneous radius of rotation is, by the law of cosines,

$$r = (r_m^2 + r_e^2 - 2r_m r_e \cos \psi)^{1/2} \tag{3.52}$$

Therefore, the corresponding centrifugal force is

$$F_{ce} = m\omega_e^2 (r_m^2 + r_e^2 - 2r_m r_c \cos \psi)^{1/2} \tag{3.53}$$

This force F_{ce} is maximum at $\psi = 180°$ and at that angle is algebraically additive to F_c. At $\psi = 0$, the total centrifugal force is $F_c - F_{ce}$. The angle between F_c and F_{ce} as derived from the law of cosines is

$$\theta = \cos^{-1} \left(\frac{r_m}{r} - \frac{r_e}{r} \cos \psi \right) \tag{3.54}$$

F_{ce} can be resolved into a radial force and a tangential force as follows:

$$F_{cer} = F_{ce} \cos \theta \tag{3.55}$$

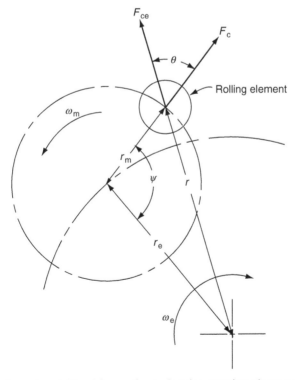

FIGURE 3.6 Rolling element centrifugal forces due to bearing rotation about an eccentric axis.

$$F_{\mathrm{cet}} = F_{\mathrm{ce}} \sin \theta \tag{3.56}$$

Hence, the total instantaneous radial centrifugal force acting on the rolling element is

$$F_{\mathrm{cr}} = m\omega_{\mathrm{m}}^2 r_{\mathrm{m}} + m\omega_{\mathrm{e}}^2 (r_{\mathrm{m}} - r_{\mathrm{e}} \cos \psi) \tag{3.57}$$

or

$$F_{\mathrm{cr}} = \frac{W}{g} [r_{\mathrm{m}}(\omega_{\mathrm{m}}^2 + \omega_{\mathrm{e}}^2) - r_{\mathrm{e}}\omega_{\mathrm{e}}^2 \cos \psi] \tag{3.58}$$

where the positive direction is that taken by the constant component F_{c}. For steel ball and roller elements, the following equations are, respectively, valid:

$$F_{\mathrm{cr}} = 2.26 \times 10^{-11} D^3 [d_{\mathrm{m}}(n_{\mathrm{m}}^2 + n_{\mathrm{e}}^2) - d_{\mathrm{e}} n_{\mathrm{e}}^2 \cos \psi] \tag{3.59}$$

$$F_{\mathrm{cr}} = 3.39 \times 10^{-11} D^2 l_{\mathrm{t}} [d_{\mathrm{m}}(n_{\mathrm{m}}^2 + n_{\mathrm{e}}^2) - d_{\mathrm{e}} n_{\mathrm{e}}^2 \cos \psi] \tag{3.60}$$

The instantaneous tangential component of eccentric centrifugal force is

$$F_{\mathrm{ct}} = m\omega_{\mathrm{e}}^2 r_{\mathrm{e}} \sin \psi \tag{3.61}$$

For steel ball and roller elements, respectively, the following equations pertain:

$$F_{\mathrm{ct}} = 2.26 \times 10^{-11} D^3 d_{\mathrm{e}} n_{\mathrm{e}}^2 \sin \psi \tag{3.62}$$

$$F_{\mathrm{ct}} = 3.39 \times 10^{-11} D^2 l_{\mathrm{t}} d_{\mathrm{e}} n_{\mathrm{e}}^2 \sin \psi \tag{3.63}$$

This tangential force alternates direction and tends to produce sliding between the rolling element and raceway. It is therefore resisted by a frictional force between the contacting surfaces.

The bearing cage also undergoes this eccentric motion and if it is supported on the rolling elements, it will impose an additional load on the individual rolling elements. This cage load may be reduced by using a material of smaller mass density.

3.2.3 GYROSCOPIC MOMENT

It can usually be assumed with minimal loss of calculational accuracy that pivotal motion due to gyroscopic moment is negligible; then, the angle β' is zero and Equation 3.22 is of no consequence. The gyroscopic moment as defined by Equation 3.21 is therefore resisted successfully by friction forces at the bearing raceways for ball bearings and by normal forces for roller bearings. Substituting Equation 3.25 into Equation 3.21, the following expression is obtained for ball bearings:

$$M_{\mathrm{g}} = \tfrac{1}{60} \rho \pi D^5 \omega_{\mathrm{R}} \omega_{\mathrm{m}} \sin \beta \tag{3.64}$$

since

$$\omega_{\mathrm{R}} = \frac{2\pi n_{\mathrm{R}}}{60} \tag{3.65}$$

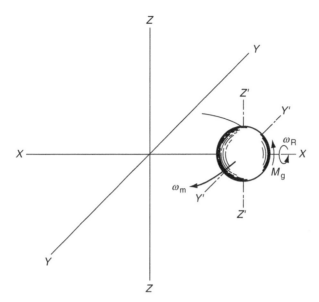

FIGURE 3.7 Gyroscopic moment due to simultaneous rotation about nonparallel axes.

and

$$\omega_m = \frac{2\pi n_m}{60} \tag{3.66}$$

The gyroscopic moment for steel ball bearings is given by

$$M_g = 4.47 \times 10^{-12} D^5 n_R n_m \sin\beta \tag{3.67}$$

Figure 3.7 shows the direction of the gyroscopic moment in a ball bearing. Accordingly, Figure 3.8 shows the ball loading due to the action of gyroscopic moment and centrifugal force on a thrust-loaded ball bearing.

See Example 3.4.

Gyroscopic moments also act on rollers in radial tapered and spherical roller bearings and on the rollers in thrust roller bearings of all types. The rollers, however, are geometrically constrained from rotating due to the induced gyroscopic moments. Therefore, a gyroscopic moment of significant magnitude tends to alter the distribution of load across the roller contour. For steel rollers, the gyroscopic moments are given by

$$M_g = 8.37 \times 10^{-12} D^4 l_t n_R n_m \sin\beta \tag{3.68}$$

3.3 HIGH-SPEED BALL BEARINGS

To determine the load distribution in a high-speed ball bearing, consider Figure 1.2, which shows the displacements of a ball bearing inner ring relative to the outer ring due to a generalized loading system, including radial, axial, and moment loads. Figure 3.9 shows the relative angular position (azimuth) of each ball in the bearing.

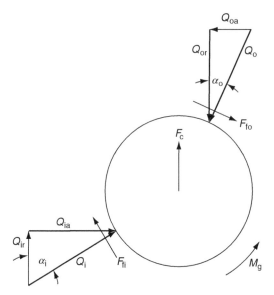

FIGURE 3.8 Forces acting on a ball in a high-speed ball bearing subjected to applied thrust load.

Under zero load, the centers of the raceway groove curvature radii are separated by a distance A as shown in Figure 1.1a. In Chapter 2 of the first volume of this handbook, it was shown that $A = BD$ where $B = f_i + f_o - 1$. Under an applied static load, the distance between the inner and outer raceway groove curvature centers will increase by the amount of the contact

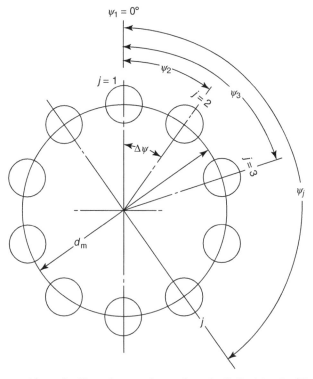

FIGURE 3.9 Angular position of rolling elements in yz plane (radial). $\Delta\psi = 2\pi/Z$, $\psi_j = 2\pi/Z(j-1)$.

deformations δ_i and δ_o as shown in Figure 1.1b. The line of action between the centers is collinear with BD (A). If, however, a centrifugal force acts on the ball, then because the ball–inner and ball–outer raceway contact angles are dissimilar, the line of action between the raceway groove curvature centers is not collinear with BD. Rather, it is discontinuous as indicated in Figure 3.10. It is assumed in Figure 3.10 that the outer raceway groove curvature center is fixed in space, and the inner raceway groove curvature center moves relative to that fixed center. Moreover, the ball center shifts by virtue of the dissimilar contact angles.

The distance between the fixed outer raceway groove curvature center and the final position of the ball center at any ball azimuth location j is

$$\Delta_{oj} = r_o - \frac{D}{2} + \delta_{oj} \tag{3.69}$$

Since

$$r_o = f_o D$$
$$\Delta_{oj} = (f_o - 0.5)D + \delta_{oj} \tag{3.70}$$

Similarly,

$$\Delta_{ij} = (f_i - 0.5)D + \delta_{ij} \tag{3.71}$$

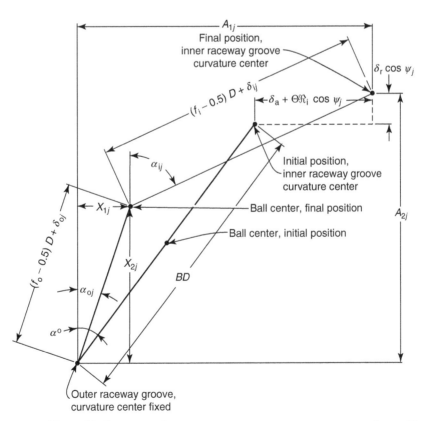

FIGURE 3.10 Positions of ball center and raceway groove curvature centers at angular position ψ_j with and without applied load.

where δ_{oj} and δ_{ij} are the normal contact deformations at the outer and inner raceway contacts, respectively.

In accordance with the relative axial displacement of the inner and outer rings δ_a and the relative angular displacements θ, the axial distance between the loci of inner and outer raceway groove curvature centers at any ball position is

$$A_{1j} = BD \sin \alpha^\circ + \delta_a + \theta \Re_i \cos \psi_j \tag{3.72}$$

where $\Re_i$ is the radius of the locus of inner raceway groove curvature centers and α° is the initial contact angle before loading. Further, in accordance with a relative radial displacement of the ring centers δ_r, the radial displacement between the loci of the groove curvature centers at each ball location is

$$A_{2j} = BD \cos \alpha^\circ + \delta_r \cos \psi_j \tag{3.73}$$

These data are intended as an explanation of Figure 3.10.

Jones [1] found it convenient to introduce new variables X_1 and X_2, as shown in Figure 3.10. It can be seen from Figure 3.10 that at any ball location

$$\cos \alpha_{oj} = \frac{X_{2j}}{(f_o - 0.5)D + \delta_{oj}} \tag{3.74}$$

$$\sin \alpha_{oj} = \frac{X_{1j}}{(f_o - 0.5)D + \delta_{oj}} \tag{3.75}$$

$$\cos \alpha_{ij} = \frac{A_{2j} - X_{2j}}{(f_i - 0.5)D + \delta_{ij}} \tag{3.76}$$

$$\sin \alpha_{ij} = \frac{A_{1j} - X_{1j}}{(f_i - 0.5)D + \delta_{ij}} \tag{3.77}$$

Using the Pythagorean Theorem, it can be seen from Figure 3.10 that

$$(A_{1j} - X_{1j})^2 + (A_{2j} - X_{2j})^2 - [(f_i - 0.5)D + \delta_{ij}]^2 = 0 \tag{3.78}$$

$$X_{1j}^2 + X_{2j}^2 - [(f_o - 0.5)D + \delta_{oj}]^2 = 0 \tag{3.79}$$

Considering the plane passing through the bearing axis and the center of a ball located at azimuth ψ_j (see Figure 3.9), the load diagram in Figure 3.11 obtains if noncoplanar friction forces are insignificant. Assuming that "outer raceway control" is approximated at a given ball location, it can also be assumed with little effect on calculational accuracy that the ball gyroscopic moment is resisted entirely by friction force at the ball–outer raceway contact. Otherwise, it is safe to assume that the ball gyroscopic moment is resisted equally at the ball–inner and ball–outer raceway contacts. In Figure 3.11, therefore, $\lambda_{ij} = 0$ and $\lambda_{oj} = 2$ for outer raceway control; otherwise, $\lambda_{ij} = \lambda_{oj} = 1$.

The normal ball loads are related to normal contact deformations as follows:

$$Q_{oj} = K_{oj}\delta_{oj}^{1.5} \tag{3.80}$$

$$Q_{ij} = K_{ij}\delta_{ij}^{1.5} \tag{3.81}$$

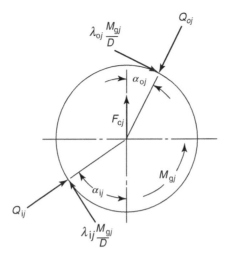

FIGURE 3.11 Ball loading at angular position ψ_j.

From Figure 3.11, considering the equilibrium of forces in the horizontal and vertical directions,

$$Q_{ij} \sin \alpha_{ij} - Q_{oj} \sin \alpha_{oj} - \frac{M_{gj}}{D}(\lambda_{ij} \cos \alpha_{ij} - \lambda_{oj} \cos \alpha_{oj}) = 0 \tag{3.82}$$

$$Q_{ij} \cos \alpha_{ij} - Q_{oj} \cos \alpha_{oj} - \frac{M_{gj}}{D}(\lambda_{ij} \sin \alpha_{ij} - \lambda_{oj} \sin \alpha_{oj}) + F_{cj} = 0 \tag{3.83}$$

Substituting Equation 3.80, Equation 3.81, and Equation 3.74 through Equation 3.77 into Equation 3.82 and Equation 3.83 yields

$$\frac{\frac{\lambda_{oj}M_{gj}X_{2j}}{D} - K_{oj}\delta_{oj}^{1.5}X_{1j}}{(f_o - 0.5)D + \delta_{oj}} + \frac{K_{ij}\delta_{ij}^{1.5}(A_{1j} - X_{1j}) - \frac{\lambda_{ij}M_{gj}}{D}(A_{2j} - X_{2j})}{(f_i - 0.5)D + \delta_{ij}} = 0 \tag{3.84}$$

$$\frac{K_{oj}\delta_{oj}^{1.5}X_{2j} + \frac{\lambda_{oj}M_{gj}X_{1j}}{D}}{(f_o - 0.5)D + \delta_{oj}} - \frac{K_{ij}\delta_{ij}^{1.5}(A_{2j} - X_{2j}) + \frac{\lambda_{ij}M_{gj}}{D}(A_{1j} - X_{1j})}{(f_i - 0.5)D + \delta_{ij}} - F_{cj} = 0 \tag{3.85}$$

Equation 3.78, Equation 3.79, Equation 3.84, and Equation 3.85 may be solved simultaneously for X_{1j}, X_{2j}, δ_{ij}, and δ_{oj} at each ball angular location once values for δ_a, δ_r, and θ are assumed. The most probable method of solution is the Newton–Raphson method for solution of simultaneous nonlinear equations.

The centrifugal force acting on a ball is calculated as follows:

$$F_c = \tfrac{1}{2}md_m\omega_m^2 \tag{3.20}$$

where ω_m is the orbital speed of the ball. Substituting the identity $\omega_m^2 = (\omega_m/\omega)^2\omega^2$ in Equation 3.20, the following equation for centrifugal force is obtained:

$$F_{cj} = \frac{1}{2}md_m\omega^2\left(\frac{\omega_m}{\omega}\right)_j^2 \tag{3.86}$$

where ω is the speed of the rotating ring and ω_m is the orbital speed of the ball at angular position ψ_j. It should be apparent that because orbital speed is a function of contact angle, it is not constant for each ball location.

Moreover, it must be kept in mind that this analysis does not consider frictional forces that tend to retard ball and hence cage motion. Therefore, in a high-speed bearing, it is to be expected that ω_m will be less than that predicted by Equation 2.47 and greater than that predicted by Equation 2.48. Unless the loading on the bearing is relatively light, however, the cage speed differential is usually insignificant in affecting the accuracy of the calculations ensuing in this chapter.

Gyroscopic moment at each ball location may be described as follows:

$$M_{gj} = J\left(\frac{\omega_R}{\omega}\right)_j \left(\frac{\omega_m}{\omega}\right)_j \omega^2 \sin\beta \tag{3.87}$$

where β is given by Equation 2.43, ω_R/ω by Equation 2.44, and ω_m/ω by Equation 2.47 or Equation 2.48.

Since K_{oj}, K_{ij}, and M_{gj} are functions of a contact angle, Equation 3.74 through Equation 3.77 may be used to establish these values during the iteration.

To find the values of δ_r, δ_a, and θ, it remains only to establish the conditions of equilibrium applying to the entire bearing. These are

$$F_a - \sum_{j=1}^{j=Z}\left(Q_{ij}\sin\alpha_{ij} - \frac{\lambda_{ij}M_{gj}}{D}\cos\alpha_{ij}\right) = 0 \tag{3.88}$$

or

$$F_a - \sum_{j=1}^{j=Z}\left[\frac{K_{ij}(A_{1j}-X_{1j})\delta_{ij}^{1.5} - \frac{\lambda_{ij}M_{gj}}{D}(A_{2j}-X_{2j})}{(f_i-0.5)D+\delta_{ij}}\right] = 0 \tag{3.89}$$

$$F_r - \sum_{j=1}^{j=Z}\left(Q_{ij}\cos\alpha_{ij} + \frac{\lambda_{ij}M_{gj}}{D}\sin\alpha_{ij}\right)\cos\psi_j = 0 \tag{3.90}$$

or

$$F_r - \sum_{j=1}^{j=Z}\left(\frac{K_{ij}(A_{2j}-X_{2j})\delta_{ij}^{1.5} - \frac{\lambda_{ij}M_{gj}}{D}(A_{1j}-X_{1j})}{(f_i-0.5)D+\delta_{ij}}\right) = 0 \tag{3.91}$$

$$M - \sum_{j=1}^{j=Z}\left[\left(Q_{ij}\sin\alpha_{ij} - \frac{\lambda_{ij}M_{gj}}{D}\cos\alpha_{ij}\right)\Re_i + \frac{\lambda_{ij}M_{gj}}{D}r_i\right]\cos\psi_j = 0 \tag{3.92}$$

or

$$M - \sum_{j=1}^{j=Z}\left[\frac{\left(K_{ij}(A_{1j}-X_{1j})\delta_{ij}^{1.5} - \frac{\lambda_{ij}M_{gj}}{D}(A_{2j}-X_{2j})\right)\Re_i}{(f_i-0.5)D+\delta_{ij}} + \lambda_{ij}f_iM_{gj}\right]\cos\psi_j = 0 \tag{3.93}$$

$$\Re_i = \tfrac{1}{2} d_m + (f_i - 0.5)D \cos \alpha° \tag{1.3}$$

Having calculated values of X_{1j}, X_{2j}, δ_{ij}, and δ_{oj} at each ball position, and knowing F_a, F_r, and M as input conditions, the values δ_a, δ_r, and θ may be determined by Equation 3.89, Equation 3.91, and Equation 3.93. After obtaining the primary unknown quantities δ_a, δ_r, and θ, it is then necessary to repeat the calculation of X_{1j}, X_{2j}, δ_{ij}, δ_{oj}, and so on, until compatible values of the primary unknown quantities δ_a, δ_r, and θ are obtained.

Solution of the system of simultaneous equations, Equation 3.78, Equation 3.79, Equation 3.84, Equation 3.85, and Equation 3.89, requires the use of a digital computer. To illustrate the results of such a calculation, the performance of a 218 angular-contact ball bearing (40° free contact angle), was evaluated over an applied thrust load range 0–44,450 N (0–10,000 lb) and shaft speed range 3,000–15,000 rpm. Figure 3.12 through Figure 3.14 show the results of the calculations.

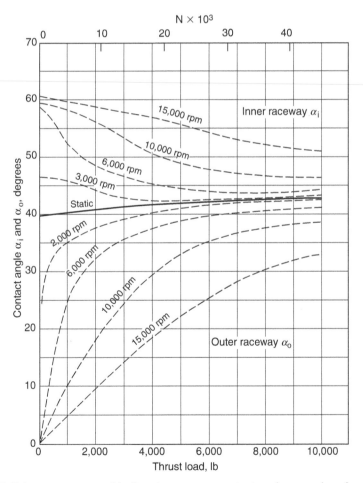

FIGURE 3.12 Ball–inner raceway and ball–outer raceway contact angles α_i and α_o for a 218 angular-contact ball bearing (free contact angle $\alpha° = 40°$).

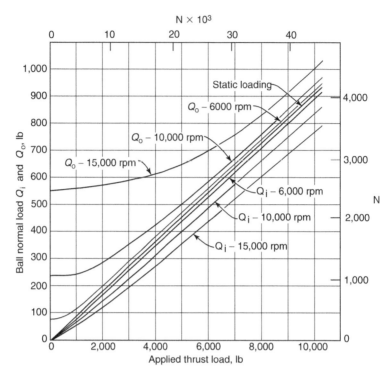

FIGURE 3.13 Ball–inner raceway and ball–outer raceway contact normal loads Q_i and Q_o for a 218 angular-contact ball bearing (free contact angle $\alpha^\circ = 40^\circ$).

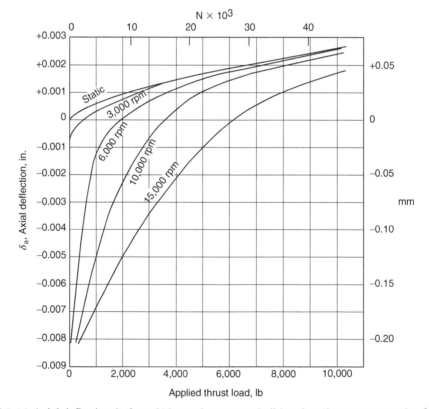

FIGURE 3.14 Axial deflection δ_a for a 218 angular-contact ball bearing (free contact angle $\alpha^\circ = 40^\circ$).

3.3.1 Ball Excursions

For an angular-contact ball bearing subjected only to thrust loading, the orbital travel of the balls occurs in a single radial plane, whose axial location is defined by X_{1j} in Figure 3.10; X_{1j} is the same at all ball azimuth angles ψ_i. For a bearing that supports combined radial and axial loads, or combined radial, axial, and moment loads, X_{1j} is different at each ball azimuth angle ψ_i. Therefore, a ball undergoes an axial movement or "excursion" as it orbits the shaft or housing center. Unless this excursion is accommodated by providing sufficient axial clearance between the ball and the cage pocket, the cage will experience nonuniform and possibly heavy loading in the axial direction. This can also cause a complex motion of the cage, that is, no longer simple rotation in a single plane, rather including an out-of-plane vibration component. Such a motion together with the aforementioned loading can lead to the rapid destruction and seizure of the bearing.

Under combined loading, because of the variation in the ball–raceway contact angles α_{ij} and α_{oj} as the ball orbits the bearing axis, there is a tendency for the ball to lead (advance ahead of) or lag (fall behind) its central position in the cage pocket. The orbital or circumferential travel of the ball relative to the cage is, however, limited by the cage pocket. Therefore, a load occurs between the ball and the cage pocket in the circumferential direction. Under steady-state cage rotation, the sum of these ball-cage pocket loads in the circumferential direction is close to zero, balanced only by friction forces. Moreover, the forces and moments acting on a ball in the bearing's plane of rotation must be in balance, including acceleration or deceleration loading and friction forces. To achieve this condition of equilibrium, the ball speeds, including orbital speed, will be different from those calculated considering only kinematic conditions, or even those indicated in Chapter 2, assuming the condition of no gyroscopic motion. This is a condition of skidding, and it will be covered in Chapter 5.

3.3.2 Lightweight Balls

To permit ball bearings to operate at higher speeds, it is possible to reduce the adverse ball inertial effects by reducing the ball mass. This is especially effective for angular-contact ball bearings as the differential between the ball–inner raceway and ball–outer raceway contact angles, $\alpha_{ij} - \alpha_{oj}$, will be reduced. To achieve this result, it was first attempted to operate bearings with complements of hollow balls [2]; however, this proved impractical because it was difficult to manufacture hollow balls that have isotropic mechanical properties. In the 1980s, hot isostatically pressed (HIP) silicon nitride ceramic was developed as an acceptable material for the manufacture of rolling elements. Bearings with balls of HIP silicon nitride, which has a density approximately 42% that of steel and an excellent compressive strength, are used in high-speed, machine tool spindle applications and are under consideration for use as aircraft gas turbine engine application mainshaft bearings. Figure 3.15 through Figure 3.17 compare the bearing performance parameters for operations at high speed of the 218 angular-contact ball bearing with steel balls and HIP silicon nitride balls.

Silicon nitride also has a modulus of elasticity of approximately 3.1×10^5 MPa $(45 \times 10^6 \text{ psi})$. In a hybrid ball bearing, that is, a bearing with steel rings and silicon nitride balls, owing to the higher elastic modulus of the ball material, the contact areas between balls and raceways will be smaller than in an all-steel bearing. This causes the contact stresses to be greater. Depending on the load magnitude, the stress level may be acceptable to the ball material, but not to the raceway steel. This situation can be ameliorated at the expense of increased contact friction by increasing the conformity of the raceways to the balls; for example, decreasing the raceway groove curvature radii. This amount of decrease is specific to each application, dependent on bearing applied loading and speed.

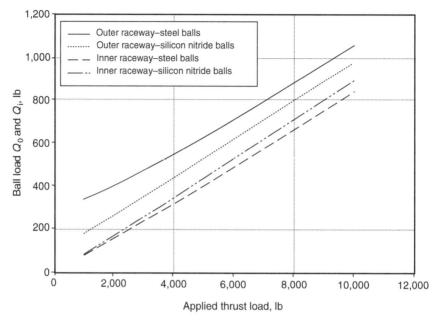

FIGURE 3.15 Outer and inner raceway–ball loads vs. bearing applied thrust load for a 218 angular-contact ball bearing operating at 15,000 rpm with steel or silicon nitride balls.

3.4   HIGH-SPEED RADIAL CYLINDRICAL ROLLER BEARINGS

Because of the high rate of heat generation accompanying relatively high friction torque, tapered roller and spherical roller bearings have not historically been employed for high-speed applications. Generally, cylindrical roller bearings have been used; however, improvements in

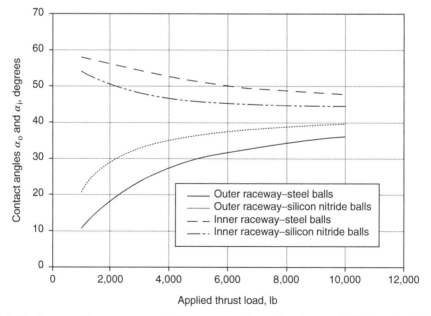

FIGURE 3.16 Outer and inner raceway–ball contact angle vs. bearing applied thrust load for a 218 angular-contact ball bearing operating at 15,000 rpm with steel or silicon nitride balls.

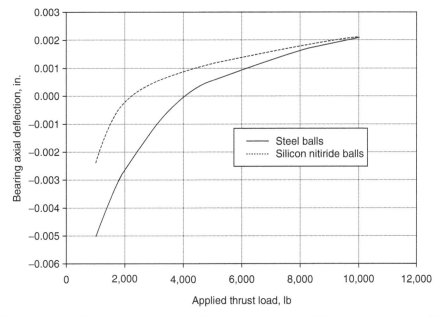

FIGURE 3.17 Axial deflection vs. bearing applied thrust load for a 218 angular-contact ball bearing operating at 15,000 rpm with steel or silicon nitride balls.

bearing internal design, accuracy of manufacture, and methods of removing generated heat via circulating oil lubrication have gradually increased the allowable operating speeds for both tapered roller and spherical roller bearings. The simplest case for analytical investigation is still a radially loaded cylindrical roller bearing and this will be considered in the following discussion.

Figure 3.18 indicates the forces acting on a roller of a high-speed cylindrical roller bearing subjected to a radial load F_r. Thus, considering equilibrium of forces,

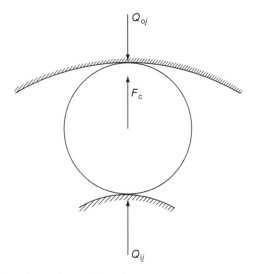

FIGURE 3.18 Roller loading at angular position ψ_j.

$$Q_{oj} - Q_{ij} - F_c = 0 \tag{3.94}$$

Rearranging Equation 1.33 yields

$$Q = K\delta^{10/9} \tag{3.95}$$

where

$$K = 8.05 \times 10^4 \, l^{8/9} \tag{3.96}$$

Therefore,

$$K\delta_{oj}^{10/9} - K\delta_{ij}^{10/9} - F_c = 0 \tag{3.97}$$

Since

$$\delta_{rj} = \delta_{ij} + \delta_{oj} \tag{3.98}$$

Equation 3.97 may be rewritten as follows:

$$\left(\delta_{rj} - \delta_{ij}\right)^{10/9} - \delta_{ij}^{10/9} - \frac{F_c}{K} = 0 \tag{3.99}$$

Equilibrium of forces in the direction of applied radial load on the bearing dictates that

$$F_r - \sum_{j=1}^{j=Z} Q_{ij} \cos \psi_j = 0 \tag{3.100}$$

or

$$\frac{F_r}{K} - \sum_{j=1}^{j=Z} \delta_{ij}^{10/9} \cos \psi_j = 0 \tag{3.101}$$

From the geometry of the loaded bearing, it can be determined that the total radial compression at any roller azimuth location ψ_j is

$$\delta_{rj} = \delta_r \cos \psi_j - \frac{P_d}{2} \tag{3.102}$$

Substitution of Equation 3.102 into Equation 3.99 yields

$$\left(\delta_r \cos \psi_j - \frac{P_d}{2} - \delta_{ij}\right)^{10/9} - \delta_{ij}^{10/9} - \frac{F_c}{K} = 0 \tag{3.103}$$

Equation 3.101 and Equations 3.103 represent a system of simultaneous nonlinear equations with unknowns δ_r and δ_{ij}. These equations may be solved for δ_r and δ_{ij} using the Newton–Raphson method. Having calculated δ_r and δ_{ij}, it is possible to calculate roller loads as follows:

$$Q_{ij} = K\delta_{ij}^{10/9} \tag{3.104}$$

$$Q_{oj} = K\delta_{ij}^{10/9} + F_c \tag{3.105}$$

Roller centrifugal force can be calculated using Equation 3.38.

These equations apply to roller bearings with line or modified line contact. Fully crowned rollers or crowned raceways may cause point contact, in which case K_i is different from K_o. These values may be determined using Equation 3.106 also given in Chapter 7 of the first volume of this handbook.

$$K_p = 2.15 \times 10^5 (\Sigma\rho)^{-1/2} (\delta^*)^{-3/2} \tag{3.106}$$

Information on high-speed roller bearings that have flexibly supported rings is given by Harris [3].

See Example 3.5.

Figure 3.19 illustrates the results of the analysis for a 209 cylindrical roller bearing with zero mounted radial clearance and subjected to applied radial load. Figure 3.20 shows the variation of bearing deflection δ_r with speed.

3.4.1 HOLLOW ROLLERS

Rollers can be made hollow to reduce roller centrifugal forces. Hollow rollers are flexible and great care must be exercised to assure that accuracy of shaft location under the applied load is maintained. Roller centrifugal force as a function of hollowness ratio D_i/D is given by

$$F_c = 3.39 \times 10^{-11} D^2 l d_m n_m^2 (1 - H^2) \tag{3.107}$$

Figure 3.21 taken from Ref. [4] shows the effect of roller hollowness in a high-speed cylindrical roller bearing on bearing radial deflection.

For the same bearing, Figure 3.22 illustrates the internal load distribution.

An added criterion for evaluation in a bearing with hollow rollers is the roller bending stress. Figure 3.23 shows the effect of roller hollowness on maximum roller bending stress. Practical limits for roller hollowness are indicated.

Great care must be given to the smooth finishing of the inside surface of a hollow roller during manufacturing as the stress raisers that occur due to a poorly finished inside surface will reduce the allowable roller hollowness ratios still further than indicated in Figure 3.22.

Lightweight rollers made from a ceramic material such as silicon nitride appear feasible to reduce roller centrifugal forces.

3.5 HIGH-SPEED TAPERED AND SPHERICAL ROLLER BEARINGS

Using digital computation and methods similar to those indicated in Chapter 1, the load distribution in other types of high-speed roller bearings can be analyzed. Harris [5] indicates all of the necessary equations. The forces acting on a generalized roller are shown in Figure 3.24. In this case, roller gyroscopic moment is given by

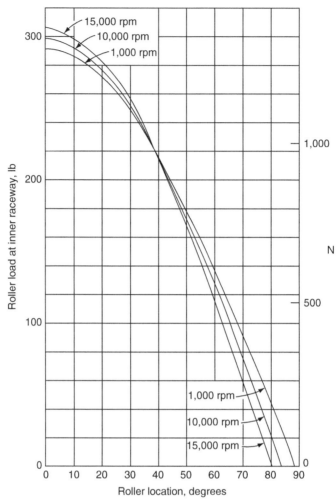

FIGURE 3.19 Distribution of load among the rollers of a 209 cylindrical roller bearing with $P_d = 0$; $F_r = 4450\,N$ (1000 lb); and operating at 1,000, 10,000, and 15,000 rpm shaft speed.

$$M_{gj} = J\omega_{mj}\omega_{Rj}\ \sin\left[\tfrac{1}{2}(\alpha_i + \alpha_o)\right] \tag{3.108}$$

3.6 FIVE DEGREES OF FREEDOM IN LOADING

Until this point, all load distribution calculation methods have been limited to, at most, three degrees of freedom in loading. This has been done in the interest of simplifying the analytical methods and the understanding thereof. Every rolling bearing applied load situation can be analyzed using a system with five degrees of freedom, considering only the applied loading. Then every specialized applied loading condition, for example, simple radial load, can be analyzed using this more complex system. Reference [5] shows the following illustrations that apply to an analytical system for a ball bearing with five degrees of freedom in applied loading (see Figure 3.25).

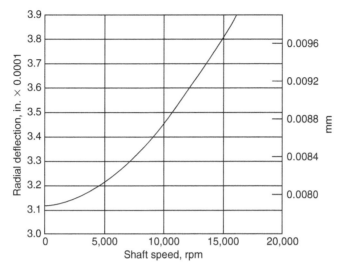

FIGURE 3.20 Radial deflection vs. speed for a 209 cylindrical roller bearing with $P_d = 0$ and $F_r = 4450$ N (1000 lb).

Note the numerical notation of applied loads, that is, $F_1, \ldots, F_5$, in lieu of F_a, F_r, and M. Figure 3.26 shows the contact angles, deformations, and displacements for the ball–raceway contacts at azimuth ψ_j. Figure 3.27 shows the ball speed vectors and inertial loading for a ball with its center at azimuth ψ_j. Note the numerical notations for raceways; $1 = o$ and $2 = i$. This is done for ease of digital programming.

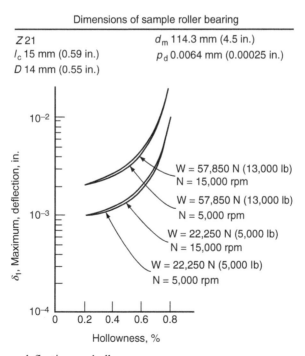

FIGURE 3.21 Maximum deflection vs. hollowness.

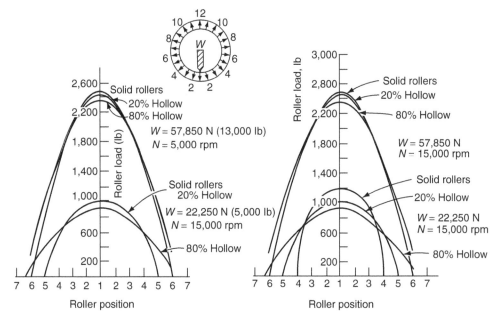

FIGURE 3.22 Roller load distribution vs. applied load, shaft speed, and hollowness.

3.7 CLOSURE

As demonstrated in the earlier discussion, analysis of the performance of high-speed roller bearings is complex and requires a computer to obtain numerical results. The complexity can

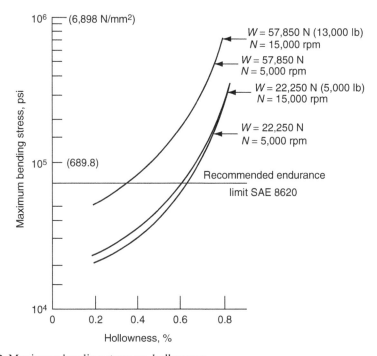

FIGURE 3.23 Maximum bending stress vs. hollowness.

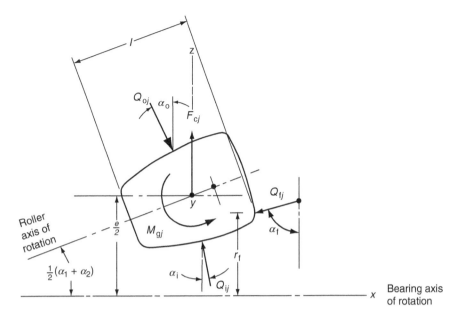

FIGURE 3.24 Roller forces and geometry.

become even greater for ball bearings. In this chapter as well as Chapter 1 and Chapter 2, for simplicity of explanation, most illustrations are confined to situations involving symmetry of loading about an axis in the radial plane of the bearing and passing through the bearing axis of rotation. The more general and complex applied loading system with five degrees of freedom is, however, discussed.

The effect of lubrication has also been neglected in this discussion. For ball bearings, it has been assumed that gyroscopic pivotal motion is minimal and can be neglected. This, of

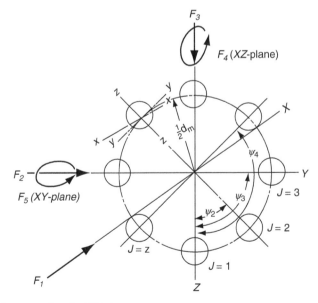

FIGURE 3.25 Bearing operating in YZ plane.

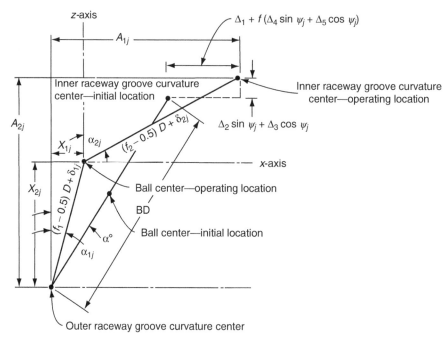

FIGURE 3.26 Contact angle, deformation, and displacement geometry.

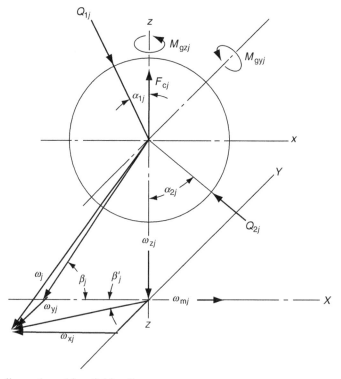

FIGURE 3.27 Ball speeds and inertial loading.

course, depends on the friction forces in the contact zones, which are affected to a great extent by lubrication. Bearing skidding is also a function of lubrication at high speeds of operation. If the bearing skids, centrifugal forces will be lower in magnitude and the performance will accordingly be different.

Notwithstanding the preceding conditions, the analytical methods presented in this chapter are extremely useful in establishing optimum bearing designs for given high-speed applications.

REFERENCES

1. Jones, A., General theory for elastically constrained ball and roller bearings under arbitrary load and speed conditions, *ASME Trans., J. Basic Eng.*, 82, 1960.
2. Harris, T., On the effectiveness of hollow balls in high-speed thrust bearings, *ASLE Trans.*, 11, 209–214, 1968.
3. Harris, T., Optimizing the fatigue life of flexibly mounted rolling bearings, *Lub. Eng.*, 420–428, October 1965.
4. Harris, T. and Aaronson, S., An analytical investigation of cylindrical roller bearings having annular rollers, *ASLE Preprint No.* 66LC-26, October 18, 1966.
5. Harris, T. and Mindel, M., Rolling element bearing dynamics, *Wear*, 23(3), 311–337, February 1973.

4 Lubricant Films in Rolling Element–Raceway Contacts

LIST OF SYMBOLS

Symbol	Description	Units
a	Semimajor axis of elliptical contact area	mm (in.)
a	Thermal expansivity	$°C^{-1}$
A	Viscosity–temperature calculation constants	
b	Semiwidth of rectangular contact area, semiminor axis of elliptical contact area	mm (in.)
B	Doolittle parameter	
C	Lubrication regime and film thickness calculation constants	
D	Roller or ball diameter	mm (in.)
d_m	Pitch diameter of bearing	mm (in.)
E	Modulus of elasticity	MPa (psi)
E'	$E/(1 - \xi^2)$	MPa (psi)
F	Force	N (lb)
F_a	Centrifugal force	N (lb)
$\bar{F}$	$F/E'\Re$	
g	Gravitational constant	mm/sec^2 ($in./sec^2$)
$\mathcal{G}$	$\lambda E'$	
G	Shear modulus	MPa (psi)
h	Lubricant film thickness	mm (in.)
h^0	Minimum lubricant film thickness	mm (in.)
H	$h/\Re$	
I	Viscous stress integral	
J	Polar moment of inertia per unit length	$N \cdot sec^2$ ($lb \cdot sec^2$)
$\bar{J}$	$J/E'\Re_i$	$mm \cdot sec^2$ ($in \cdot sec^2$)
k_b	Lubricant thermal conductivity	W/m · °C (Btu/hr · in. · °F)
K_0	Bulk modulus parameter	Pa · °K
K_∞	Bulk modulus parameter	Pa
l	Roller effective length	mm (in.)
L	Factor for calculating film thickness reduction due to thermal effects	
M	Moment	N · mm (in. · lb)
n	Speed	rpm
p	Pressure	MPa (psi)
Q	Force acting on roller or ball	N (lb)
$\bar{Q}$	$Q/E'\Re$	

r	relative occupied volume expansion factor	
R	relative occupied volume	m^3
R$_o$	relative occupied volume at 20°C	m^3
R	Cylinder radius	mm (in.)
$\Re$	Equivalent radius	mm (in.)
s	rms surface finish (height)	mm (in.)
SSU	Saybolt university viscosity	sec
t	Time	sec
T	Lubricant temperature	°C,°K (°F, °R)
u	Fluid velocity	mm/sec (in./sec)
U	Entrainment velocity ($U_1 - U_1$)	mm/sec (in./sec)
$\overline{U}$	$\eta_0 U/2E'\Re$	
v	Fluid velocity, displacement in y direction	mm/sec, mm (in./sec, in.)
V	Volume	mm^3
V$_o$	Volume at 20°C	mm^3
V	Sliding velocity ($U_1 - U_1$)	mm/sec (in./sec)
$\overline{V}$	$\eta_0 V/E'\Re$	
w	Deformation in z direction	mm (in.)
y	Distance in y direction	mm (in.)
z	Distance in z direction	mm (in.)
β'	Coefficient for calculating viscosity as a function of temperature	
γ	($D \cos \alpha)/d_m$	
$\dot{\gamma}$	Lubricant shear rate	sec^{-1}
ε	Strain	mm/mm (in./in.)
ε	occupied volume expansivity	°C^{-1}
η	Lubricant viscosity	cp (lb · sec/in.^{2})
η_b	Base oil viscosity (grease)	cp (lb · sec/in.^{2})
η_{eff}	Effective viscosity (grease)	cp (lb · sec/in.^{2})
η_0	Fluid viscosity at atmospheric pressure	cp (lb · sec/in.^{2})
κ	Ellipticity ratio a/b	
λ	Pressure coefficient of viscosity	mm^2/N (in.^{2}/lb)
Λ	Lubricant film parameter	
v_b	Kinematic viscosity	stokes (cm^2/sec)
ξ	Poisson's ratio	
ρ	Weight density	g/mm^3/(lb/in.^{3})
σ	Normal stress	MPa (psi)
τ	Shear stress	MPa (psi)
θ	Angle	rad
$\overline{Y}$	Factor to calculate φ_{TS}	
φ	Film thickness reduction factor	
Φ	Factor to calculate φ_S	
ψ	Angular location of roller	rad
ω	Rotational speed	rad/sec

Subscripts

b	Entrance to contact zone
e	Exit from contact zone
G	Grease

i	Inner raceway film
j	Roller location
m	Orbital motion
NN	Non-Newtonian lubricant
o	Outer raceway film
R	Roller
S	Lubricant starvation
SF	Surface roughness (finish)
T	Temperature
TS	Temperature and lubricant starvation
x	*x* Direction, that is, transverse to rolling
y	*y* Direction, that is, direction of rolling
z	*z* Direction
μ	Rotating raceway
v	Nonrotating raceway
0	Minimum lubricant film
1, 2	Contacting bodies

4.1 GENERAL

Ball and roller bearings require fluid lubrication if they are to perform satisfactorily for long periods of time. Although modern rolling bearings in extreme temperature, pressure, and vacuum environment aerospace applications have been adequately protected by dry film lubricants, such as molybdenum disulfide among many others, these bearings have not been subjected to severe demands regarding heavy load and longevity of operation without fatigue. It is further recognized that in the absence of a high-temperature environment only a small amount of lubricant is required for excellent performance. Thus, many rolling bearings can be packed with greases containing only small amounts of oil and then be mechanically sealed to retain the lubricant. Such rolling bearings usually perform their required functions for indefinitely long periods of time. Bearings that are lubricated with excessive quantities of oil or grease tend to overheat and burn up.

The mechanism of the lubrication of rolling elements operating in concentrated contact with a raceway was not established mathematically until the late 1940s; it was not proven experimentally until the early 1960s. This is to be compared with the existence of hydrodynamic lubrication in journal bearings, which was established by Reynolds in the 1880s. It is known, for instance, that a fluid film completely separates the bearing surface from the journal or slider surface in a properly designed bearing. Moreover, the lubricant can be oil, water, gas, or some other fluid that exhibits adequate viscous properties for the intended application. In rolling bearings, however, it was only relatively recently established that fluid films could, in fact, separate rolling surfaces subjected to extremely high pressures in the zones of contact. Today, the existence of lubricating fluid films in rolling bearings is substantiated in many successful applications where these films are effective in completely separating the rolling surfaces. In this chapter, methods will be presented for the calculation of the thickness of lubricating films in rolling bearing applications.

4.2 HYDRODYNAMIC LUBRICATION

4.2.1 REYNOLDS EQUATION

Because it appeared possible that lubricant films of significant proportions do occur in the contact zones between rolling elements and raceways under certain conditions of load and speed, several investigators have examined the hydrodynamic action of lubricants on rolling

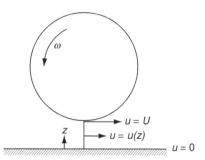

FIGURE 4.1 Cylinder rolling on a plane with lubricant between cylinder and plane.

bearings according to classical hydrodynamic theory. Martin [1] presented a solution for rigid rolling cylinders as early as 1916. In 1959, Osterle [2] considered the hydrodynamic lubrication of a roller bearing assembly.

It is of interest at this stage to examine the mechanism of hydrodynamic lubrication at least in two dimensions. Accordingly, consider an infinitely long roller rolling on an infinite plane and lubricated by an incompressible isoviscous Newtonian fluid with viscosity η. For a Newtonian fluid, the shear stress τ at any point obeys the relationship

$$\tau = \eta \frac{\partial u}{\partial z} \tag{4.1}$$

where $\partial u/\partial z$ is the local fluid velocity gradient in the z direction (see Figure 4.1). Because the fluid is viscous, fluid inertia forces are small compared with the viscous fluid forces. Hence, a particle of fluid is subjected only to fluid pressure and shear stresses as shown in Figure 4.2.

Noting the stresses of Figure 4.2 and recognizing that static equilibrium exists, the sum of the forces in any direction must equal zero. Therefore,

$$\sum F_y = 0$$

$$p\,dz \;-\; \left(p + \frac{\partial p}{\partial y}\right) dz + \tau\,dy - \left(\tau + \frac{\partial \tau}{\partial z}\right) dy = 0$$

and

$$\frac{\partial p}{\partial y} = -\frac{\partial \tau}{\partial z} \tag{4.2}$$

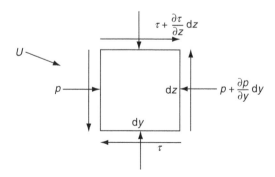

FIGURE 4.2 Stresses on a fluid particle in a two-dimensional flow field.

Differentiating Equation 4.1 once with respect to z yields

$$\frac{\partial \tau}{\partial z} = -\eta \frac{\partial^2 u}{\partial z^2}$$ (4.3)

Substituting Equation 4.3 into Equation 4.2, one obtains

$$\frac{\partial p}{\partial y} = \eta \frac{\partial^2 u}{\partial z^2}$$ (4.4)

Assuming for the moment that $\partial p/\partial y$ is constant, Equation 4.4 may be integrated twice with respect to z. This procedure gives the following expression for local fluid velocity u:

$$u = \frac{1}{2\eta} \frac{\partial p}{\partial y} z^2 + c_1 z + c_2$$ (4.5)

The velocity U may be ascribed to the fluid adjacent to the plane that translates relative to a roller. At a point on the opposing surface, it is proper to assume that $u = 0$, that is, at $z = 0$, $u = U$ and at $z = h$, $u = 0$. Substituting these boundary conditions into Equation 4.5, it can be determined that

$$u = \frac{1}{2\eta} \frac{\partial p}{\partial y} z(z - h) + U\left(1 - \frac{z}{h}\right)$$ (4.6)

where h is the film thickness.

Considering the fluid velocities surrounding the fluid particle as shown in Figure 4.3, one can apply the law of continuity of flow in steady state, that is, mass influx equals mass efflux. Hence, as density is constant for an incompressible fluid

$$u\,dz - \left(u + \frac{\partial u}{\partial y} dy\right) dz + v\,dy - \left(v + \frac{\partial v}{\partial z} dz\right) dy = 0$$ (4.7)

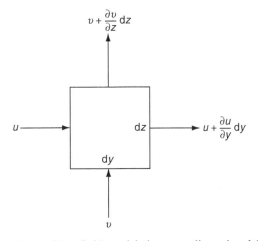

FIGURE 4.3 Velocities associated with a fluid particle in a two-dimensional flow field.

Therefore,

$$\frac{\partial u}{\partial y} = -\frac{\partial v}{\partial z} \qquad (4.8)$$

Differentiating Equation 4.6 with respect to y and equating this to Equation 4.8 yields

$$\frac{\partial v}{\partial z} = -\frac{\partial}{\partial y}\left[\frac{1}{2\eta}\frac{\partial p}{\partial y}z(z-h) + U\left(1 - \frac{z}{h}\right)\right] \qquad (4.9)$$

Integrating Equation 4.9 with respect to z gives

$$\int \frac{\partial v}{\partial z}dz = -\int_0^h dv = 0 = \int_0^h \frac{\partial}{\partial y}\left[\frac{1}{2\eta}\frac{\partial p}{\partial y}z(z-h) + U\left(1 - \frac{z}{h}\right)\right]dz \qquad (4.10)$$

and

$$\frac{\partial}{\partial y}\left(h^2 \frac{\partial p}{\partial y}\right) = 6\eta U \frac{\partial h}{\partial y} \qquad (4.11)$$

Equation 4.11 is commonly called the Reynolds equation in two dimensions.

4.2.2 FILM THICKNESS

To solve the Reynolds equation, it is only necessary to evaluate film thickness as a function of y, that is, $h = h(y)$. For a cylindrical roller near a plane as shown in Figure 4.4, it can be seen that

$$h = h^0 + \frac{y^2}{2R} \qquad (4.12)$$

where h^0 is the minimum lubricant film thickness. Substituting Equation 4.12 into Equation 4.11 gives

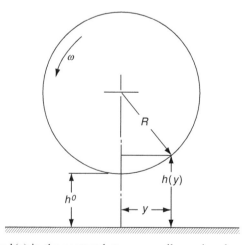

FIGURE 4.4 Film thickness $h(y)$ in the contact between a roller and a plane.

$$\frac{\partial}{\partial y}\left[\left(h^0 + \frac{y^2}{2R}\right)^3 \frac{\partial p}{\partial y}\right] = \frac{6\eta U y}{R} \qquad (4.13)$$

Equation 4.13 varies only in y; hence,

$$\frac{\mathrm{d}}{\mathrm{d}y}\left[\left(h^0 + \frac{y^2}{2R}\right)^3 \frac{\mathrm{d}p}{\mathrm{d}y}\right] = \frac{6\eta U y}{R} \qquad (4.14)$$

4.2.3 LOAD SUPPORTED BY THE LUBRICANT FILM

Integration of Equation 4.14 yields pressure over the lubricant film as a function of distance y. If both contact surfaces are considered portions of rotating cylinders, then

$$U = U_1 + U_2 \qquad (4.15)$$

where subscripts 1 and 2 refer to the respective cylinders. Moreover, an equivalent radius $\Re$ is defined as

$$\Re = (R_1^{-1} + R_2^{-1})^{-1} \qquad (4.16)$$

Note that for an outer raceway R^{-1} is negative. The load per unit axial length of contact carried by the lubricant film is given by

$$q = \int p(y)\,\mathrm{d}y \qquad (4.17)$$

Considering hydrodynamic lubrication with a constant viscosity (isoviscous) fluid permits the solution of these equations for relatively lightly loaded contacts such as those that occur in fluid-lubricated journal bearings.

4.3 ISOTHERMAL ELASTOHYDRODYNAMIC LUBRICATION

4.3.1 VISCOSITY VARIATION WITH PRESSURE

The normal pressure between contacting rolling bodies in ball and roller bearings tends to be of magnitude 700 MPa (100,000 psi) and higher. Figure 4.5 shows some experimental data on viscosity variation with pressure for a few bearing lubricants. It is seen that, at a given temperature, viscosity is an exponential function of pressure. Therefore, between the contacting surfaces in a normal rolling bearing application, lubricant viscosity can be several orders of magnitude higher than its value at atmospheric pressure.

In 1893, Barus [3] established an empirical equation for the variation of viscosity with pressure, an isothermal relationship. The Barus equation is

$$\eta = \eta_0 e^{\lambda p} \qquad (4.18)$$

In Equation 4.18, λ, the pressure–viscosity coefficient, is a constant at a given temperature. In 1953, an ASME [4] study published viscosity vs. pressure curves for various fluid lubricants. On the basis of the ASME data, it is apparent that the Barus equation is a crude approximation

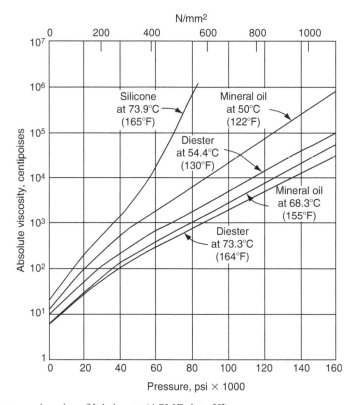

FIGURE 4.5 Pressure viscosity of lubricants (ASME data [5]).

because the pressure–viscosity coefficient decreases with both pressure and temperature for most fluid lubricants. The lubricant film thickness obtained in a concentrated contact has been established as a function of the viscosity of the lubricant entering the contact. Therefore, for the purpose of determining the thickness of the lubricant film, the viscosity–pressure coefficient at atmospheric pressure is utilized.

Roelands [5] later established an equation defining the viscosity–pressure relationship for given fluids; however, including the influence of temperature on viscosity as well:

$$\frac{\log_{10} \eta + 1.2}{\log_{10} \eta_0 + 1.2} = \left(\frac{T_0 + 135}{T + 135}\right)^{S_0} \left(1 + \frac{p}{2000}\right)^z \qquad (4.19)$$

In Equation 4.19, pressure is expressed in kgf/cm^2 and temperature in °K; exponents S_0 and z are determined empirically for each lubricant. At high pressures, Equation 4.19 indicates viscosities substantially lower than those produced using the Barus Equation 4.18.

Sorab and VanArsdale [6] developed an expression for viscosity vs. pressure and temperature, which can be applied to several of the lubricants employed in the ASME [4] study:

$$\ln\frac{\eta}{\eta_0} = A_1\left(\frac{p}{p_0} - 1\right) + A_2\left(\frac{T_0}{T} - 1\right) + A_3\left(\frac{p}{p_0} - 1\right)^2 + A_4\left(\frac{T_0}{T} - 1\right)^2$$
$$+ A_5\left(\frac{p}{p_0} - 1\right)\left(\frac{T_0}{T} - 1\right) \qquad (4.20)$$

In Equation 4.20, temperature is stated in °K. Ref. [6] provides values of the coefficients A_i for the various lubricants tested in Ref. [4]. As an example, the coefficients for the diester fluid viscosity vs. pressure curve of Figure 4.5 are

A_1 1.48×10^{-3}
A_2 11.78
A_3 -7.7×10^{-8}
A_4 14.31
A_5 2.17×10^{-3}

This fluid may be considered representative of an aircraft power transmission fluid lubricant. Sorab and VanArsdale [6] demonstrate that Equation 4.20 is superior to the Roelands equation in approximating the ASME viscosity–pressure–temperature data. Nevertheless each of the approximations has only been demonstrated over the 0–1034 MPa (0–150 kpsi) pressure range and 25–218°C (77–425°F) temperature range of the ASME data. Contact pressures and temperatures in many ball and roller bearing applications are apt to exceed these ranges; therefore, it becomes necessary to extrapolate these data substantially beyond the range of the experimentation. This is not critical for the determination of lubricant film thicknesses. In the estimation of bearing friction, however, lubricant viscosity at pressures higher than 1034 MPa and at temperatures greater than 218°C has a great influence on the magnitudes of friction forces calculated and hence on the accuracy of the calculations.

Bair and Kottke [7], based on experimental studies of lubricants at high pressures (for example, up to 2000 MPa), developed the following equation to describe absolute viscosity as a function of pressure and temperature:

$$\eta = \eta_0 \exp\left[B\left(\frac{R_0 r}{V/V_0 - R_0 r} - \frac{R_0}{1 - R} \right) \right] \tag{4.21}$$

where η_0 is the viscosity at atmospheric pressure and 20°C. Parameter R_0, relative occupied volume at 20°C, and B according to Doolittle [8] are given in Table 4.1.

The occupied volume, assumed to vary linearly with temperature, is given by

$$r = 1 + \varepsilon(T - T_0) \tag{4.22}$$

where ε is the occupied volume expansivity; it tends to be negative. The variation of volume with pressure and temperature is determined from

$$\frac{V}{V_0} = [1 + a(T - T_0)]\left\{ 1 - \frac{1}{1 + K_0'} \ln\left[1 + \frac{p}{K_0}(1 + K_0') \right] \right\} \tag{4.23}$$

TABLE 4.1
Doolittle–Tait Parameters for $T_0 = 20°C$

Lubricant	η_0 (Pa · sec)	a (1/C · 10^{-4})	ε (1/C · 10^{-3})	R_0	B	K_0	K_∞ (GPa)	K_0 (GPa · °K)
SAE 20	0.1089	8	−1.034	0.6980	3.520	10.40	−0.9282	580.7
PAO ISO 68	0.0819	8	−1.035	0.6622	3.966	11.38	−0.9881	580.8
Mil-L-23699	0.04667	7.42	−1.28	0.6641	3.382	10.741	−1.0149	570.8

where a is the thermal expansivity, K_0' is the assumed constant, and the bulk modulus varies with temperature according to

$$K_0 = K_\infty + \frac{K_0}{T} \qquad (4.24)$$

where T is in °K. Equation 4.21 through Equation 4.24 tend to give better predictions of viscosity at elevated pressures than does Roelands [5]; however, they still tend to predict viscosities higher than that experienced in ball and roller bearing applications.

Harris [9] introduced the use of a sigmoid curve as defined by Equation 4.25 to fit the ASME [4] data.

$$\eta = C_1 + \frac{C_2}{1 + e^{-(p-C_3)/C_4}} \qquad (4.25)$$

In Equation 4.25, $C_1, \ldots, C_4$ are constants determined from the curve-fitting procedure for a given lubricant at a given temperature. Figure 4.6 illustrates the sigmoid curves for the ASME data for a Mil-L-7808 ester-type lubricant at 37.8, 98.9, and 218.3°C (100, 210, and 425°F). The salient feature of the sigmoid viscosity vs. pressure curve is the virtually constant viscosity value at extremely high pressures. As noted by Bair and Winer [10,11], the fluid in a high-pressure, concentrated contact undergoes transformation to a glassy state; that is, the fluid essentially becomes a solid during its time in the contact. It therefore appears reasonable to assume that fluid viscosity becomes essentially constant with pressure during the fluid's time in the contact. To accurately predict bearing friction torque, this becomes an important consideration for the use of a sigmoid curve to describe lubricant viscosity in the contact. Conversely, using a sigmoid curve to approximate lubricant viscosity at atmospheric and low pressures does not provide the accuracy of either the Roelands [5] or Sorab and VanArsdale

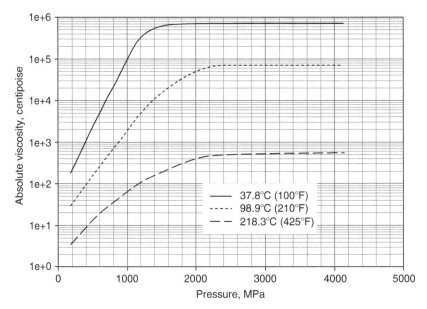

FIGURE 4.6 Viscosity vs. pressure and temperature for Mil-L-7808 ester-type lubricant (sigmoid curve fit to ASME data [4]—extrapolation from 1000 to 4000 MPa). (ASME Research Committee on Lubrication, Pressure–viscosity report—Vol. 11, ASME, 1953.)

[6] model. Either of these models may be used in the estimation of lubricant viscosity to calculate lubricant film thickness.

4.3.2 DEFORMATION OF CONTACT SURFACES

Because of the fluid pressures present between contacting rolling bodies causing the increases in viscosity noted in Figure 4.5, the rolling surfaces deform appreciably in proportion to the thickness of a fluid film between the surfaces. The combination of the deformable surface with the hydrodynamic lubricating action constitutes the "elastohydrodynamic" (EHD) problem. The solution of this problem established the first feasible analytical means of estimating the thickness of fluid films, the local pressures, and the tractive forces that occur in rolling bearings.

Dowson and Higginson [12], for the model in Figure 4.7, used the following formulation for film thickness at any point in the contact:

$$h = h^0 + \frac{y^2}{2R_1} + \frac{y^2}{2R_2} + w_1 + w_2 \tag{4.26}$$

Solid displacements w are calculated for a semi-infinite solid in a condition of plane strain. As the width of the loaded zone is extremely small compared with the dimensions of the contacting bodies, an approximation that $w_1 = w_2$ is valid. Hence, for the equivalent cylinder radius,

$$\Re = (R_1^{-1} + R_2^{-1})^{-1} \tag{4.16}$$

and the film thickness is given by

$$h = h^0 + \frac{y^2}{2}R + w \tag{4.27}$$

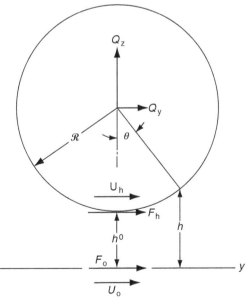

FIGURE 4.7 Forces and velocities pertaining to the equivalent roller.

To solve the plane strain problem, the following stress function was assumed:

$$\Phi = -\frac{Q}{\pi} y \, \tan^{-1} \frac{y}{z} \tag{4.28}$$

Using this stress function, the stresses due to a narrow strip of pressure over the width $\mathrm{d}s$ in the y direction are determined as follows:

$$\sigma_y = -\frac{2y^2 zp \, \mathrm{d}s}{\pi(y^2 + z^2)^2} \tag{4.29}$$

$$\sigma_z = -\frac{2z^3 p \, \mathrm{d}s}{\pi(y^2 + z^2)^2} \tag{4.30}$$

$$\tau_{yz} = -\frac{2yz^2 p \, \mathrm{d}s}{\pi(y^2 + z^2)^2} \tag{4.31}$$

σ_y and σ_z are normal stresses and τ_{yz} is the shear stress. By Hooke's law, the strains are given by

$$\varepsilon_y = \frac{(1 - \xi^2)\sigma_y}{E} - \frac{\xi(1 + \xi)\sigma_z}{E} \tag{4.32}$$

$$\varepsilon_z = \frac{(1 - \xi^2)\sigma_z}{E} - \frac{\xi(1 + \xi)\sigma_y}{E} \tag{4.33}$$

$$\varepsilon_{yz} = \frac{2(1 + \xi)\tau_{yz}}{E} = \frac{\tau_{yz}}{G} \tag{4.34}$$

where G is the shear modulus of elasticity and ξ is Poisson's ratio. In plane strain,

$$\varepsilon_y = \frac{\partial v}{\partial y}, \qquad \varepsilon_z = \frac{\partial w}{\partial z}, \qquad \text{and} \qquad \varepsilon_{yz} = \frac{\partial v}{\partial z} + \frac{\partial w}{\partial y}$$

Using these relationships, and Equation 4.29 through Equation 4.34, it can be established that at the surface, that is, at $z = 0$,

$$w = -\frac{2(1 - \xi^2)}{\pi E} \int_{S_1}^{S_2} p \, \ln \, (y - S) \, \mathrm{d}S + \text{constant} \tag{4.35}$$

To solve for w, Dowson and Higginson [12] divided the pressure curve into segments and represented the pressure thereunder by

$$p = \zeta_1 + \zeta_2 S + \zeta_3 S^2 \tag{4.36}$$

where ζ_1, ζ_2, and ζ_3 are constants for that segment. Using p in this form, Equation 4.35 can be integrated to obtain the surface deformation. This procedure, of course, is used for an assumed pressure distribution.

To obtain h^0, the Reynolds equation is used in accordance with the pressure variation of viscosity:

$$\frac{d}{dy}\left(h^2 e^{-\lambda p}\frac{dp}{dy}\right) = 6\eta_0 U \frac{dh}{dy} \tag{4.37}$$

Performing the indicated differentiation and rearranging yields

$$h^3 e^{-\lambda p}\left[\frac{d^2 p}{dy^2} - \lambda\left(\frac{dp}{dy}\right)^2\right] + \frac{dh}{dy}\left(6u\eta_0 + 3h^2 e^{-\lambda p}\frac{dp}{dy}\right) = 0 \tag{4.38}$$

At the inlet and at the outlet of the contact,

$$\frac{d^2 p}{dy^2} - \lambda\left(\frac{dp}{dy}\right)^2 = 0 \tag{4.39}$$

such that Equation 4.38 becomes

$$\frac{dh}{dy}\left(6U\eta_0 + 3h^2 e^{-\lambda p}\frac{dp}{dy}\right) = 0 \tag{4.40}$$

At the outlet end of the pressure curve, $dh/dy = 0$. This condition applies to the point of minimum film thickness. At the inlet, Equation 4.40 is solved by

$$\frac{dp}{dy} = -\frac{2\eta_0 e^{\lambda p} U}{h^2} \tag{4.41}$$

Thus, if viscosity and speed are known, the value of h for the point at which Equation 4.40 is satisfied in the inlet region can be evaluated for a given pressure curve. Solving Equation 4.41 for h_b (at inlet) gives

$$h_b = \left[-\frac{2\eta_0 e^{\lambda p} U}{(dp/dy)_b}\right]^{1/2} \tag{4.42}$$

Once h_b has been determined, the entire film shape can be estimated by using the integrated form of the Reynolds equation, that is,

$$\frac{dp}{dy} = -6\eta_0 e^{\lambda p} U \left(\frac{1}{h^2} - \frac{h_e}{h^3}\right) \tag{4.43}$$

Substitution of dp/dy from Equation 4.41 for the point at which $h = h_b$ determines that $h_e = 2h_b/3$. At other positions y, film thickness h may be determined from the following cubic equation developed from Equation 4.43:

$$\frac{dp/dy}{6\eta_0 e^{\lambda p} U}h^3 + h - h_e = 0 \tag{4.44}$$

At the point of maximum pressure, $dp/dy = 0$ and Equation 4.38 becomes

$$\frac{\mathrm{d}h}{\mathrm{d}y} = -\frac{h^3}{6\eta_0 e^{\lambda p} U} \frac{\mathrm{d}^2 p}{\mathrm{d}y^2} \tag{4.45}$$

In cases of most interest, the pressure curve is predominantly Hertzian such that

$$p = p_0 \left[1 - \left(\frac{y}{b}\right)^2\right]^{1/2} \tag{4.46}$$

where p_0 is the maximum pressure and b is the semiwidth of the contact zone. Thus, at $y = 0$, $p = p_0$, Equation 4.45 becomes

$$\frac{\mathrm{d}h}{\mathrm{d}y} = \frac{p_0 h^3}{3\eta_0 e^{\lambda p_0} U b^2} \tag{4.47}$$

Consequently, if h is small (as it must be in a rolling bearing under load) and the viscosity is high (as it will become because of high pressure), $\mathrm{d}h/\mathrm{d}y$ is very small and the film is essentially of uniform thickness. This result is shown by Dowson and Higginson [12], and also by Grubin [13].

4.3.3 PRESSURE AND STRESS DISTRIBUTION

In a later presentation Dowson and Higginson [12] and Grubin [13] indicated that dimensionless film thickness $H = h/\Re$ could be expressed as follows:

$$H = f\left(\bar{Q}_z, \bar{U}, \mathcal{G}\right) \tag{4.48}$$

where

$$\bar{Q} = \frac{Q_z}{lE'\Re} \tag{4.49}$$

$$\bar{U} = \frac{\eta_0 U}{2E'\Re} \tag{4.50}$$

$$\mathcal{G} = \lambda E' \tag{4.51}$$

$$E' = \frac{E}{1 - \xi^2} \tag{4.52}$$

In the expression for H and in Equation 4.47 and Equation 4.48, the equivalent radius in the direction of rolling for a ball or roller bearing is given by

$$\Re_m = \frac{D}{2}(1 \mp \gamma) \tag{4.53}$$

In Equation 4.53, the upper sign refers to the inner raceway contact and the lower sign to the outer raceway contact. The velocities with which fluid is swept into the rolling element–raceway contacts are given by Equation 4.54 and Equation 4.55 for the inner and outer raceway contacts, respectively.

$$U_i = \frac{d_m}{2}\left[(1 - \gamma)(\omega - \omega_m) + \gamma\omega_R\right] \tag{4.54}$$

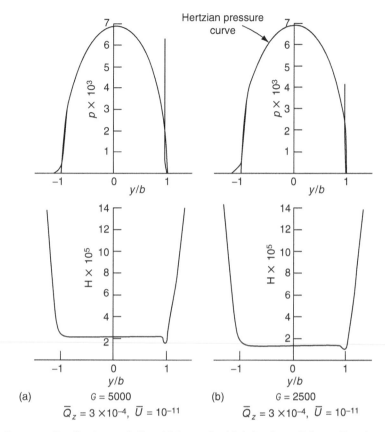

FIGURE 4.8 Pressure distribution and film thickness for high-load conditions. (Reprinted from Dowson, D. and Higginson, G., *J. Mech. Eng. Sci.*, 2(3), 1960. With permission.)

$$U_o = \frac{d_m}{2}[(1 + \gamma)\omega_m + \gamma\omega_R] \qquad (4.55)$$

Dowson and Higginson [14] presented the results shown in Figure 4.8 and Figure 4.9 for $\mathcal{G} = 2500$ and 5000 corresponding to bronze rollers and steel rollers, respectively, lubricated by a mineral oil. The load $\bar{Q}_z = 0.00003$ corresponds approximately to 483 MPa (70,000 psi) and $\bar{Q}_z = 0.0003$ corresponds approximately to 1380 MPa (200,000 psi). Dimensionless speed $\bar{U} = 10^{-11}$ corresponds to surface velocities in the order of 1524 mm/sec (5 ft/sec) for an equivalent roller radius of 25.4 mm (1 in.) operating in mineral oil.

Note from Figure 4.8 and Figure 4.9 that the departure from the Hertzian pressure distribution is less significant as the load increases. The second pressure peak at the outlet end of the contact corresponds to a local decrease in the film thickness at that point. Otherwise, the film is essentially of uniform thickness. The latter condition was confirmed by tests conduced by Sibley and Orcutt [15].

Additionally, Dowson and Higginson [14] demonstrated the effect of distorted pressure distribution on maximum subsurface shear stress. Figure 4.10 shows contours of τ_{yzmax}/p_{max}. Note that the shear stress increases in the vicinity of the second pressure peak and tends toward the surface.

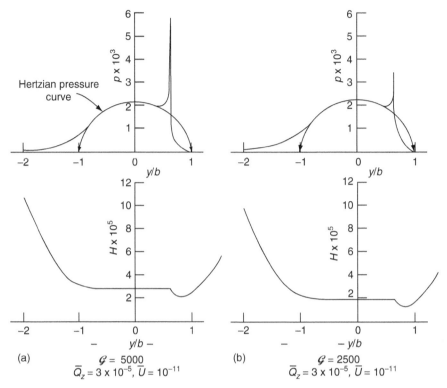

FIGURE 4.9 Pressure distribution and film thickness for light-load conditions. (Reprinted from Dowson, D. and Higginson, G., *J. Mech. Eng. Sci.*, 2(3), 1960. With permission.)

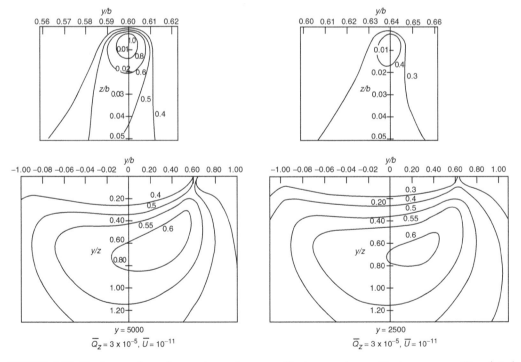

FIGURE 4.10 Contours of maximum shear stress amplitude—maximum Hertz pressure. (Reprinted from *J. Mech. Eng. Sci.*, 2(3). 1960. With permission.)

4.3.4 Lubricant Film Thickness

Grubin [13] developed a formula for minimum film thickness in line contact, that is, the thickness of the lubricant film between the protuberance at the trailing edge of the contact on the equivalent roller surface and the opposing surface of the relative flat. The Grubin formula is based on the assumption that the rolling surfaces deform as if dry contact occurs and is given in a dimensionless format:

$$H^0 = \frac{1.95(\mathcal{G}\bar{U})^{0.727}}{\bar{Q}_z^{0.091}} \tag{4.56}$$

where $H^0 = h^0/\Re_y$.

Based on analytical studies and experimental results, Dowson and Higginson [16] established the following formula to calculate the minimum film thickness:

$$H^0 = \frac{2.65\,\bar{U}^{0.7}\mathcal{G}^{0.54}}{\bar{Q}_z^{0.13}} \tag{4.57}$$

A significant feature of both equations is the relatively large dependency of film thickness on speed and lubricant viscosity and the comparative insensitivity to load. Testing conducted by Sibley and Orcutt [15] using radiation techniques seemed to confirm the Grubin equation; however, the agreement between the Dowson and Grubin formulas is apparent. Today, the Dowson equation is recommended as representative of line contact lubrication conditions.

Equation 4.56 and Equation 4.57 describe the minimum lubricant film thickness. The film thickness at the center of the contact, plateau film thickness, is approximated by

$$H_c = \tfrac{4}{3}H^0 \tag{4.58}$$

Archard and Kirk [17] described the minimum film thickness between two spheres as

$$H^0 = \frac{0.84(\mathcal{G}\bar{U})^{0.741}}{\bar{Q}_z^{0.074}} \tag{4.59}$$

Using a ball–disk test rig with a clear sapphire disk and interferometry, it is possible to obtain photographs of the lubricant film thickness distribution in a moving ball–disk contact. Figure 4.11 shows the horseshoe pattern corresponding to the high-pressure ridge associated with the minimum lubricant film thickness. The central or plateau film thickness is enclosed by the horseshoe.

A more generalized formula for minimum lubricant film thickness in an elliptical area point contact was subsequently developed by Hamrock and Dowson [18]:

$$H^0 = \frac{3.63\,\bar{U}^{0.68}\mathcal{G}^{0.49}\left(1 - e^{-0.68\kappa}\right)}{\bar{Q}_z^{0.073}} \tag{4.60}$$

where $\bar{Q}_z$ for point contact is given by

$$\bar{Q}_z = \frac{Q}{E'\Re^2} \tag{4.61}$$

Sometimes for elliptical point contact, an equivalent line contact load is considered as follows:

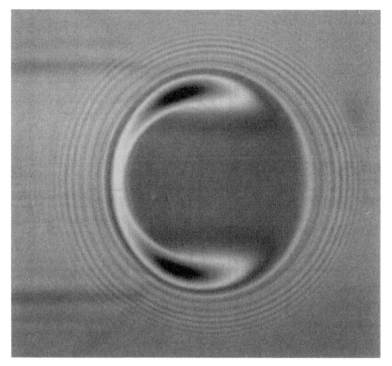

FIGURE 4.11 Photograph of fluid-lubricated steel ball–sapphire disk contact. Interferometric fringes indicate variation of film thickness and hence pressure. (Wedeven,L., Optical Measurements in Elasto-hydrodynamic rolling contact bearings, Ph.D. Thesis, University of London, 1917.)

$$\bar{Q}_{ez} = \frac{3Q}{4E'\Re_y a} \tag{4.62}$$

In Equation 4.60, κ is the ellipticity ratio a/b. The central or plateau lubricant film thickness is given by

$$H^0 = \frac{2.69\,\bar{U}^{0.67}\mathcal{G}^{0.53}\left(1 - 0.61\mathrm{e}^{-0.73\kappa}\right)}{\bar{Q}_z^{0.067}} \tag{4.63}$$

Kotzalas [20] conducted a study of lubricant film formation using both Roelands equation (Equation 4.19) and a fitted sigmoid curve (Equation 4.25) to define lubricant viscosity vs. pressure at a given temperature. He established that the calculated lubricant film thickness distributions are substantially identical irrespective of which of the two models for viscosity vs. pressure is used.

See Example 4.1 and Example 4.2.

4.4   VERY-HIGH-PRESSURE EFFECTS

Maximum Hertz pressures occurring in the rolling element–raceway contacts typically fall in the range of 1000–2000 MPa (approximately 150–300 kpsi); however, in modern bearing applications, particularly endurance tests, it is not unusual for maximum Hertz pressure to reach 4000 MPa. To prevent damage to laboratory test equipment and the materials under test, experiments used to confirm the lubricant film thickness equations provided here have typically been confined to pressures not exceeding 1500 MPa. Venner [22] conducted EHL analyses at high pressures and concluded that lubricant films predicted by the equations, both minimum and central lubricant film thicknesses, are somewhat thinner than calculated by these equations. Using a tungsten carbide ball on a sapphire disk and ultrathin film interferometry and digital techniques, Smeeth and Spikes [23] measured lubricant film thicknesses at maximum Hertz pressures up to 3500 MPa. They confirmed Venner's conclusions, finding that, above contact loading of 2000 MPa, the minimum lubricant film thickness varies inversely as dimensionless load to the 0.3 power as compared with the 0.073 power indicated in Equation 4.60. The data shown by Smeeth and Spikes [23] might further be represented by Equation 4.64 and Equation 4.65:

$$\left(\frac{h_{0hp}}{h_0}\right)^{1/2} = 1.0943 - 4.597 \times 10^{-12} p_{max}^3 \tag{4.64}$$

$$\frac{h_{chp}}{h_{cen}} = 0.8736 - 8.543 \times 10^{-9} p_{max}^2 \tag{4.65}$$

These equations define the ratio of film thickness resulting from very high pressure to that calculated using Equation 4.60 and Equation 4.63 for minimum and central film thicknesses, respectively.

4.5 INLET LUBRICANT FRICTIONAL HEATING EFFECTS

At high bearing operating speeds, some of the frictional heat generated in each concentrated contact is dissipated in the lubricant momentarily residing in the inlet zone of the contact. This effect, examined first by Cheng [24], tends to increase the temperature of the lubricant in the contact. Vogels [25] gives the following expression for viscosity:

$$\eta_b = A_1 e^{\beta'/(T_b + A_2)} \tag{4.66}$$

where T_b is in °C and A_1, A_2, and β' are parameters to be defined for each lubricant. Three temperature–viscosity data points are required to determine A_1, A_2, and β' as follows:

$$A_1 = \eta_1 e^{-\beta'/(T_b + A_2)} \tag{4.67}$$

$$A_2 = \frac{A_3 T_1 - T_3}{1 - A_3} \tag{4.68}$$

$$\beta' = \frac{(T_2 + A_2)(T_1 + A_2)}{(T_2 - T_1)} \ln\left(\frac{\eta_1}{\eta_2}\right) \tag{4.69}$$

$$A_3 = \frac{(T_3 - T_2)}{(T_2 - T_1)} \frac{\ln(\eta_1/\eta_2)}{\ln(\eta_2/\eta_3)} \tag{4.70}$$

If only two temperature–viscosity data points are known and A_2 can be fixed to 273, Equation 4.66 can be simplified to:

$$\eta_b = \eta_{ref}e^{\beta(1/T_b - 1/T_{ref})} \qquad (4.71)$$

where T is now in $°K$ and η_{ref} is the absolute viscosity at reference temperature T_{ref}. As T_{ref} is generally room temperature and as T_b is usually higher than room temperature, Equation 4.71 generally takes the form:

$$\eta_b = \eta_{ref}e^{-A_4\beta} \qquad (4.72)$$

showing that as temperature increases, lubricant viscosity decreases.

In accordance with this, it is clear that the lubricant film thickness will be reduced as a result of temperature increase in the contact. Cheng [26] and subsequently, Murch and Wilson [27], Wilson [28], and Wilson and Sheu [29] developed thermal reduction factors for lubricant film thickness from numerical solutions of the thermal EHL problem for rolling–sliding contacts. Gupta et al. [30] recommended the film thickness reduction factor in Equation 4.73.

$$\phi_t = \frac{1 - 13.2\left(\frac{p_0}{E}\right)L^{0.42}}{1 + 0.213(1 + 2.23S^{0.83})L^{0.64}} \qquad (4.73)$$

where p_0 is the Hertzian pressure and dimensionless parameters L and S are defined as follows:

$$L = -\left(\frac{\partial \eta}{\partial T}\right)_b \frac{(u_1 + u_2)^2}{4k_b} \qquad (4.74)$$

$$S = 2\frac{(u_1 - u_2)}{(u_1 + u_2)} \qquad (4.75)$$

Particularly for line contacts, Hsu and Lee [31] provided Equation 4.76.

$$\phi_T = \frac{1}{1 + 0.0766\mathcal{G}^{0.687}Q_L^{0.447}L^{0.527}e^{0.875S}} \qquad (4.76)$$

See Example 4.3.

4.6   STARVATION OF LUBRICANT

The basic formulas for calculation of lubricant film thickness assume an adequate supply of lubricant to the contact zones. The condition in which the volume of lubricant on the surfaces entering the contact is insufficient to develop a full lubricant film is called starvation. Factors to determine the reduction of the apparent lubricant film thickness have been developed as functions of the distance of the lubricant meniscus in the inlet zone from the center of the contact. As yet, no definitive equations have been developed to accurately calculate the aforementioned distance; therefore, the meniscus distance has to be determined experimentally. Figure 4.12 illustrates the concept of meniscus distance. References [33–37] give further details about this concept.

In consideration of the meniscus distance problem, a condition of zero reverse flow is defined. Under this condition, the minimum velocity of the point situated at the meniscus distance from the contact center is, by definition, zero. If the meniscus distance is greater, the latter point will have a negative velocity, that is, reverse flow. The zero reverse flow condition

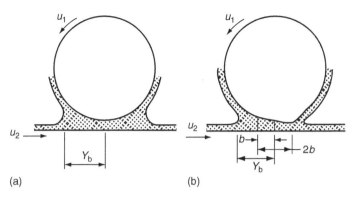

FIGURE 4.12 Meniscus distance in (a) hydrodynamic and (b) elastohydrodynamic lubrication.

is therefore a quasistable situation, because no lubricant is lost to the contact owing to reverse flow. In the case of a minimum quantity of lubricant supplied, for example, oil mist or grease lubrication, the lubricant film thickness reduction factor owing to starvation effects, according to Refs. [33,36], lies between 0.71 (in pure rolling) and 0.46 (in pure sliding). Castle and Dowson [36] give the following equation for line contact:

$$\varphi_s = 1 - e^{-1.347\Phi^{0.69}\phi^{0.13}} \tag{4.77}$$

where

$$\Phi = \frac{\frac{y_b}{b} - 1}{\left[2\left(\frac{\Re_y}{b}\right)^2 H_c\right]^{2/3}} \tag{4.78}$$

It is clear that Φ is zero if the meniscus distance should equal b and in that case $\varphi_s = 0$. Accordingly, an accurate estimation of the meniscus distance is necessary to the effective employment of a lubricant starvation factor. In the absence of this value, the condition of zero reverse flow provides a practical limitation and a starvation factor of $\varphi_s = 0.70$.

Thermal effects on lubricant film formation under conditions approaching lubricant starvation are extremely significant owing to the absence of excess lubricant to help dissipate frictional heat generation in the contacts. Accordingly, the lubricant film reduction factors for thermal effects and starvation are not multiplicative and a combined factor is required. Goksem et al. [33] derived the following expression for elastohydrodynamic line contact:

$$\varphi_{TS} = \varphi_T\left(1 - \frac{1}{(4.6 + 1.15L^{0.6})^{(0.67\bar{Q}_z \bar{Y}/\varphi_T H_c)^{(0.52/(1+0.001L))}}}\right) \tag{4.79}$$

where L is given by Equation 4.74 and

$$\bar{Y} = y_b\left(y_b^2 - 1\right)^{1/2} - \ln\left[y_b + \left(y_b^2 - 1\right)^{1/2}\right] \tag{4.80}$$

For the zero reverse flow condition, the combined reduction factor for the central lubricant film thickness is

$$\varphi_{TS} = \varphi_T \left(1 - \frac{1}{(4.6 + 1.15 L^{0.6})^{(0.6345/\varphi_T)^{(0.52/(1+0.001L))}}} \right) \qquad (4.81)$$

For point contact, Equation 4.79 through Equation 4.81 can be used in conjunction with Equation 4.62 for equivalent line contact loading.

See Example 4.4.

4.7 SURFACE TOPOGRAPHY EFFECTS

In the methods and equations used in the calculation of lubricant film thickness thus far in this chapter, only the macrogeometries of the rolling components have been considered; that is, the surfaces of the components have been assumed to be smooth. In practice, each ball, roller, or raceway surface has a roughness superimposed upon the principal geometry. This roughness, or more correctly surface topography similar to the earth's surface superimposed upon the spherical surface of the planet, is introduced by the surface finishing processes during component manufacture. In recent history, substantial manufacturing development efforts have been expended to produce ultrasmooth rolling component surfaces. Figure 4.13 schematically illustrates a rough rolling component surface.

For a given surface, the roughness is most commonly defined by the arithmetic average (AA) peak-to-valley distance. This is easily measurable using stylus devices such as the Talysurf machine. Using surface-measuring devices, more extensive properties of surface microgeometry can also be measured; see Ref. [38]. To date, AA surface roughnesses, R_A, as fine as 0.05 μm (2 μin.) have been produced on ball bearing raceways approaching 600 mm (24 in.) diameter. Balls larger than 25 mm (1 in.) diameter are routinely produced with R_A values of 0.005 μm (0.2 μin.). It is, however, not certain that $R_A = 0$ is an ideal microgeometry from a lubrication effectiveness or surface fatigue endurance standpoint.

Ground M.S. RMS Surface Roughness 1.5 μm
3 mm x 9 mm

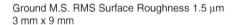
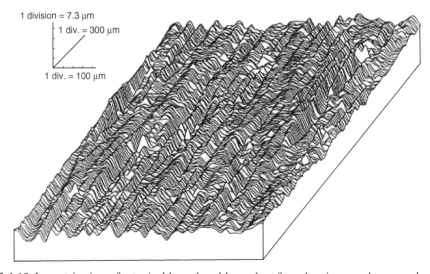

1 division = 7.3 μm

1 div. = 300 μm

1 div. = 100 μm

FIGURE 4.13 Isometric view of a typical honed and lapped surface showing roughness peaks.

Depending on the thickness of the lubricant film relative to the roughnesses of the rolling contact surfaces, the direction of the roughness pattern can affect the film-building capability of the lubricant. If the surface roughness has a pattern wherein the microgrooves are transverse to the direction of motion, this could result in a beneficial lubricant film-building effect. Conversely, if the lay of the roughness is parallel to the direction of motion, the effect can be to produce a thinner lubricant film. The most successful applications of rolling bearings are those in which fluid lubricant films over the rolling element–raceway contacts are sufficiently thick to completely separate those components. This is generally defined by the parameter Λ as follows:

$$\Lambda = \frac{h^0}{\sqrt{s_r^2 + s_{RE}^2}} \tag{4.82}$$

In Equation 4.82, h^0 is the minimum lubricant film thickness, s_r the root mean square (rms) roughness of the raceway surface, and s_{RE} is the rms roughness of the ball or roller surface. In general, the rms roughness value is taken as $1.25 \, R_A$.

Patir and Cheng [39] first investigated the effect of the lay of surface topography on the lubricant film thickness generated. They developed a correction factor for lubricant film thickness based on the distances between contact surface "hills" and "valleys" in directions transverse and parallel to rolling motion. Tønder and Jakobsen [40] using a ball-on-disk test rig and optical interferometry confirmed the general conclusion of Patir and Cheng that transverse lay tends to generate thicker films than does longitudinal lay. Kaneta et al. [41] in a similar experimental effort determined that, in the thin film region ($\Lambda < 1$), film thickness for surfaces with transverse lay tends to increase with slide/roll ratio due to deformation of asperities. When $\Lambda > 3$, however, deformation of asperities can be neglected.

Chang et al. [42] analytically investigated the effects of surface roughness considering the effects of lubricant shear thinning due to frictional heating. They determined that these effects serve to mitigate the pressure rippling influence on lubricant film thickness. Ai and Cheng [43], considering the randomized surface roughness of Figure 4.14, conducted an extensive analysis revisiting the influences of surface topographical lay. They generated three-dimensional plots of point contact pressure and film thickness distribution for transverse, longitudinal, and oblique topographical lays. Figure 4.15 through Figure 4.17 illustrate the effects for the randomized surface roughness. They indicated that roughness orientation has a noticeable effect on pressure fluctuation. They further noted that oblique roughness lay induces localized three-dimensional pressure fluctuations in which the maximum pressure may be greater than that produced by transverse roughness lay. It is to be noted that the oblique roughness lay more likely is representative of the surfaces generated during bearing component manufacture. Oblique surface roughness lay may also result in the minimum lubricant film thicknesses compared with transverse or longitudinal roughness lays. Ai and Cheng [43] further noted, however, that when Λ is sufficiently large such that the surfaces are effectively separated, the effect of lay on film thickness and contact pressure is minimal.

Guangteng and Spikes [44], using ultrathin film, optical interferometry, managed to measure the mean EHL film thickness of very thin film, isotropically rough surfaces occurring in rolling balls on flat contacts. They found that, for $\Lambda < 2$, the mean EHL film thicknesses were less than those for smooth surfaces. Subsequently, using the spacer layer imaging method developed by Cann et al. [45] to map EHL contacts, Guangteng et al. [46] indicated that rolling elements having real, random, rough surfaces; for example, rolling bearing components. The mean film thicknesses tend to be less than those calculated for rolling elements that have smooth surfaces. This implies that, in the mixed EHL regime, for example,

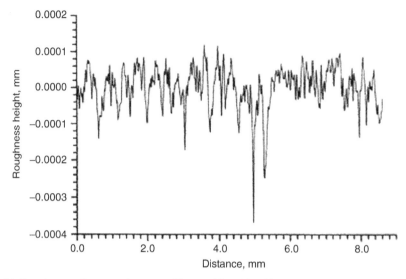

FIGURE 4.14 Random surface roughness profile considered by Ai and Cheng. (From Ai, X. and Cheng, H., *Trans. ASME, J. Tribol.*, 118, 59–66, January 1996. With permission.)

$\Lambda < 1.5$, the mean lubricant film thicknesses will tend to be less than those predicted by the equations given for rolling contacts with smooth surfaces. The amount of the reduction may only be determined by testing; empirical relationships need to be developed.

4.8 GREASE LUBRICATION

When grease is used as a lubricant, the lubricant film thickness is generally estimated using the properties of the base oil of the grease while ignoring the effect of the thickener. It has been determined, however, by several researchers [47–50] that in a given application, owing to a

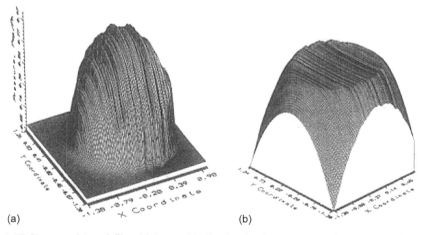

FIGURE 4.15 Pressure (a) and film thickness (b) distribution in an EHL point contact with transverse topographical lay, random surface roughness. Motion is in the x direction. (From Ai, X. and Cheng, H., *Trans. ASME, J. Tribol.*, 118, 59–66, January 1996. With permission.)

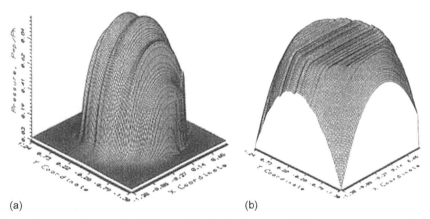

(a)   (b)

FIGURE 4.16 Pressure (a) and film thickness (b) distribution in an EHL point contact with longitudinal topographical lay, random surface roughness. Motion is in the *x* direction. (From Ai, X. and Cheng, H., *Trans. ASME, J. Tribol.*, 118, 59–66, January 1996. With permission.)

contribution by the thickener, grease may form a thicker lubricant film than that determined using only the properties of the base oil. Kauzlarich and Greenwood [51] developed an expression for the thickness of the film formed by greases in line contact under a Herschel–Bulkley constitutive law in which shear stress τ and shear rate $\dot{\gamma}$ are related by the equation

$$\tau = \tau_y + \alpha \dot{\gamma}^{\beta} \tag{4.83}$$

where τ_y is the yield stress and α and β are considered physical properties of the grease.

For a Newtonian fluid,

$$\tau = \eta \dot{\gamma} \tag{4.84}$$

where η is the viscosity.

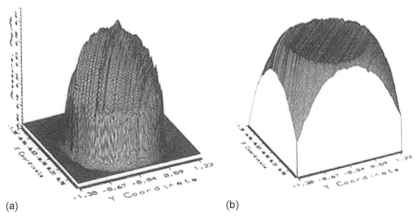

(a)   (b)

FIGURE 4.17 Pressure (a) and film thickness (b) distribution in an EHL point contact with oblique topographical lay, random surface roughness. Motion is in the *x* direction. (From Ai, X. and Cheng, H., *Trans. ASME, J. Tribol.*, 118, 59–66, January 1996. With permission.)

The effective viscosity under a Herschel–Bulkley law is thus found by equating τ from Equation 4.83 and Equation 4.84 so that

$$\eta_{\text{eff}} = \frac{\tau_y + \alpha \dot{\gamma}^\beta}{\dot{\gamma}} \qquad (4.85)$$

In this form, it is seen that for $\alpha > 1$, η_{eff} increases indefinitely with the shear rate, and for $\alpha < 1$, η_{eff} approaches zero as the strain rate increases. Palacios et al. [49] argued that it is more reasonable to assume that at high shear rates greases will behave like their base oils. They accordingly proposed a modification of the Herschel Bulkley law to the form

$$\tau = \tau_y + \alpha \dot{\gamma}^\beta + \eta_b \dot{\gamma} \qquad (4.86)$$

where η_b is the base oil viscosity. In this form, provided $\alpha < 1$, η_{eff} approaches η_b as the strain rate approaches ∞. Values of τ_y, α, β, and η_b are given in Ref. [52] for three greases from 35 to 80°C (95 to 176°F).

Since viscosity appears raised to the 0.67 power in Equation 4.63, Palacios and Palacios [52] proposed that h_G, the film thickness of a grease, and h_b, the film thickness of the base oil, will be in the proportion

$$\frac{h_G}{h_b} = \left(\frac{\eta_{\text{eff}}}{\eta_b} \right)^{0.67} \qquad (4.87)$$

They proposed that this evaluation be made at a shear rate equal to $0.68u/h_G$, which requires iteration to determine h_G. Their suggested approach is to calculate h_b from Equation 4.63, determine $\dot{\gamma} = 0.68u/h_b$, and then h_G from Equation 4.87. The shear rate is then recalculated using h_G. The process is repeated until convergence occurs. The analysis was applied to line contact, but it should also be valid for elliptical contacts with a/b in the range of 8–10 (typical for ball bearing point contacts).

In her investigations, Cann [53,54] notes that the portion of the film associated with the grease thickener is a residual film composed of the degraded thickener deposited in the bearing raceways. The hydrodynamic component is generated by the relative motion of the surfaces due to oil, both in the raceways and supplied by the reservoirs of grease adjacent to the raceways. She further notes that at low temperatures grease films are generally thinner than those for the fully flooded, base fluid lubricant. This is due to the predominant bulk grease starvation and the inability of the high viscosity, bled lubricant to resupply the contact. At higher temperatures of operation, grease forms films considerably thicker than those considering only the base oil. This is attributed to the increased local supply of lubricant to the contact area due to the lower oil viscosity at the elevated temperature producing a partially flooded EHL film augmented by a boundary film of deposited thickener.

Therefore, it can be stated that with grease lubrication the degree of starvation tends to increase with increasing base oil viscosity, thickener content, and speed of rotation. It tends to decrease with increasing temperature. For rolling bearing applications, the film thickness may only be a fraction of that calculated for fully flooded, oil lubrication conditions. A most likely saving factor is that as lubricant films become thinner, friction and hence temperature increase. This tends to reduce viscosity permitting increased return flow to the rolling element–raceway contacts. Nevertheless, depending on the aforementioned operating conditions of grease base oil viscosity, grease thickener content, and rotational speed, lubricant film thicknesses may be expected to be only a fraction of those calculated using Equation 4.57,

Equation 4.60, and Equation 4.63. According to data shown by Cann [54], fractional values might range from 0.9 down to 0.2.

4.9 LUBRICATION REGIMES

Although this chapter has concentrated on elastohydrodynamic lubrication in rolling contacts, the general solution presented for the Reynolds equation covers a gamut of lubrication regimes; for example:

- Isoviscous hydrodynamic (IHD) or classical hydrodynamic lubrication
- Piezoviscous hydrodynamic (PHD) lubrication, in which lubricant viscosity is a function of pressure in the contact
- Elastohydrodynamic (EHD) lubrication, in which both the increase in viscosity with pressure and the deformations of the rolling component surfaces are considered in the solution

Dowson and Higginson [55] created Figure 4.18 to define these regimes for line contact in terms of the dimensionless quantities for film thickness, load, and rolling velocity; Equation 4.48 through Equation 4.50.

Markho and Clegg [56] established a parameter, called C_1 herein, for a fixed value of $\mathcal{G}$; This factor was used to define the lubrication regime. Dalmaz [57] subsequently established Equation 4.88 to cover all practical values of $\mathcal{G}$.

$$C_1 = \log_{10}\left[1.5 \times 10^6 \left(\frac{\mathcal{G}}{5000}\right)^2 \frac{\bar{Q}_z^3}{\bar{U}}\right] \tag{4.88}$$

Table 4.2 shows the relationship of parameter C_1 to the operating lubrication regimes.

For calculation of the lubricant film thicknesses in rolling element–raceway contacts, only the PHD and EHD regimes need to be considered. For calculations associated with the cage–rolling element contacts, probably a consideration of the hydrodynamic regime is sufficient. In this case, Martin [1] gave the following equation for film thickness in line contact:

$$H = 4.9 \frac{\bar{U}}{\bar{Q}_z} \tag{4.89}$$

For point contact, Brewe and Hamrock [58] give

$$H = \left\{\frac{\frac{\bar{Q}_z}{\bar{U}}\left(1 + \frac{2\mathcal{R}_x}{3\mathcal{R}_y}\right)}{\left(128 \frac{\mathcal{R}_y}{\mathcal{R}_x}\right)^{1/2}\left[0.131 \tan^{-1}\left(\frac{\mathcal{R}_y}{2\mathcal{R}_x}\right) + 1.163\right]} + 2.6511\right\}^{-2} \tag{4.90}$$

For the PHD regime in line contacts, data from Ref. [56] have been used to establish the following expression for minimum film thickness:

$$H = 10^{C_4} \times \left(\frac{\mathcal{G}}{5000}\right)^{0.35(1+C_1)} \tag{4.91}$$

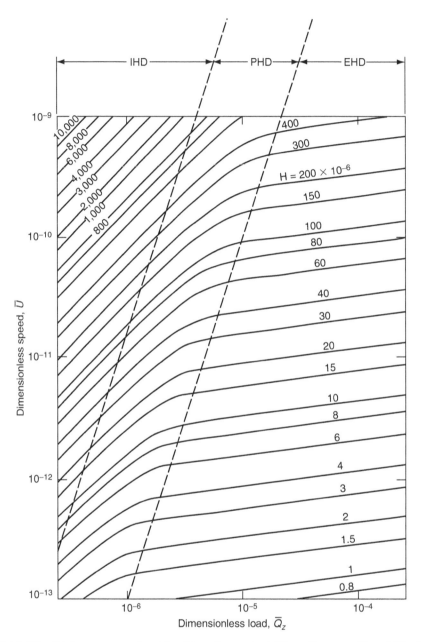

FIGURE 4.18 Film thickness vs. speed and load for a line contact. (From Dowson, D. and Higginson, G., *Proc. Inst. Mech. Eng.*, 117, 1963.)

where

$$C_2 = \log_{10}\left(618\,\bar{U}^{0.6617}\right) \tag{4.92}$$

$$C_3 = \log_{10}\left(1.285\,\bar{U}^{0.0025}\right) \tag{4.93}$$

and C_1 is given by Equation 4.88. In Equation 4.91, C_4 is given by

TABLE 4.2
Lubrication Regimes

Parameter Limits	Lubrication Regime	Characteristics
$C_1 \leq -1$	IHD	Low contact pressure, no significant surface deformation
$-1 < C_1 < 1$	PHD	No significant surface deformation, lubricant viscosity increases with pressure
$C_1 \geq 1$	EHD	Surface deformation and lubricant viscosity increase with pressure

$$C_4 = C_2 + C_1 C_3 (C_1^2 - 3) - 0.094 C_1 \ (C_1^2 - 0.77 C_1 - 1) \tag{4.94}$$

Dalmaz [57] also developed numerical results for point contact film thicknesses in the PHD regime; an analytical relationship was not then established.

4.10 CLOSURE

In the earlier discussion, it has been demonstrated analytically that a lubricant film can separate the rolling elements from the contacting raceways. Moreover, the fluid friction forces developed in the contact zones between the rolling elements and raceways can significantly alter the bearing's mode of operation. It is desirable from the standpoint of preventing increased stresses caused by metal-to-metal contact that the minimum film thickness should be sufficient to completely separate the rolling surfaces. The effect of film thickness on bearing endurance is discussed in Chapter 8.

A substantial amount of analytical and experimental research from the 1960s into the 21st century has contributed greatly to the understanding of the lubrication mechanics of concentrated contacts in rolling bearings. Perhaps the original work of Grubin [13] will prove to be as significant as that conducted by Reynolds during the 1880s.

Apart from acting to separate rolling surfaces, the lubricant is frequently used as a medium to dissipate the heat generated by bearing friction as well as to remove heat that would otherwise be transferred to the bearing from the surroundings at elevated temperatures. This topic is discussed in Chapter 7.

REFERENCES

1. Martin, H., Lubrication of gear teeth, *Engineering*, 102, 199, 1916.
2. Osterle, J., On the hydrodynamic lubrication of roller bearings, *Wear*, 2, 195, 1959.
3. Barus, C., Isothermals, isopiestics, and isometrics relative to viscosity, *Am. J. Sci.*, 45, 87–96, 1893.
4. ASME Research Committee on Lubrication, Pressure–viscosity report—Vol. 11, *ASME*, 1953.
5. Roelands, C., *Correlation Aspects of Viscosity–Temperature–Pressure Relationship of Lubricating Oils*, Ph.D. Thesis, Delft University of Technology, 1966.
6. Sorab, J. and VanArsdale, W., A correlation for the pressure and temperature dependence of viscosity, *Tribol. Trans.*, 34(4), 604–610, 1991.
7. Bair, S. and Kottke, P., Pressure–viscosity relationships for elastohydrodynamics, Preprint AM03-1, STLE Annual Meeting, New York, 2003.
8. Doolittle, A., Studies in Newtonian flow II, the dependence of the viscosity of liquids on free-space, *J. Appl. Phys.*, 22, 1471–1475, 1951.
9. Harris, T., Establishment of a new rolling bearing life calculation method, Final Report, U.S. Navy Contract N68335-93-C-0111, January 15, 1994.
10. Bair, S. and Winer, W., Shear strength measurements of lubricants at high pressure, *Trans. ASME, J. Lubr. Technol., Ser. F*, 101, 251–257, 1979.

11. Bair, S. and Winer, W., Some observations in high pressure rheology of lubricants, *Trans. ASME, J. Lubr. Technol., Ser. F*, 104, 357–364, 1982.

12. Dowson, D. and Higginson, G., A numerical solution to the elastohydrodynamic problem, *J. Mech. Eng. Sci.*, 1(1), 6, 1959.

13. Grubin, A., Fundamentals of the hydrodynamic theory of lubrication of heavily loaded cylindrical surfaces, *Investigation of the Contact Machine Components*, Kh. F. Ketova (ed.) [Translation of Russian Book No. 30, Chapter 2], Central Scientific Institute of Technology and Mechanical Engineering, Moscow, 1949.

14. Dowson, D. and Higginson, G., The effect of material properties on the lubrication of elastic rollers, *J. Mech. Eng. Sci.*, 2(3), 1960.

15. Sibley, L. and Orcutt, F., Elastohydrodynamic lubrication of rolling contact surfaces, *ASLE Trans.*, 4, 234–249, 1961.

16. Dowson, D. and Higginson, G., *Proc. Inst. Mech. Eng.*, 182(Part 3A), 151–167, 1968.

17. Archard, G. and Kirk, M., Lubrication at point contacts, *Proc. R. Soc. Ser. A*, 261, 532–550, 1961.

18. Hamrock, B. and Dowson, D., Isothermal elastohydrodynamic lubrication of point contacts—Part III—fully flooded results, *Trans. ASME, J. Lubr. Technol.*, 99, 264–276, 1977.

19. Wedeven, L., *Optical Measurements in Elastohydrodynamic Rolling Contact Bearings*, Ph.D. Thesis, University of London, 1971.

20. Kotzalas, M., *Power Transmission Component Failure and Rolling Contact Fatigue*, Ph.D. Thesis, Pennsylvania State University, 1999.

21. Avallone, E. and Baumeister, T., *Standard Handbook for Mechanical Engineers*, 9th ed., McGraw-Hill, New York, 1987.

22. Venner, C., Higher order mutlilevel solvers for the EHL line and point contact problems, *ASME Trans., J. Tribol.*, 116, 741–750, 1994.

23. Smeeth, S. and Spikes, H., Central and minimum elastohydrodynamic film thickness at high contact pressure, *ASME Trans., J. Tribol.*, 119, 291–296, 1997.

24. Cheng, H., A numerical solution to the elastohydrodynamic film thickness in an elliptical contact, *Trans. ASME, J. Lubr. Technol.*, 92, 155–162, 1970.

25. Vogels, H., Das Temperaturabhängigkeitsgesetz der Viscosität von Flíssigkeiten, *Phys. Z.*, 22, 645–646, 1921.

26. Cheng, H., A refined solution to the thermal-elastohydrodynamic lubrication of rolling and sliding cylinders, *ASLE Trans.*, 8(4), 397–410, 1965.

27. Murch, L. and Wilson, W., A thermal elastohydrodynamic inlet zone analysis, *Trans. ASME, J. Lubr. Technol., Ser. F*, 97(2), 212–216, 1975.

28. Wilson, A., An experimental thermal correction for predicted oil film thickness in elastohydrodynamic contacts, *Proc. 6th Leeds–Lyon Symp. Tribol.*, 1979.

29. Wilson, W. and Sheu, S., Effect of inlet shear heating due to sliding on elastohydrodynamic film thickness, *Trans. ASME, J. Lubr. Technol., Ser. F*, 105(2), 187–188, 1983.

30. Gupta, P., et al., Viscoelastic effects in Mil-L-7808 type lubricant, Part I: Analytical formulation, *Tribol. Trans.*, 35(2), 269–274, 1992.

31. Hsu, C. and Lee, R., An efficient algorithm for thermal elastohydrodynamic lubrication under rolling/sliding line contacts, *J. Vibr. Acoust. Reliab. Des.*, 116(4), 762–768, 1994.

32. MacAdams, W., *Heat Transmission*, 3rd ed., McGraw-Hill, New York, 1954.

33. Goksem, P. and Hargreaves, R., The effect of viscous shear heating in both film thickness and rolling traction in an EHL line contact—Part II: Starved condition, *Trans. ASME, J. Lubr. Technol.*, 100, 353–358, 1978.

34. Dowson, D., Inlet boundary conditions, *Leeds–Lyon Symp.*, 1974.

35. Wolveridge, P., Baglin, K., and Archard, J., The starved lubrication of cylinders in line contact, *Proc. Inst. Mech. Eng.*, 185, 1159–1169, 1970–1971.

36. Castle, P. and Dowson, D., A theoretical analysis of the starved elastohydrodynamic lubrication problem, *Proc. Inst. Mech. Eng.*, 131, 131–137, 1972.

37. Hamrock, B. and Dowson, D., Isothermal elastohydrodynamic lubrication of point contact—Part IV: Starvation results, *Trans. ASME, J. Lubr. Technol.*, 99, 15–23, 1977.

38. McCool, J., Relating profile instrument measurements to the functional performance of rough surfaces, *Trans. ASME, J. Tribol.*, 109, 271–275, April 1987.
39. Patir, N. and Cheng, H., Effect of surface roughness orientation on the central film thickness in EHD contacts, *Proc. 5th Leeds–Lyon Symp. Tribol.*, 15–21, 1978.
40. Tønder, P. and Jakobsen, J., Interferometric studies of effects of striated roughness on lubricant film thickness under elastohydrodynamic conditions, *Trans. ASME, J. Tribol.*, 114, 52–56, January 1992.
41. Kaneta, M., Sakai, T., and Nishikawa, H., Effects of surface roughness on point contact EHL, *Tribol. Trans.*, 36(4), 605–612, 1993.
42. Chang, L., Webster, M., and Jackson, A., On the pressure rippling and roughness deformation in elastohydrodynamic lubrication of rough surfaces, *Trans. ASME, J. Tribol.*, 115, 439–444, July 1993.
43. Ai, X. and Cheng, H., The effects of surface texture on EHL point contacts, *Trans. ASME, J. Tribol.*, 118, 59–66, January 1996.
44. Guangteng, G. and Spikes, H., An experimental study of film thickness in the mixed lubrication regime, *Proc. 24th Leeds–Lyon Symp., Elastohydrodynamics*, 159–166, September 1996.
45. Cann, P., Hutchinson, J., and Spikes, H., The development of a spacer layer imaging method (SLIM) for mapping elastohydrodynamic contacts, *Tribol. Trans.*, 39, 915–921, 1996.
46. Guangteng, G., et al., Lubricant film thickness in rough surface, mixed elastohydrodynamic contact, ASME Paper 99-TRIB-40, October 1999.
47. Wilson, A., The relative thickness of grease and oil films in rolling bearings, *Proc. Inst. Mech. Eng.*, 193, 185–192, 1979.
48. Mïnnich, H. and Glöckner, H., Elastohydrodynamic lubrication of grease-lubricated rolling bearings, *ASLE Trans.*, 23, 45–52, 1980.
49. Palacios, J., Cameron, A., and Arizmendi, L., Film thickness of grease in rolling contacts, *ASLE Trans.*, 24, 474–478, 1981.
50. Palacios, J., Elastohydrodynamic films in mixed lubrication: an experimental investigation, *Wear*, 89, 303–312, 1983.
51. Kauzlarich, J. and Greenwood, J., Elastohydrodynamic lubrication with Herschel–Bulkley model reases, *ASLE Trans.*, 15, 269–277, 1972.
52. Palacios, J. and Palacios, M., Rheological properties of greases in EHD contacts, *Tribol. Int.*, 17, 167–171, 1984.
53. Cann, P., Starvation and reflow in a grease-lubricated elastohydrodynamic contact, *Tribol. Trans.*, 39(3), 698–704, 1996.
54. Cann, P., Starved grease lubrication of rolling contacts, *Tribol. Trans.*, 42(4), 867–873, 1999.
55. Dowson, D. and Higginson, G., Theory of roller bearing lubrication and deformation, *Proc. Inst. Mech. Eng.*, 117, 1963.
56. Markho, P. and Clegg, D., Reflections on some aspects of lubrication of concentrated line contacts, *Trans. ASME, J. Lubr. Technol.*, 101, 528–531, 1979.
57. Dalmaz, G., Le Film Mince Visquex dans les Contacts Hertziens en Regimes Hydrodynamique et Elastohydrodynamique, Docteur d'Etat Es Sciences Thesis, I.N.S.A. Lyon, 1979.
58. Brewe, D. and Hamrock, B., Analysis of starvation on hydrodynamic lubrication in non-conforming contacts, ASME Paper 81-LUB-52, 1981.

5 Friction in Rolling Element–Raceway Contacts

LIST OF SYMBOLS

Symbol	Description	Units
a	Semimajor axis of contact ellipse	mm (in.)
A_c	True average contact area	mm^2 ($in.^2$)
A_0	Apparent contact area	mm^2 ($in.^2$)
b	Semiminor axis of contact ellipse	mm (in.)
d	Separation of mean plane of summits and smooth plane	mm (in.)
d_i	Raceway track diameter	mm (in.)
D	Rolling element diameter	mm (in.)
D_{SUM}	Summit density	mm^{-2} ($in.^{-2}$)
E_1, E_2	Elastic moduli of bodies 1 and 2	MPa (psi)
E'	Reduced elastic modulus	MPa (psi)
F	Contact friction force	N (lb)
$F_0(\ ), F_1(\ ),$		
$F_{3/2}(\ )$	Tabular functions for the Greenwood–Williamson model	
h	Lubricant film thickness	mm (in.)
h_c	Central or plateau lubricant film thickness	mm (in.)
L	Roller length end-to-end	mm (in.)
l_{eff}	Roller effective length	mm (in.)
l_s	Roller straight length	mm (in.)
m_0	Zeroth-order spectral moment, $\equiv R_q^2 \equiv s^2$	μm^2 ($\mu in.^2$)
m_2	Second-order spectral moment	
m_4	Fourth-order spectral moment	mm^{-2} ($in.^{-2}$)
n	Contact density	mm^{-2} ($in.^{-2}$)
n_p	Plastic contact density	mm^{-2} ($in.^{-2}$)
q	x/a	
Q	Contact load	N (lb)
Q_a	Asperity-supported load	N (lb)
Q_f	Fluid-supported load	N (lb)
R	Radius of deformed surface	mm (in.)
R	Summit sphere radius	mm (in.)
R_q	Root mean square (rms) value of surface profile	μm ($\mu in.$)
S	Composite rms surface roughness for bodies 1 and 2	μm ($\mu in.$)
S_s	Standard deviation of summit heights for bodies 1 and 2	mm (in.)
s_1, s_2	Surface rms roughnesses for bodies 1 and 2	μm ($\mu in.$)
t	y/a	
T	Temperature	°C (°F)
u	Surface velocity	mm/sec (in./sec)

u_m	Raceway surface velocity	mm/sec (in./sec)
u_{RE}	Rolling element surface velocity	mm/sec (in./sec)
U	Rolling velocity $= 1/2\,(u_{RE} + u_m)$	mm/sec (in./sec)
v	Sliding velocity	mm/sec (in./sec)
w	Deflection of summit	μm (μin.)
w_p	Variable governing asperity density	μm (μin.)
Y	Yield strength in simple tension	MPa (psi)
z_s	Summit height relative to summit mean plane	mm (in.)
$\bar{z}_s$	Distance between surface and summit mean plane	mm (in.)
$z(x)$	Surface profile	mm (in.)
α	Bandwidth parameter	
γ	Shear rate	$\sec^{-1}$
η	Absolute viscosity	N-sec/m^2
		(lb-sec/in.^{2})
Λ	Lubricant film parameter, h/s	
μ	Friction or traction coefficient	
μ_a	Asperity–asperity friction coefficient	
ν_1, ν_2	Poisson's ratio for bodies 1 and 2	
σ	Normal contact stress or pressure	MPa (psi)
σ_0	Maximum normal contact stress or pressure	MPa (psi)
Φ_o	Maximum normal contact stress or pressure MPa(psi)	
τ	Shear stress	MPa (psi)
τ_f	Shear stress due to fluid	MPa (psi)
τ_{lim}	Limiting shear stress in fluid	MPa (psi)
τ_N	Shear stress in Newtonian fluid lubrication	MPa (psi)
$\phi(\ )$	Gaussian probability density function	mm^{-1} (in.$^{-1}$)

5.1 GENERAL

Ball and roller bearings were historically called antifriction bearings because of the low friction properties associated with them. Actually, the major portion of friction associated with rolling bearings is caused by sliding motions in the contacts between components such as rolling elements and raceways, rolling elements and cage, roller ends and roller guide flanges, and cage rails and inner or outer ring lands. This excludes the friction due to sliding between bearing seals and inner or outer ring lands; this friction is generally greater than that produced by all of the other sources of friction combined. In this chapter, the friction between rolling elements and raceways will be investigated.

Rolling bearings are generally operated with oil lubrication; this can be accomplished using circulating oil, bath oil, air–oil mist, or grease. Grease lubricant is an organic or inorganic thickener containing oil that exudes from the thickener to become the predominant lubricant. In Chapter 4, it was shown that the lubricant film acts to separate the rolling elements from the raceways. This separation can be complete or partial. With complete separation, friction depends wholly on the properties of the lubricant at the contact temperatures and pressures. In the latter case, peaks or asperities from the rolling/sliding surfaces come into contact under boundary lubrication conditions, resulting in increased friction. Thus, it is important to establish the lubricant film thickness in each contact.

In Chapter 4, it was shown that lubricant film thickness occurring in a fluid-lubricated (oil-lubricated), rolling element–raceway contact depends on contact geometry and load, rolling speed, and lubricant properties. The lubricant properties, in turn, depend on the temperature of the lubricant both within and on entering the contact. The temperatures depend on the friction heat generated and on the heat dissipation paths available to the bearing. Methods to determine bearing temperatures will be discussed in Chapter 7; in this chapter, it will be assumed that temperatures are known.

Under conditions where fluid or grease lubrication is precluded, rolling bearings may also be operated with solid-film lubricants; for example, graphite, molybdenum disulfide, or other compounds. These lubricants generally cause rolling bearings to operate with higher friction and temperatures than do fluid lubricants. This form of lubrication is similar to boundary lubrication, resulting in less friction than direct rolling component contact; however, heat dissipation capability is greatly reduced.

5.2 ROLLING FRICTION

5.2.1 DEFORMATION

The balls or rollers in a bearing are mainly subjected to loads perpendicular to the tangent plane at each contact surface. Because of these normal loads, the rolling elements and raceways are deformed at each contact, producing according to Hertz, a radius of curvature of the common contacting surfaces equal to the harmonic mean of the radii of the contacting bodies. For a roller of diameter D, bearing on a cylindrical raceway of diameter d_i, the radius of curvature of a contact surface is

$$R = \frac{d_i D}{d_i + D} \tag{5.1}$$

Because of the deformation indicated above and because of the rolling motion of the roller over the raceway, which requires a tangential force to overcome rolling resistance, raceway material is squeezed up to form a bulge in the forward portion of the contact as shown in Figure 5.1. A depression is subsequently formed in the rear of the contact area. Thus, an additional tangential force is required to overcome the resisting force of the bulge. The bulge is very small and the friction force is insignificant.

5.2.2 ELASTIC HYSTERESIS

As may be observed in the discussion, as a rolling element under compressive load travels over a raceway, the material in the forward portion of the contact in the direction of rolling undergoes compression while the material in the rear of the contact is relieved of stress. It is recognized that as load is increasing, a given stress corresponds to a smaller deflection than when load is decreasing (see Figure 5.2). The area between the curves in Figure 5.2 is called the hysteresis loop, and it represents an energy loss (friction power loss). Generally, friction due to elastic hysteresis is very small compared with other types of friction occurring in rolling

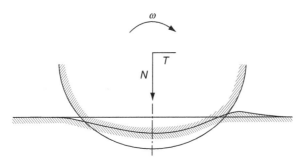

FIGURE 5.1 Roller–raceway contact showing bulge due to rolling deformation.

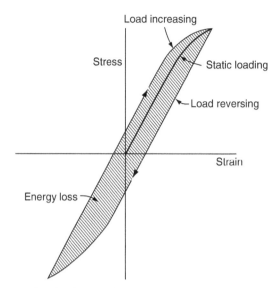

FIGURE 5.2 Hysteresis loop for elastic material subjected to reversing stresses.

bearings. Drutowski [1] verified this by experimenting with balls rolling between flat plates. Friction coefficients as low as 0.0001 can be determined from the data of Ref. [1] for 12.7 mm (0.5 in.) chrome steel balls rolling on chrome steel plates under normal loads of 356 N (80 lb). Greenwood and Tabor [2] evaluated the rolling resistance due to elastic hysteresis. They found that the frictional resistance is substantially less than that due to sliding if the normal load is sufficiently large.

Drutowski [3] also demonstrated the linear dependence of rolling friction on the volume of stressed material. In both Refs. [1,3], he further showed the dependence of elastic hysteresis on the material under stress and the specific load on the contact area.

5.3 SLIDING FRICTION

5.3.1 MICROSLIP

If a radial cylindrical roller bearing had rollers and raceways of exactly the same lengths, if the rollers were accurately guided by frictionless flanges, and if the bearing operated with zero misalignment under moderate speed, then gross sliding in the roller–raceway contacts would not occur. Gross sliding refers to the total slip of one surface over another. Depending on the elastic properties of the contacting bodies and the coefficient of friction between the contacting surfaces, microslip could occur. Using Figure 5.3, the coefficient of friction is defined as the ratio of the tangential force F to the normal force Q. Microslip is defined as the partial sliding of one surface relative to the other:

$$\mu = \frac{F}{Q} \tag{5.2}$$

Reynolds [4] first referred to microslip when, in his experiments involving rolling of an elastically stiff cylinder on rubber, he observed that since the rubber stretched in the contact zone, the cylinder rolled forward a distance less than its circumference in one complete revolution about its axis. This experiment was conducted in the absence of a lubricating medium, that is, dry contact.

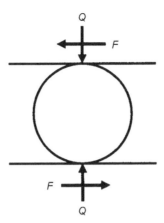

FIGURE 5.3 Roller between two plane surfaces—loaded by normal forces Q and tangential forces F.

Poritsky [5] demonstrated the microslip or creep phenomenon in two dimensions considering the action of a locomotive driving wheel, also dry contact. The normal load between contacting cylinders was assumed to generate a parabolic stress distribution, similar to a Hertzian stress distribution, over the contact surfaces as illustrated in Figure 5.4. Superimposed on this stress distribution with stresses σ_z was a tangential stress τ_x. In this case, the local coefficient of friction in the contact is

$$\mu_x = \frac{\tau_x}{\sigma_z} \tag{5.3}$$

Using this model, Poritsky demonstrated the existence of a "locked" region over which no slip occurs and a region of relative movement or slip over a contact area for which it was historically assumed that only rolling occurred. This is illustrated in Figure 5.5.

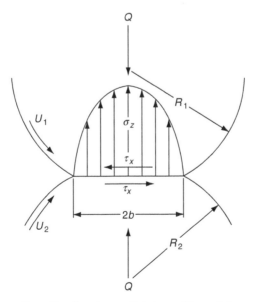

FIGURE 5.4 Rolling under action of surface tangential stress. (From Johnson, K., Tangential tractions and micro-slip, *Rolling Contact Phenomena*, Elsevier, Amsterdam, 1962, pp. 6–28. Reprinted with permission from American Elsevier Publishing Company.)

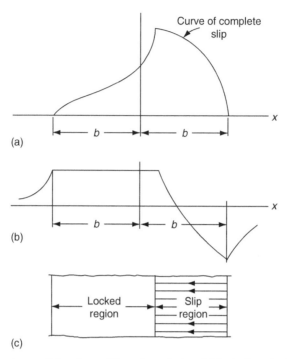

FIGURE 5.5 (a) Surface tangential actions; (b) surface strains; (c) locked and microslip regions. (From Cain, B., *J. Appl. Mech.*, 72, 465, 1950. Reprinted with permission from American Elsevier Publishing Company.)

Cain [6] further determined that in pure rolling the locked region coincided with the leading edge of the contact area. It must be emphasized that the locked region can only occur when the friction coefficient is very high as between two unlubricated surfaces.

Heathcote [7] determined that a "hard" ball "rolling" in a closely conforming groove can roll without sliding only on two narrow bands. Ultimately, Heathcote obtained a formula for the rolling friction in this situation. While Heathcote slip is very similar to that which occurs because of rolling element–raceway deformation, Heathcote's analysis takes no account of the ability of the surfaces to elastically deform and accommodate the difference in surface velocities by differential expansion. Johnson [8] expanded on the Heathcote analysis by slicing an elliptical contact area, such as that in a ball–raceway contact, into differential slabs of area as shown in Figure 5.6 and thereafter applying the Poritsky analysis for each slab. Johnson's analysis using elastic tangential compliance demonstrates a lower coefficient of friction; this assumes sliding rather than microslip. Figure 5.7 shows the locked and slip regions that obtain within the contact ellipse.

5.3.2 SLIDING DUE TO ROLLING MOTION: SOLID-FILM OR BOUNDARY LUBRICATION

5.3.2.1 Direction of Sliding

Even though called rolling bearings, the major source of friction during their operation is sliding. In Chapter 2, it was demonstrated that sliding occurs in most ball and roller bearings due to the macrogeometry, that is, basic internal geometry of the bearing. For a radial ball bearing subjected to a simple radial load, Figure 2.7 demonstrates that in a single contact pure rolling can only occur at two points, designated "A." At all other points along the contact, sliding must occur in a direction parallel to rolling motion. Outside of points A, sliding occurs

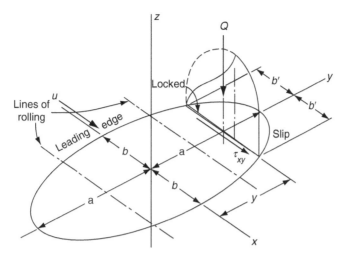

FIGURE 5.6 Ball–raceway contact ellipse showing locked region and microslip region—radial ball bearing. (From Johnson, K., Tangential tractions and micro-slip, *Rolling Contact Phenomena*, Elsevier, Amsterdam, 1962, pp. 6–28. Reprinted with permission from American Elsevier Publishing Company.)

in one direction; between points A sliding occurs in the opposite direction. The elliptical contact area showing sliding velocity directions may be characterized as shown in Figure 5.8; it assumes that the coefficient of friction is not sufficiently great to cause the possibility of a locked region. This is always the case for oil-lubricated bearings, and it is usually the case for bearings operating effectively with solid-film lubricants such as molybdenum disulfide and graphite.

5.3.2.2 Sliding Friction

In Chapter 6, the first volume of this handbook, the normal stress at any point (x, y) in the contact was given by the equation below:

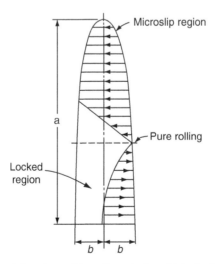

FIGURE 5.7 (a) Surface tangential actions; (b) surface strains; (c) locked and microslip regions. (From Johnson, K., Tangential tractions and micro-slip, *Rolling Contact Phenomena*, Elsevier, Amsterdam, 1962, pp. 6–28. Reprinted with permission from American Elsevier Publishing Company.)

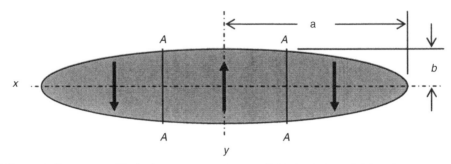

FIGURE 5.8 Ball–raceway elliptical contact area in a radially loaded, radial bearing. Arrows show sliding direction.

$$\sigma = \frac{3Q}{2\pi ab}\left[1 - \left(\frac{x}{a}\right)^2 - \left(\frac{y}{b}\right)^2\right]^{1/2} \tag{5.4}$$

According to Equation 5.3 then at any point (x, y), surface friction shear stress parallel to the rolling direction is given by

$$\tau_y = \frac{3\mu Q}{2\pi ab}\left[1 - \left(\frac{x}{a}\right)^2 - \left(\frac{y}{b}\right)^2\right]^{1/2} \tag{5.5}$$

Friction force parallel to the rolling direction is calculated by integrating over the contact area from $-a$ to $+a$ and $-b$ to $+b$. Letting $q = x/a$ and $t = y/b$,

$$F_y = \frac{3\mu Q}{2\pi ab}\int\limits_{-1}^{+1}\int\limits_{-\sqrt{1-q^2}}^{+\sqrt{1-q^2}}\left(1 - q^2 - t^2\right)^{1/2}dt\,dq = \frac{3\mu Q}{2\pi ab}\,I \tag{5.6}$$

where the integral I is calculated in three parts as follows:

$$I_1 = c_{v1}\int\limits_{-1}^{-A/a}\int\limits_{-\sqrt{1-t^2}}^{+\sqrt{1-t^2}}\left(1 - q^2 - t^2\right)^{1/2}dt\,dq$$

$$I_2 = c_{v2}\int\limits_{-A/a}^{+A/a}\int\limits_{-\sqrt{1-t^2}}^{+\sqrt{1-t^2}}\left(1 - q^2 - t^2\right)^{1/2}dt\,dq \tag{5.7}$$

$$I_3 = c_{v3}\int\limits_{+A/a}^{+1}\int\limits_{-\sqrt{1-t^2}}^{+\sqrt{1-t^2}}\left(1 - q^2 - t^2\right)^{1/2}dt\,dq$$

where c_{vn}, the sliding velocity direction coefficient, is $+1$ or -1 depending on the direction of sliding.

Equation 5.6 and Equation 5.7 are valid for operating conditions involving solid-film lubrication and boundary lubrication where friction coefficient μ can be characterized as a constant.

5.3.3 SLIDING DUE TO ROLLING MOTION: FULL OIL-FILM LUBRICATION

5.3.3.1 Newtonian Lubricant

When the lubricant film completely separates the rolling surfaces, Newtonian fluid lubrication is assumed, giving as stated in Chapter 4 the following relationship for surface friction shear stress:

$$\tau = \eta \frac{\partial u}{\partial z} \qquad (4.1)$$

where η is the fluid viscosity, u the fluid velocity in the direction of rolling motion, and z is the distance into the gap between the rolling contact surfaces. Since the gap is very small compared with the dimensions of the rolling components, Equation 4.1 can be simplified to

$$\tau = \eta \frac{v}{h} \qquad (5.8)$$

where v is the sliding velocity and h is the plateau lubricant film thickness. This equation assumes constant viscosity. Recall from Chapter 4 that h is a function of the viscosity of the lubricant entering the contact. For a given lubricant, this viscosity is mainly dependent on temperature. To calculate surface friction shear stress, however, the viscosity of the lubricant in the contact must be used. Since this viscosity is not constant, the use of simple Newtonian lubrication in rolling contact is limited to very low load applications.

5.3.3.2 Lubricant Film Parameter

The parameter Λ was established during the 1960s to indicate the degree to which a lubricant film separates the surfaces in rolling "contact":

$$\Lambda = \frac{h}{\left(s_m^2 + s_R^2\right)^{1/2}} \qquad (5.9)$$

where s_m is the root mean square (rms) roughness of the raceway and s_R is the rms roughness of the rolling element. These values are usually obtained as R_a in arithmetic average units; rms $= 1.25 \times R_a$. Full-film separation can be assumed for $\Lambda \geq 3$.

5.3.3.3 Non-Newtonian Lubricant in an Elastohydrodynamic Lubrication Contact

The friction shear stress for a non-Newtonian lubricant does not occur according to Equation 4.1. Several investigators [9–12] examined the effects of non-Newtonian lubricant behavior in the elastohydrodynamic lubrication (EHL) model. Bell [10] studied the effects of a Ree–Eyring fluid for which the shear rate is described by Equation 5.10:

$$\dot{\gamma} = \frac{\tau_0}{\eta} \sin\ h\left(\frac{\tau}{\tau_0}\right) \qquad (5.10)$$

where Eyring stress τ_0 and viscosity η are functions of temperature and pressure. Houpert [13] and Evans and Johnson [14] used the Ree–Eyring model for the analysis of EHL traction. When τ is small, Equation 5.10 describes a linear viscous behavior approaching that of a Newtonian lubricant. It has been established, however, that at high lubricant shear rates, the

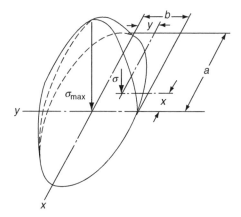

FIGURE 5.9 Ellipsoidal surface compressive stress distribution of point contact.

non-Newtonian characteristics tend to cause decreases in viscosity. As indicated, this occurs under conditions involving substantial sliding in addition to rolling. Since the film thickness that obtains is primarily a function of the lubricant properties at the inlet to the contact, a non-Newtonian lubricant will not significantly influence lubricant film thickness.

Non-Newtonian lubrication does, however, significantly influence friction in the contact. Because of friction, lubricant temperature in the contact rises causing viscosity to decrease. Since pressure increases greatly in, and varies over, the contact, it is evident that Equation 4.1 becomes

$$\tau = \eta(T,p)\frac{\partial u}{\partial z} \tag{5.11}$$

Assuming that the contact areas and pressure distributions are represented in Figure 5.9 for point contact and Figure 5.10 for line contact (as shown in Chapter 6 in the first volume of this handbook), Equation 5.11 defines the localized shear stress τ at any point (x, y) on the contact surface. As EHL films are very thin compared with the macrogeometrical dimensions of the rolling components, it is appropriate to approximate Equation 5.11 as follows:

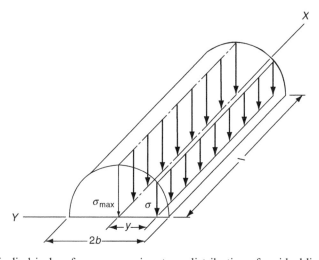

FIGURE 5.10 Semicylindrical surface compressive stress distribution of an ideal line contact.

$$\tau = \eta(T,p)\frac{v}{h} \tag{5.12}$$

where v is sliding velocity and h is the plateau lubricant film thickness. In Chapter 4, several equations were presented describing lubricant viscosity *vs* temperature and pressure. Of these, Equation 4.21 by Bair and Kottke (Ref. [7] of Chapter 4) or Equation 4.25 recommended by Harris (Ref. [9] of Chapter 4) may be substituted in Equation 5.12 for $\eta(T,p)$ to help calculate τ with satisfactory results.

5.3.3.4 Limiting Shear Stress

Gecim and Winer [12] and Bair and Winer [15] suggested alternative expressions for the relationship between shear stress and strain rate incorporating a limiting shear stress. They proposed that for a given pressure, temperature, and degree of sliding, there is a maximum shear stress that can be sustained. Based on experimental data using a disk machine, Figure 5.11 from Ref. [16] shows curves of traction coefficient *vs* pressure and slide–roll ratio that illustrate this phenomenon. Traction coefficient is defined as the ratio of average shear stress to average normal stress. From experiments, Schipper et al. [17] indicated a range of values for limiting shear stress; for example, $0.07 < \tau_{\mathrm{lim}}/p_{\mathrm{ave}} < 0.11$.

5.3.3.5 Fluid Shear Stress for Full-Film Lubrication

Trachman and Cheng [18] and Tevaarwerk and Johnson [19] investigated traction in rolling–sliding contacts and determined that Equation 4.1 pertains only to a situation involving relatively low slide-to-roll ratio; for example, less than 0.003 and shown in Figure 5.11. Following the method of Trachman and Cheng, at a given temperature and pressure it is possible to define local contact friction as follows:

$$\tau_{\mathrm{f}} = \left(\tau_{\mathrm{N}}^{-1} + \tau_{\mathrm{lim}}^{-1}\right)^{-1} \tag{5.13}$$

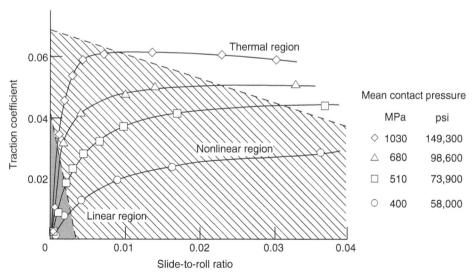

FIGURE 5.11 Curves of traction measured using a disk machine operating in line contact. (From Schipper, D., et al., *ASME Trans., J. Tribol.*, 112, 392–397, 1990. With permission.)

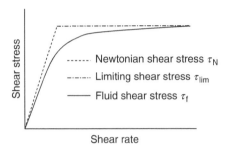

FIGURE 5.12 Schematic illustration of Equation 5.13. (From Houpert, L., *ASME Trans., J. Lubr. Technol.*, 107(2), 241, 1985. With permission.)

where τ_N is the Newtonian portion of the friction shear stress as defined by Equation 4.1 and τ_{lim} is the maximum shear stress that can be sustained at the contact pressure. Figure 5.12 schematically demonstrates Equation 5.13.

5.3.4 SLIDING DUE TO ROLLING MOTION: PARTIAL OIL-FILM LUBRICATION

5.3.4.1 Overall Surface Friction Shear Stress

When the lubricant film is insufficient to completely separate the surfaces in rolling contact, that is for $\Lambda < 3$, some of the surface peaks, also called asperities, as illustrated in Figure 5.13, break through the lubricant film and contact each other. The sliding friction shear stress during this asperity–asperity interaction occurs in the regime of boundary lubrication and may be calculated using Equation 5.5 for a ball–raceway or point contact. Only a portion of

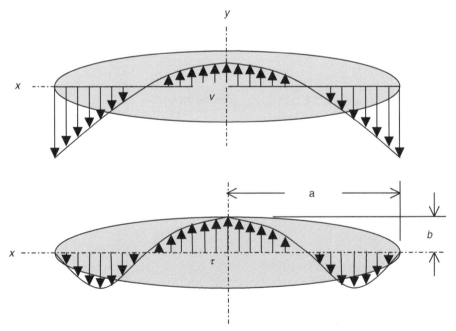

FIGURE 5.13 Distributions of sliding velocity and surface friction shear stress over an elliptical area of rolling element–raceway contact in a radially loaded, radial ball bearing.

the contact, however, operates in this manner; the remainder of the contact surface operates according to fluid-film lubrication; that is, Equation 5.13. Therefore, as given by Harris and Barnsby [20], the friction shear stress acting at any point (x, y) in the contact may be described by Equation 5.14:

$$\tau = c_v \frac{A_c}{A_0} \mu_a \sigma + \left(1 - \frac{A_c}{A_0}\right) \left(\tau_N^{-1} + \tau_{lim}^{-1}\right)^{-1} \tag{5.14}$$

where A_c is the area associated with asperity–asperity contact, A_0 is the total contact area, and σ is the normal stress or contact pressure. Coefficient of sliding $c_v = +1$ or -1 depending on the direction of sliding velocity. In Equation 5.14, it is necessary to define values for τ_{lim} and μ. These values can only be determined through full-scale bearing testing. Based on comparison of predicted to tested bearing heat generation rates, τ_{lim} can be estimated as $0.1 p_{ave}$ and $\mu_a \approx 0.1$ for oil-lubricated bearings.

For an oil-lubricated, elliptical area contact, operating mainly in rolling motion, the sliding velocity and surface friction shear stress distributions are illustrated in Figure 5.13.

5.3.4.2 Friction Force

It can be observed from Figure 5.13 that friction shear stress τ is a strong function of sliding velocity v notwithstanding the microcontact portion of Equation 5.14. The friction force acting over the contact surface is obtained by integration.

$$F_y = \int \tau dA = ab \int_{-1}^{+1} \int_{-\sqrt{1-q^2}}^{+\sqrt{1-q^2}} c_v \frac{A_c}{A_0} \mu_a \sigma + \left(1 - \frac{A_c}{A_0}\right) \left(\tau_N^{-1} + \tau_{lim}^{-1}\right)^{-1} dt \, dq \tag{5.15}$$

Contact pressure σ (or p) at any point (x, y) is determined from Equation 5.4.

At a given temperature, lubricant viscosity in the contact might be calculated using Equation 4.25:

$$\eta = C_1 + \frac{C_2}{1 + e^{-(\sigma - C_3)/C_4}} \tag{4.25}$$

5.4 REAL SURFACES, MICROGEOMETRY, AND MICROCONTACTS

5.4.1 Real Surfaces

To calculate friction force F using Equation 5.15, it is also necessary to determine the ratio A_c/A_0. Therefore, the microgeometry of the rolling contact surfaces must be considered. In calculating the lubricant film thickness in Chapter 4, the rolling contact surfaces are considered perfectly smooth. The assumption is now made that the lubricant film thickness calculated using that assumption separates the mean planes of the "rough" surfaces as illustrated in Figure 5.14.

The surfaces fluctuate randomly about their mean planes in accordance with a probability distribution. The rms value of this distribution is denoted σ_1 for the upper surface and σ_2 for the lower surface. When the combined surface fluctuations at a given position exceed the gap h due to the lubricant film, a microcontact occurs. At the microcontacts, the surfaces deform elastically and possibly plastically. The aggregate of the microcontact areas is generally a small fraction ($<5\%$) of the nominal area of contact for $1 \leq \Lambda \leq 3$.

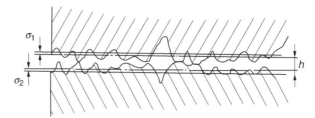

FIGURE 5.14 Asperity contacts through partial oil film.

A microcontact model uses surface microgeometry data to predict, at a minimum, the density of microcontacts, the real area of contact, and the elastically supported mean load. One of the earliest and simplest microcontact models is that of Greenwood and Williamson (GW) [21]. Generalizations of this model applicable to isotropic surfaces have been developed by Bush et al. [22] and by O'Callaghan and Cameron [23]. Bush et al. [24] also treated a strongly anisotropic surface. One of the most comprehensive models yet developed is ASPERSIM [25], which requires a nine-parameter microgeometry description and accounts for anisotropic as well as isotropic surfaces. A comparison of various microcontact models conducted by McCool [26] has shown that the GW model, despite its simplicity, compares favorably with the other models. Because it is much easier to implement than the other models, the GW model is the microcontact model considered here.

5.4.2 GW MODEL

For the contact of real surfaces, Greenwood and Williamson [21] developed one of the first models that specifically accounted for the random nature of interfacial phenomena. The model applies to the contact of two flat plastic planes, one rough and the other smooth. It is readily adapted to the case of two rough surfaces as discussed further. In the GW model, the rough surface is presumed to be covered with local high spots or asperities whose summits are spherical. The summits are presumed to have the same radius R, but randomly variable heights, and to be uniformly distributed over the rough surface with a known density D_{SUM} of summits/unit area.

The mean height of the summits lies above the mean height of the surface as a whole by the amount $\bar{z}_s$ indicated in Figure 5.15. The summit heights z_s are assumed to follow a Gaussian probability law with a standard deviation σ_s. Figure 5.16 shows the assumed form for the summit height distribution or probability density function (pdf) $f(z_s)$. It is symmetrical about the mean summit height. The probability that a summit has a height, measured relative to the summit mean plane in the interval $(z_s, z_s + dz_s)$, is expressed in terms of the pdf as $f(z_s) \, dz_s$. The probability that a randomly selected summit has a height in excess of some value d is the area under the pdf to the right of d. The equation of the pdf is

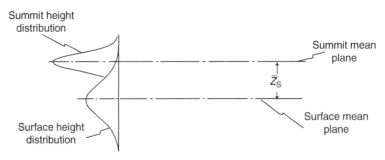

FIGURE 5.15 Surface and summit mean planes and distributions.

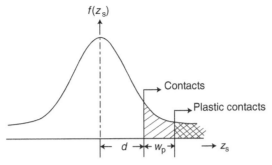

Summit height distribution

FIGURE 5.16 Spherical capped asperity in contact.

$$f(z_s) = \frac{e^{-(z_s/2S_s)^2}}{S_s\sqrt{2\pi}} \tag{5.16}$$

Therefore, the probability that a randomly selected summit has height in excess of d is

$$P[z_s > d] = \int_d^\infty f(z_s)\, ds \tag{5.17}$$

This integration must be performed numerically. Fortunately, however, the calculation can be related to tabulated areas under the standard normal curve for which the mean is 0 and the standard deviation is 1.0.

Using the standard normal density function $\phi(x)$, the probability that a summit has a height greater than d above the summit mean plane is calculated.

$$P[z_s > d] = \int_{d/S_s}^\infty \phi(x)\, dx = F_0\left(\frac{d}{S_s}\right) \tag{5.18}$$

where $F_0(t)$ is the area under the standard normal curve to the right of the value t. Values $F_0(t)$ for t ranging from 1.0 to 4.0 are given in column 2 of Table CD5.1.

It is assumed that when large flat surfaces are pressed together, their mean planes remain parallel. Thus, if a rough surface and a smooth surface are pressed against each other until the summit mean plane of the rough surface and the mean plane of the smooth surface are separated by an amount d, the probability that a randomly selected summit will be a microcontact is

$$P[\text{summit is a contact}] = P[z_s > d] = F_0\left(d/S_s\right) \tag{5.19}$$

As the number of summits per unit area is D_{SUM}, the average expected number of contacts in any unit area is

$$n = D_{SUM}F_0(d/S_s) \tag{5.20}$$

Given that a summit is in contact because its height z_s exceeds d, the summit must deflect by the amount $w = z_s - d$, as shown in Figure 5.16.

For notational simplicity, the subscript in z_s is henceforth deleted. For a sphere of radius R elastically deflecting by the amount w, the Hertzian solution gives the contact area:

$$A = \pi R w = \pi R(z - d) = \pi a^2 \qquad z > d \tag{5.21}$$

where a is the contact radius.

The corresponding asperity load is

$$Q_a = \tfrac{4}{3} E' R^{1/2} w^{3/2} = \tfrac{4}{3} E' R^{1/2} (z - d)^{3/2} \qquad z > d \tag{5.22}$$

where $E' = [(1 - v_1^2)/E_1 + (1 - v_2^2)/E_2]^{-1}$ and E_i, v_i ($i = 1, 2$) are Young's moduli and Poisson's ratios for the two bodies. The maximum Hertzian pressure in the microcontact is

$$\sigma = 1.5 \frac{P}{A} = \frac{2E' w^{1/2}}{\pi R^{1/2}} = \left(\frac{2E'}{\pi R^{1/2}} \right)(z - d)^{1/2} \tag{5.23}$$

Both A and Q_a are functions of the random variable z. The average or expected values of functions of random variables are obtained by integrating the function and the probability density of the random variable over the space of possible values of the random variable. The expected summit contact area is thus

$$A = \int_d^\infty \pi R(z - d) f(z) \, \mathrm{d}z \tag{5.24}$$

which transforms to

$$A = \pi R \sigma_s \int_{d/\sigma_s}^\infty \left(x - \frac{d}{S_s} \right) \phi_x \, \mathrm{d}x = \pi R S_s F_1 \left(\frac{d}{\sigma_s} \right) \tag{5.25}$$

where

$$F_1(t) = \int_t^\infty (x - t) \, \phi_x \, \mathrm{d}x \tag{5.26}$$

$F_1(t)$ is also given in Table CD5.1.

The expected total contact area as a fraction of the apparent area is obtained as the product of the average asperity contact area contributed by a single randomly selected summit and the density of summits. Thus, the ratio of contact to apparent area, A_c/A_0, is

$$\frac{A_c}{A_0} = \pi R S_s D_{\text{SUM}} F_1 \left(\frac{d}{S_s} \right) \tag{5.27}$$

By the same argument, the total load per unit area supported by asperities is

$$\frac{Q_a}{A_0} = \frac{4}{3} E' R^{1/2} S_s^{3/2} D_{\text{SUM}} F_{3/2} \left(\frac{d}{S_s} \right) \tag{5.28}$$

where

$$F_{3/2}(t) = \int_t^\infty (x - t)^{3/2} \, \phi(x) \, \mathrm{d}x \tag{5.29}$$

$F_{3/2}(t)$ is also given in Table CD5.1.

5.4.3 PLASTIC CONTACTS

A contacting summit will experience some degree of plastic flow when the maximum shear stress exceeds half the yield stress in simple tension. In the contact of a sphere and a flat, the maximum shear stress is related to the maximum Hertzian stress σ_0 by

$$\tau_{\max} = 0.31\sigma_0 \tag{5.30}$$

Thus, some degree of plastic deformation is present at a contact if $\tau_{\max} > Y/2$. Using the expression for σ_0 Equation 5.23 gives

$$\frac{0.31 \times 2E'\,(z - d)^{1/2}}{\pi R^{1/2}} > \frac{Y}{2} \tag{5.31}$$

or

$$z - d > 6.4R\left(\frac{Y}{E'}\right)^2 \equiv w_p \tag{5.32}$$

$$z > d + w_p \tag{5.33}$$

Thus, any summit whose height exceeds $d + w_p$ will have some degree of plastic deformation. The probability of a plastic summit is given by the shaded area in Figure 5.16 to the right of $d + w_p$. The expected number of plastic contacts per unit area becomes

$$n_p = D_{\text{SUM}}F_0\left(\frac{d}{S_s} + w_p^*\right) \tag{5.34}$$

where

$$w_p^* \equiv \frac{w_p}{S_s} \equiv 6.4\left(\frac{R}{S_s}\right)\left(\frac{Y}{E'}\right)^2 \tag{5.35}$$

For fixed d/σ_s the degree of plastic asperity interaction is determined by the value of w_p^*: the higher is w_p^*, the fewer the plastic contacts. Accordingly, GW use the inverse, $1/w_p^*$, as a measure of the plasticity of an interface. For a given nominal pressure Q/A_0, d/S_s is found by solving Equation 5.28 assuming that most of the load is elastically supported.

5.4.4 APPLICATION OF THE GW MODEL

To use the GW model for a lubricated contact, (1) the height d relative to the mean plane of the summit heights to h, the thickness of the lubricant film between the contact surfaces, and (2) values of the GW parameters R, D_{SUM}, and σ_s must be established. For (1), the first step is to calculate the composite roughness rms value of the two surfaces as

$$s = \left(s_1^2 + s_2^2\right)^{1/2} \tag{5.36}$$

When the mean plane of a rough surface with this rms value is held at height h above a smooth plane, the rms value of the gap width is the same as shown in Figure 5.17, where both surfaces are rough. It is in this sense that the surface contact of two rough surfaces may be translated into the equivalent contact of a rough surface and a smooth surface. As shown in Figure 5.15, the summit and surface mean planes are separated by an amount z_s.

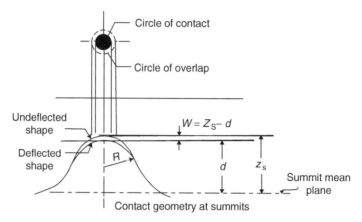

FIGURE 5.17 Distribution of summit heights.

For an isotropic surface with normally distributed height fluctuations, the value of z_s has been found by Bush et al. [22] to be

$$\bar{z}_s = \frac{4s}{\sqrt{\pi\alpha}} \tag{5.37}$$

The quantity α, known as the bandwidth parameter, is defined by

$$\alpha = \frac{m_0 m_4}{m_2^2} \tag{5.38}$$

where m_0, m_2, and m_4 are known as the zeroth, second, and fourth spectral moments of a profile. They are equivalent to the mean square height, slope, and second derivative of a profile in an arbitrary direction; that is

$$m_0 = E(z^2) = s^2 \tag{5.39}$$

$$m_2 = E\left[\left(\frac{dz}{dx}\right)^2\right] \tag{5.40}$$

$$m_2 = E\left[\left(\frac{d^2 z}{dx^2}\right)^2\right] \tag{5.41}$$

where $z(x)$ is a profile in an arbitrary direction x, $E[\,]$ denotes statistical expectation, and m_0 is simply the mean square surface height. The square root of m_0 or rms is sometimes referred to as S or R_q and forms part of the usual output of a stylus-measuring device. Some profile-measuring devices also give the rms slope, which is the same as $(m_2)^{1/2}$ converted from radians to degrees. No commercial equipment is yet available to measure m_4. Measurements of m_4 made so far have used custom computer processing of the signal output of profile measurement equipment.

Bush et al. [24] also show that the variance S_s^2 of the surface summit height distribution is related to S^2, the variance of the composite surfaces, by

$$S_s^2 = \left(1 - \frac{0.8968}{\alpha}\right) S^2 \tag{5.42}$$

A summit located a distance d from the summit height mean plane is at a distance $h = d + \bar{z}_s$, from the surface mean plane. Thus,

$$d = h - \bar{z}_s \tag{5.43}$$

Using Equation 5.37 for $\bar{z}_s$ and Equation 5.42 for s_s gives

$$\frac{d}{s_s} = \frac{h/s - 4/\sqrt{\pi\alpha}}{((1 - 0.8968)/\alpha)^{1/2}} \tag{5.44}$$

Equation 5.44 shows that d/s_s is linearly related to the lubricant film parameter Λ.

For a specified value of Λ, d/s_s is calculated from Equation 5.44. For an isotropic surface, the two parameters D_{SUM} and R may be expressed as (from Ref. [27])

$$D_{\text{SUM}} = \frac{m_4}{6\pi m_2 \sqrt{3}} \tag{5.45}$$

$$R = \frac{3}{8}\sqrt{\frac{\pi}{m_4}} \tag{5.46}$$

For an anisotropic surface, the value of m_2 will vary with the direction in which the profile is taken on the surface. The maximum and minimum values occur in two orthogonal "principal" directions. Sayles and Thomas [28] recommend the use of an equivalent isotropic surface for which m_2 is calculated as the harmonic mean of the m_2 values found along the principal directions. The value of m_4 is similarly taken as the harmonic mean of the m_4 values in these two directions.

5.4.5 ASPERITY-SUPPORTED AND FLUID-SUPPORTED LOADS

For a specified contact with semiaxes a and b, under a load Q, with plateau lubricant film thickness h and given values of m_0, m_2, and m_4, the load Q_a carried by the asperities is determined by first calculating Q/A_0 from Equation 5.28 and using

$$Q_a = \pi ab \left(\frac{Q}{A_0}\right) \tag{5.47}$$

The fluid-supported load is then

$$Q_f = Q - Q_a \tag{5.48}$$

If $Q_a > Q$, the implication is that the lubricant film thickness is larger than that calculated using smooth surface theory. In this case, Equation 5.28 must be solved iteratively until $Q_a = Q$.

See Example 5.1.

5.4.6 SLIDING DUE TO ROLLING MOTION: ROLLER BEARINGS

5.4.6.1 Sliding Velocities and Friction Shear Stresses

For roller bearings operating with predominantly rolling motion, the roller–raceway contact friction analyses are very similar to those described for ball–raceway contacts. As indicated in

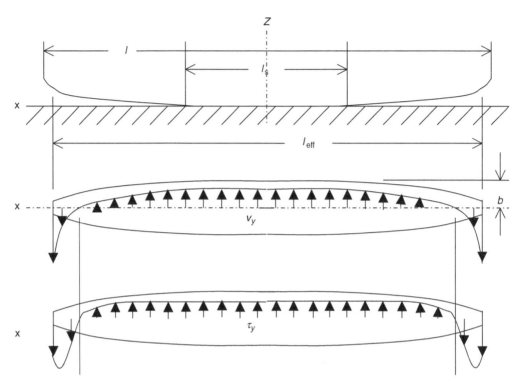

FIGURE 5.18 Distributions of sliding velocity and surface friction shear stress over an area of crowned roller–raceway contact in a cylindrical roller bearing under load. Roller crowning is illustrated in the uppermost drawing. In this contact, ideal normal stress distribution is not achieved.

Chapter 6 in the first volume of this handbook, rollers and raceways are crowned to avoid or minimize edge loading, and under applied load the contact surface is curved in the plane passing through the bearing axis of rotation and the center of rolling contact. Pure rolling is defined by instant centers at which no relative motion of the contacting elements occurs; that is, the surfaces have the same velocities at such points. Therefore, in a radial, cylindrical roller bearing having crowned components, only two points of pure rolling can exist on the major axis of each contact surface. At all other points sliding must occur. The same is basically true for the roller–raceway contacts in radial, spherical, and tapered roller bearings. Figure 5.18 schematically depicts sliding velocities and surface friction shear stresses in a crowned cylindrical roller–raceway contact.

5.4.6.2 Contact Friction Force

As demonstrated in Chapter 1 and Chapter 3, the friction force over the contact is calculated by dividing the contact into n laminae; then,

- Establishing the normal stress distribution over each lamina k
- Determining the average lubricant viscosity η_k using a pressure–viscosity relationship at contact temperature
- Calculating the plateau lubricant film thickness and subsequently A_c/A_0 using the GW method
- Determining sliding velocities v_k based on contact deformation criteria

- Calculating the surface friction shear stress τ_k for each lamina k using Equation 5.14
- Using Simpson's rule, numerically integrating $\tau_k \times A_k$ across the contact, where $A_k = 2 b_k \times w$, the width of a lamina

Depending on the geometries of the rolling components and the amount of normal loading between them, sliding motions that accompany the essential rolling motion can vary in significance with regard to the friction generated due to rolling. Generally, for mainly rolling motion, the amount of rolling contact friction tends to be small.

5.4.7 SLIDING DUE TO SPINNING AND GYROSCOPIC MOTIONS

5.4.7.1 Sliding Velocities and Friction Shear Stresses

Ball bearings that operate with nonzero contact angles; for example, angular-contact and thrust ball bearings, experience spinning contact motions, and gyroscopic moments that cause gyroscopic motions. Nonzero contact angle roller bearings also experience spinning motions; however, gyroscopic moments are resisted by nonuniform roller–raceway loading per unit length. Spinning motions and gyroscopic motions in ball bearings were discussed in Chapter 2. The sliding velocity distribution and surface friction shear stress distribution over a loaded angular-contact ball bearing contact that experiences rolling, spinning and gyroscopic motions is illustrated in Figure 5.19. In Figure 5.19, v_y is the sliding velocity in the direction of rolling, and v_x is the sliding velocity in the direction transverse to rolling, caused by gyroscopic motion; v_y gives rise to friction shear stress component τ_y, and v_x gives rise to friction shear stress component τ_x. This is shown by expanding Equation 5.14 as follows:

$$\tau_y = c_v \frac{A_c}{A_0} \mu_a \sigma + \left(1 - \frac{A_c}{A_0}\right) \left(\frac{h}{\eta v_y} + \frac{1}{\tau_{\text{lim}}}\right)^{-1} \tag{5.49}$$

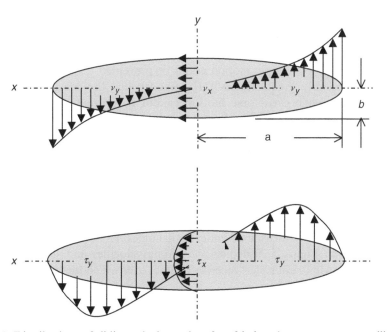

FIGURE 5.19 Distributions of sliding velocity and surface friction shear stress over an elliptical area of rolling element–raceway contact in an angular-contact ball bearing.

$$\tau_x = c_v \frac{A_c}{A_0} \mu_a \sigma + \left(1 - \frac{A_c}{A_0}\right) \left(\frac{h}{\eta v_x} + \frac{1}{\tau_{\text{lim}}}\right)^{-1} \tag{5.50}$$

As an alternative to Equation 5.49 and Equation 5.50, Harris [29] used 37 sets of data of traction coefficient *vs* slide-to-roll ratio and Λ collected on a v-ring-single ball test rig to generate the following empirical relationship:

$$\mu = -2.066 \times 10^{-3} + 2.612 \times 10^{-6} \left[\frac{1}{\Lambda} \ln\left(\frac{\eta}{\eta_0}\right)\right]^2 - 5.605 \times 10^{-2} \left[\frac{v}{U} \ln\left(\frac{v}{U}\right)\right] \tag{5.51}$$

where η is the lubricant viscosity at contact pressure, η_0 is the lubricant viscosity at atmospheric pressure, and U is the rolling velocity. Traction coefficient μ is directional; that is, μ_y or μ_x and was developed considering average normal stress over the contact. It might, however, be considered as occurring at a point in a contact such that $\tau_y = \mu_y \sigma$ and $\tau_x = \mu_x \sigma$. The lubricant used during the v-ring-ball testing was a Mil-L-23699 polyolester.

5.4.7.2 Contact Friction Force Components

The friction force components in the rolling direction F_y and in the gyroscopic direction F_x may be determined by integration over the contact area. Accordingly,

$$F_y = \int \tau_y \, dA = ab \int_{-1}^{+1} \int_{-\sqrt{1-q^2}}^{+\sqrt{1-q^2}} c_v \frac{A_c}{A_0} \mu_a \sigma + \left(1 - \frac{A_c}{A_0}\right) \left(\frac{h}{\eta v_y} + \frac{1}{\tau_{\text{lim}}}\right)^{-1} dt \, dq \tag{5.52}$$

$$F_x = \int \tau_x \, dA = ab \int_{-1}^{+1} \int_{-\sqrt{1-q^2}}^{+\sqrt{1-q^2}} c_v \frac{A_c}{A_0} \mu_a \sigma + \left(1 - \frac{A_c}{A_0}\right) \left(\frac{h}{\eta v_x} + \frac{1}{\tau_{\text{lim}}}\right)^{-1} dt \, dq \tag{5.53}$$

Jones [30] assumed that gyroscopic motion could be prevented if the ball–raceway friction coefficient was sufficiently great. Harris [31] demonstrated the inaccuracy of the Jones assumption; but, that while gyroscopic motion cannot be prevented in the presence of a gyroscopic moment, its speed is nevertheless very small compared with ball speeds about the two orthogonal axes.

5.4.8 Sliding in a Tilted Roller–Raceway Contact

In Chapter 1, it was shown that rollers in cylindrical roller or tapered roller bearings subjected to moment loading or misalignment that causes moment loading undergo tilt angles ζ_j to accommodate the applied load; the subscript refers to the roller azimuth location. Similarly, cylindrical rollers subjected to thrust load undergo tilt angles. Thus, the normal loading on each contact is nonuniform. Figure 5.20 depicts the sliding velocities and surface friction shear stresses in a crowned cylindrical roller–raceway contact over which the loaded roller is tilted.

5.5 CLOSURE

This chapter contains a generalized approach to predicting surface friction stresses and forces for rolling element–raceway contacts; that is, both solid-film lubrication and oil-lubrication

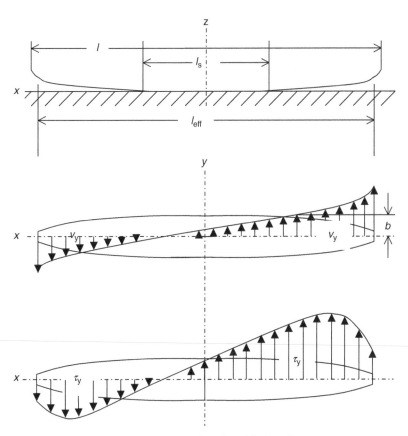

FIGURE 5.20 Distributions of sliding velocity and surface friction shear stress over an area of crowned roller–raceway contact in a cylindrical roller bearing under load. The roller is tilted over the contact to accommodate bearing misalignment or applied thrust load. Roller crowning is illustrated in the uppermost drawing.

conditions are considered. In the former case, Coulomb friction is assumed and the direction of friction shear stress at a given surface point is dictated by the direction of sliding motion at that point. With regard to oil-lubrication, the approach is taken to predicting key performance-related parameters descriptive of real EHL contacts. These parameters include true contact area, plastic contact area, fluid and asperity load sharing, and the relative contributions of the fluid and asperities to overall friction. It is recognized that, using more elegant and complex analytical methods such as very fine mesh, multithousand node, finite-element analysis together with solutions of the Reynolds and energy equations in three dimensions, it is possible to obtain a more generalized solution with perhaps increased accuracy. Unfortunately, using currently available computing equipment, such solutions would require several hours of computational time to enable the performance analysis of a single operating condition for a rolling bearing containing only a small complement of rolling elements.

The equations provided in this chapter for frictional shear stress are based on the assumption of Hertz pressure (normal stress) applied over the contact. In the case of oil-lubricated bearings, the Hertzian stress distribution is assumed to be unmodified by EHL conditions. This assumption is sufficiently accurate for most rolling element–raceway contacts in that such loading is reasonably heavy; for example, generally at least several hundred MPa. Furthermore, the assumption is made that Equation 5.14 can be applied at every point in the contact. With respect to the Coulomb friction component of surface shear stress, it is recognized that surface roughness peaks cause local pressures in excess of Hertzian values and these will cause localized shear stresses in excess of those predicted by Equation 5.14. Accommodation of these variations tends to increase the computational time beyond current engineering practicality. Therefore, for engineering purposes, frictional shear stress may be calculated according to the average condition in each contact.

REFERENCES

1. Drutowski, R., Energy losses of balls rolling on plates, *Friction and Wear*, Elsevier, Amsterdam, 1959, pp. 16–35.
2. Greenwood, J. and Tabor, D., *Proc. Phys. Soc. London*, 71, 989, 1958.
3. Drutowski, R., Linear dependence of rolling friction on stressed volume, *Rolling Contact Phenomena*, Elsevier, Amsterdam, 1962.
4. Reynolds, O., *Philos. Trans. R. Soc. London*, 166, 155, 1875.
5. Poritsky, H., *J. Appl. Mech.*, 72, 191, 1950.
6. Cain, B., *J. Appl. Mech.*, 72, 465, 1950.
7. Heathcote, H., *Proc. Inst. Automob. Eng., London*, 15, 569, 1921.
8. Johnson, K., Tangential tractions and micro-slip, *Rolling Contact Phenomena*, Elsevier, Amsterdam, 1962, pp. 6–28.
9. Sasaki, T., Mori, H., and Okino, N., Fluid lubrication theory of rolling bearings parts I and II, *ASME Trans., J. Basic Eng.*, 166, 175, 1963.
10. Bell, J., Lubrication of rolling surfaces by a Ree–Eyring fluid, *ASLE Trans.*, 5, 160–171, 1963.
11. Smith, F., Rolling contact lubrication—the application of elastohydrodynamic theory, ASME Paper 64-Lubs-2, April 1964.
12. Gecim, B. and Winer, W., A film thickness analysis for line contacts under pure rolling conditions with a non-Newtonian rheological model, ASME Paper 80C2/LUB 26, August 8, 1980.
13. Houpert, L., New results of traction force calculations in EHD contacts, *ASME Trans., J. Lubr. Technol.*, 107(2), 241, 1985.
14. Evans, C. and Johnson, K., The rheological properties of EHD lubricants, *Proc. Inst. Mech. Eng.*, 200(C5), 303–312, 1986.
15. Bair, S. and Winer, W., A rheological model for elastohydrodynamic contacts based on primary laboratory data, *ASME Trans., J. Lubr. Technol.*, 101(3), 258–265, 1979.
16. Johnson, K. and Cameron, A., *Proc. Inst. Mech. Eng.*, 182(1), 307, 1967.
17. Schipper, D., et al., Micro-EHL in lubricated concentrated contacts, *ASME Trans., J. Tribol.*, 112, 392–397, 1990.
18. Trachman, E. and Cheng, H., Thermal and non-Newtonian effects on traction in elastohydrodynamic contacts, *Proc. Inst. Mech. Eng.*, 2nd Symposium on Elastohydrodynamic Lubrication, Leeds, 1972, pp. 142–148.
19. Tevaarwerk, J. and Johnson, K., A simple non-linear constitutive equation for EHD oil films, *Wear*, 35, 345–356, 1975.
20. Harris, T. and Barnsby, R., Tribological performance prediction of aircraft turbine mainshaft ball bearings, *Tribol. Trans.*, 41(1), 60–68, 1998.
21. Greenwood, J. and Williamson, J., Contact of nominally flat surfaces, *Proc. R. Soc. London, Ser. A.*, 295, 300–319, 1966.
22. Bush, A., Gibson, R., and Thomas, T., The elastic contact of a rough surface, *Wear*, 35, 87–111, 1975.
23. O'Callaghan, M. and Cameron, M., Static contact under load between nominally flat surfaces, *Wear*, 36, 79–97, 1976.
24. Bush, A., Gibson, R., and Keogh, G., Strongly anisotropic rough surfaces, ASME paper 78-LUB-16, 1978.
25. McCool, J. and Gassel, S., The contact of two surfaces having anisotropic roughness geometry, ASLE Special Publication (SP-7), 29–38, 1981.

26. McCool, J., Comparison of models for the contact of two surfaces having anisotropic roughness geometry, *Wear*, 107, 7–60, 1986.
27. Nayak, P., Random process model of rough surfaces, *ASME Trans., J. Tribol.*, 93F, 398–407, 1971.
28. Sayles, R. and Thomas, T., Thermal conductances of a rough elastic contact, *Appl. Energy*, 2, 249–267, 1976.
29. Harris, T., Establishment of a new rolling bearing fatigue life calculation model, Final Report U.S. Navy Contract N00421-97-C-1069, February 23, 2002.
30. Jones, A., Motions in loaded rolling element bearings, *ASME Trans., J. Basic Eng.*, 1–12, 1959.
31. Harris, T., An analytical method to predict skidding in thrust-loaded, angular-contact ball bearings, *ASME Trans., J. Lubr. Techol.*, 93, 17–24, 1971.

6 Friction Effects in Rolling Bearings

LIST OF SYMBOLS

Symbol	Description	Units
a	Semimajor axis of projected contact ellipse	mm (in.)
A_c	True average contact area	mm^2 (in.^{2})
A_0	Apparent contact area	mm^2 (in.^{2})
A_1	Ball center axial position variable	mm (in.)
A_2	Ball center radial position variable	mm (in.)
b	Semiminor axis of projected contact ellipse	mm (in.)
B	$f_i + f_o - 1$	
D	Roller or ball diameter	mm (in.)
E_1, E_2	Elastic moduli of bodies 1 and 2	MPa (psi)
E'	Reduced elastic modulus	MPa (psi)
f	r/D	
F	Contact friction force	N (lb)
F_c	Centrifugal force	N (lb)
F_{CL}	Friction force between cage rail and ring land	N (lb)
g	Gravitational constant	mm/sec^2 (in./sec^2)
h	Lubricant film thickness	mm (in.)
h_c	Central or plateau lubricant film thickness	mm (in.)
J	Mass moment of inertia	kg. $\cdot$ mm^2 (in. $\cdot$ lb $\cdot$ sec^2)
l	Roller length end-to-end	mm (in.)
l_{eff}	Effective roller length	mm (in.)
l_s	Roller straight length	mm (in.)
M	Moment	N $\cdot$ mm (in. $\cdot$ lb)
M_g	Gyroscopic moment	N $\cdot$ mm (in. $\cdot$ lb)
q	x/a	
Q	Roller or ball load	N (lb)
Q_a	Roller end–guide flange load	N (lb)
Q_{CG}	Cage web–rolling element load	N (lb)
R	Radius of deformed contact surface	mm (in.)
t	y/b	
T	Temperature	°C (°F)
u	Surface velocity	mm/sec (in./sec)

v	Sliding velocity	mm/sec (in./sec)
X_1	Ball center axial position variable	mm (in.)
X_2	Ball center radial position variable	mm (in.)
w	Width of a lamina, width	mm (in.)
W	Lubricant flow rate through bearing	cm^3/mm (gal/min.)
Z	Number of rolling elements	
γ	Shear rate	sec^{-1}
δ_a	Bearing axial deflection	mm (in.)
δ	Contact deformation	mm (in.)
ζ	$2f/(2f+1)$	
ζ	Roller tilting angle	°, rad
η	Lubricant viscosity	cp (lb · sec/in.^{2})
μ	Coefficient of friction for boundary or solid-film lubrication	
ν_1, ν_2	Poisson's ratio for bodies 1 and 2	
ρ	Radius	mm (in.)
ξ	Lubricant effective density	g/mm^3 (lb/in.^{3})
ξ_1	Lubricant density	g/mm^3 (lb/in.^{3})
ξ	Roller skewing angle	°, rad
σ	Normal contact stress or pressure	MPa (psi)
σ_0	Maximum normal contact stress or pressure	MPa (psi)
τ	Shear stress	MPa (psi)
ω	Rotational speed	rad/sec
Ω	Ring rotational speed	rad/sec

Subscripts

CG	Cage
CL	Cage land
CP	Cage pocket
CR	Cage rail
g	Gyroscopic motion
i	Inner raceway
j	Rolling element location
n	Outer or inner raceway or ring, o or i
m	Cage or orbital motion
o	Outer raceway
R	Roller
x'	x' Direction
y'	y' Direction
z'	z' Direction
λ	Lamina

6.1   GENERAL

In Chapter 5, the sources and magnitudes of friction in ball–raceway and roller–raceway contacts were defined. While these are the salient considerations in the study of effects of friction on rolling bearing performance, other sources of friction in the bearing can have significant and even overriding effects on bearing performance. For example, the type of oil lubrication and the amount of lubricant in the bearing, and the interaction of the cage with the rolling elements and with piloting surfaces on the bearing rings are important sources of friction. Also, the interaction of integral contact seals with bearing rings will generally have a friction effect substantially greater than all of the other sources heretofore indicated. Seal friction, however, is not a topic explored in detail in this text.

Rolling element speeds can be significantly influenced by friction, affecting rolling element centrifugal forces, gyroscopic moments, and bearing endurance. Excessive friction at high speeds can cause rolling elements to undergo gross sliding over the raceways. This motion called skidding can reduce bearing endurance. Friction can have ancillary, but important, effects on bearing performance. In roller bearings, friction between roller ends and guide ring flanges can cause rollers to skew, shortening bearing endurance. All of these effects will be discussed in this chapter.

6.2 BEARING FRICTION SOURCES

6.2.1 SLIDING IN ROLLING ELEMENT–RACEWAY CONTACTS

As indicated above, this salient feature of rolling bearing performance is discussed in detail in Chapter 5.

6.2.2 VISCOUS DRAG ON ROLLING ELEMENTS

In fluid-lubricated rolling bearings, during operation a certain amount of lubricant occupies the free space within the boundaries of the bearing. Because of their orbital motion, the balls or rollers must force their way through this fluid; the viscous fluid creates a drag force that retards the orbital motion. The fluid within the bearing free space is a mixture of gas (usually air) and lubricant. It is assumed that the drag caused by the gaseous atmosphere is insignificant; rather, the drag force depends on the quantity of the lubricant dispersed in the gas–lubricant mixture. Therefore, the mixture has an effective viscosity and an effective specific gravity. The viscous drag force acting on a ball as indicated in Ref. [1] can be approximated by

$$F_v = \frac{c_v \pi \xi D^2 (d_m \omega_m)^{1.95}}{32g} \tag{6.1}$$

where ξ is the weight of the lubricant in the bearing free space divided by the volume of the free space. Similarly, for an orbiting roller,

$$F_v = \frac{c_v \xi l D (d_m \omega_m)^{1.95}}{16g} \tag{6.2}$$

The drag coefficients c_v in Equation 6.1 and Equation 6.2 can be obtained from Ref. [2] among many others.

From the testing of ball bearings operating with circulating oil lubrication, Parker [3] established an empirical formula to estimate the percentage of the bearing free space occupied by the fluid lubricant. Using Parker's formula, it is possible to calculate the effective fluid density ξ as indicated in the following equation:

$$\xi = \frac{\xi_l W^{0.37}}{n d_m^{1.7}} \times 10^5 \tag{6.3}$$

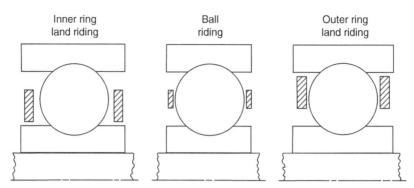

FIGURE 6.1 Cage types.

6.2.3 SLIDING BETWEEN THE CAGE AND THE BEARING RINGS

Three basic cage types are used in ball and roller bearings: (1) ball riding (BR) or roller riding (RR), (2) inner ring land riding (IRLR), and (3) outer ring land riding (ORLR). These are illustrated schematically in Figure 6.1.

BR and RR cages are usually of relatively inexpensive manufacture and are usually not used in critical applications. The choice of an IRLR or ORLR cage depends largely on the application and designer preference. An IRLR cage is driven by a force between the cage rail and inner ring land as well as by the rolling elements. ORLR cage speed is retarded by cage rail/outer ring land drag force. The magnitude of the drag or drive force between the cage rail and ring land depends on the resultant cage/rolling element loading, the eccentricity of the cage axis of rotation, and the speed of the cage relative to the ring on which it is piloted. If the cage rail/ring land normal force is substantial, hydrodynamic short bearing theory [4] might be used to establish the friction force F_{CL}. For a properly balanced cage and a very small resultant cage/rolling element load, Petroff's law can be applied; for example,

$$F_{CL} = \frac{\eta \pi w_{CR} c_n d_{CR} (\omega_c - \omega_n)}{1 - (d_1/d_2)} \quad \begin{array}{l} c_o = 1 \\ c_i = -1 \end{array} \quad (6.4)$$

where d_2 is the larger of the cage rail and ring land diameters and d_1 is the smaller.

6.2.4 SLIDING BETWEEN ROLLING ELEMENTS AND CAGE POCKETS

At any given azimuth location, there is generally a normal force acting between the rolling element and its cage pocket. This force can be positive or negative depending on whether the rolling element is driving the cage or *vice versa*. It is also possible for a rolling element to be free in the pocket with no normal force exerted; however, this situation will be of less usual occurrence. Insofar as rotation of the rolling element about its own axes is concerned, the cage is stationary. Therefore, pure sliding occurs between rolling elements and cage pockets. The amount of friction that occurs thereby depends on the rolling element–cage normal loading, lubricant properties, rolling element speeds, and cage pocket geometry. The last variable is substantial in variety. Generally, application of simplified elastohydrodynamic theory should suffice to analyze the friction forces.

6.2.5 SLIDING BETWEEN ROLLER ENDS AND RING FLANGES

In a tapered roller bearing and in a spherical roller bearing with asymmetrical rollers, concentrated contact always occurs between the roller ends and the inner (or outer) ring

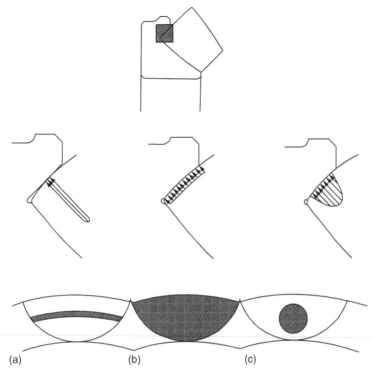

FIGURE 6.2 Contact types and pressure profiles between sphere end rollers and flanges in a spherical roller thrust bearing.

flange owing to a force component that drives the rollers against the flange. Also, in a radial cylindrical roller bearing, which can support thrust load in addition to the predominant radial load by virtue of having flanges on both inner and outer rings, sliding occurs simultaneously between the roller ends and both inner and outer rings. In these cases, the geometries of the flanges and roller ends are extremely influential in determining the sliding friction between those contacting elements.

The most general case for roller end–flange contact occurs, as shown in Figure 6.2, in a spherical roller thrust bearing. The different types of contact are illustrated in Table 6.1 for rollers having sphere ends.

Rydell [5] indicates that optimal frictional characteristics are achieved with point contacts between roller ends and flanges. Additionally, Brown et al. [6] studied roller end wear criteria for high-speed cylindrical roller bearings. They found that increasing roller

TABLE 6.1
Roller End–Flange Contact vs. Geometry[a]

	Flange Geometry	Type of Contact
a	Portion of a cone	Line
b	Portion of sphere, $R_f = R_{re}$	Entire surface
c	Portion of sphere, $R_f > T_{re}$	Point

[a] R_f is the flange surface radius of curvature; R_{re} is the roller end radius of curvature.

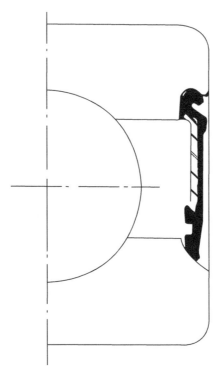

FIGURE 6.3 Deep-groove ball bearing with integral seal.

corner radius runout tends to increase wear. Increasing roller end clearance and l/D ratio also tend toward increased roller wear, but are of lesser consequence than roller corner radius runout.

6.2.6 SLIDING FRICTION IN SEALS

Many rolling bearings, particularly grease-lubricated, deep-groove ball bearings, are assembled with integral seals. As illustrated in Figure 6.3, such seals generally consist of an elastomeric material partially encased in a steel or plastic carrier. The elastomeric sealing bears (rides) either on a ring land or on a special recess or groove cut into the inner ring as shown in Figure 6.3. In any case, the seal friction due to sliding between the elastomer and bearing ring surface normally exceeds the total of all other sources of friction in the bearing unit. The technology of seal friction depends frequently on the specific mechanical structure of the seal and on the properties of the elastomeric material. Analysis of seal friction is not covered in this text.

6.3 BEARING OPERATION WITH SOLID-FILM LUBRICATION: EFFECTS OF FRICTION FORCES AND MOMENTS

6.3.1 BALL BEARINGS

In Chapter 5, it was shown that friction in solid-film lubricated ball–raceway contacts could be analyzed considering Coulomb friction; that is, surface friction shear stress τ at a given point (x, y) in the contact surface to be represented as $\mu\sigma$, μ is a coefficient of friction and σ is

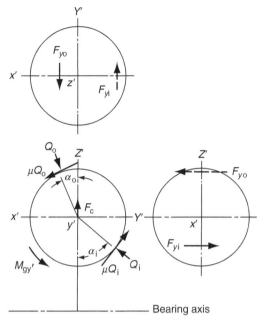

FIGURE 6.4 Forces and moments acting on a ball.

the normal stress at point (x, y). With this assumption, Harris [7] achieved a general solution entailing equilibrium of forces and moments for a thrust-loaded angular-contact ball bearing. In this case, the forces and moments acting on a bearing ball were as shown in Figure 6.4. It was also assumed that the gyroscopic motion about the y' axis is negligible, and the elliptical areas of contact could be divided into two or three zones of sliding as illustrated in Figure 6.5.

Now, for the ball–raceway contacts as shown in Figure 6.5,

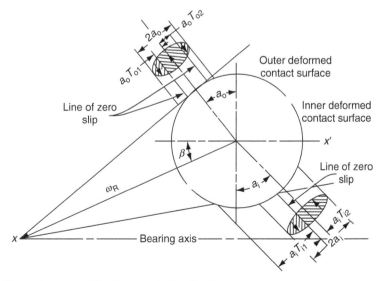

FIGURE 6.5 Contact areas, rolling lines, and slip directions.

$$F_{y'n} = 2\mu a_n b_n c_n \left(\int_{-1}^{T_{n1}} \int_{0}^{\sqrt{1-q^2}} \sigma_n \, dt \, dq - \int_{T_{n1}}^{T_{n2}} \int_{0}^{\sqrt{1-q^2}} \sigma_n \, dt \, dq - \int_{T_{n2}}^{+1} \int_{0}^{\sqrt{1-q^2}} \sigma_n \, dt \, dq \right) \quad (6.5)$$

where $q = x'/a_n$, $t = y'/b_n$, T_{n1} and T_{n2} define lines of rolling motion, n refers to inner or outer ball–raceway contact, that is $n = i$ or o, and σ_n the normal stress or pressure at any point in the contact surface, in accordance with the following equation, is given by

$$\sigma_n = \frac{3Q_n}{2\pi a_n b_n} \left(1 - q^2 - t^2\right)^{1/2} \quad (5.4)$$

Substituting Equation 5.4 in Equation 6.5 and integrating yields

$$F_{y'n} = 3\mu Q_n c_n \left[\frac{2}{3} + \sum_{k=1}^{k=2} c_k T_{nk} \left(1 - \frac{T_{nk}^2}{3}\right) \right] \quad (6.6)$$

$$n = o, i; \; c_o = 1; \; c_i = -1; \; c_1 = 1, c_2 = -1$$

From Figure 2.13 and Figure 2.14, radii r_n from the ball center to points on the contact areas are given by

$$r_n = \left(R_n^2 - x_n^2\right)^{1/2} - \left(R_n^2 - a_n^2\right)^{1/2} + \left[\left(\frac{D}{2}\right)^4 - a_n^2 \right]^{1/2} \qquad n = o, i \quad (6.7)$$

Using Equation 6.7 and Equation 5.4, the equation for friction moments is

$$M_{x'n} = 2\mu a_n b_n c_n \left[\int_{-1}^{T_{n1}} \int_{0}^{\sqrt{1-q^2}} \sigma_n r_n \cos(\alpha_n + \theta_n) \, dt \, dq - \int_{T_{n1}}^{T_{n2}} \int_{0}^{\sqrt{1-q^2}} \sigma_n r_n \cos(\alpha_n + \theta_n) \, dt \, dq \right]$$

$$+ 2\mu a_n b_n c_n \left[\int_{T_{n2}}^{1} \int_{0}^{\sqrt{1-q^2}} \sigma_n r_n \cos(\alpha_n + \theta_n) \, dt \, dq \right] \quad (6.8)$$

In Equation 6.8, $\sin \theta_n = x_n'/r_n$. Using the trigonometric identity,

$$\cos(\alpha_n + \theta_n) = \cos \alpha_n \cos \theta_n - \sin \alpha_n \sin \theta_n \quad (6.9)$$

As θ_n is small, $\cos \theta_n \Rightarrow 1$. Substituting into Equation 6.8 and integrating yields

$$M_{x'n} = 3\mu Q_n D c_n \left\{ \frac{2}{3} \cos \alpha_n + \sum_{k=1}^{k=2} c_k T_{nk} \left[\left(1 - \frac{T_{nk}^2}{3}\right) \cos \alpha_n - \frac{a_n T_{nk}}{D} \left(1 - \frac{T_{nk}^2}{2}\right) \sin \alpha_n \right] \right\} \quad (6.10)$$

$$n = o, i; c_o = 1, c_i = -1; c_1 = 1, c_2 = -1$$

Similarly,

$$M_{z'n} = 3\mu Q_n D c_n$$
$$\times \left\{ \frac{2}{3} \sin \alpha_n + \sum_{k=1}^{k=2} c_k T_{nk} \right.$$
$$\times \left. \left[\left(1 - \frac{T_{nk}^2}{3} \right) \sin \alpha_n - \frac{a T_{nk}}{D} \left(1 - \frac{T_{nk}^2}{2} \right) \cos \alpha_n \right] \right\} \qquad (6.11)$$
$$n = 0, i; \quad c_0 = 1; \quad c_i = -1$$
$$k = 1, 2; \quad c_1 = 1; \quad c_2 = -1$$

Using Figure 6.4, it can be established that four conditions of force and moment equilibrium about the x', y', and z' axes must be satisfied together with four ball position equations determined in Chapter 3. These eight equations must be solved for two position variables, two contact deformations, bearing axial deflection, and speed ω_m, $\omega_{x'}$, and $\omega_{z'}$.

Thus, there are eight equations and eight unknowns; however, the rolling lines T_{nk}, of which there are three as shown in Figure 6.5, are functions of speed ω_m, $\omega_{x'}$, and $\omega_{z'}$. To establish the required relationship, the major axes of the deformed contact surfaces as shown in Figure 2.13 and Figure 2.14 are considered arcs of great circles defined by

$$(x_n' - X)^2 + (z_n' - Z)^2 - (\zeta_n D)^2 = 0 \qquad (6.12)$$

where $\zeta = 2f/(2f+1)$ and $f = r/D$. From Figure 2.13 and Figure 2.14, it can be determined that the offset of the ball center from the circle center is given by the coordinates

$$X = \frac{D}{2} [(4\zeta_n^2 - k_n^2)^{1/2} - (1 - k_n^2)^{1/2}] \sin \alpha_n \qquad (6.13)$$

$$Z = \frac{D}{2} [(4\zeta_n^2 - k_n^2)^{1/2} - (1 - k_n^2)^{1/2}] \cos \alpha_n \qquad (6.14)$$

where $k_n = 2a_n/D$. Zero sliding velocity is determined from the equations

$$(\Omega_o - \omega_m) \left(\frac{d_m}{2} + z' \right) + \omega_x z' + \omega_z x' = 0 \qquad (6.15)$$

$$(\omega_m - \Omega_i) \left(\frac{d_m}{2} + z' \right) + \omega_x z' + \omega_z x' = 0 \qquad (6.16)$$

Equation 6.12, Equation 6.15, and Equation 6.16 can be solved simultaneously to yield x'_{nk}, z'_{nk} locations at which zero sliding velocity occurs on the deformed surface circle. It can be shown that

$$T_{nk} = \frac{(x'_{nk}{}^2 + z'_{nk}{}^2)}{a_n} \sin \left[\frac{\pi}{2} - \alpha_n - \tan^{-1} \left(\frac{z'_{nk}}{x'_{nk}} \right) \right], \qquad k = 1, 2 \qquad (6.17)$$

Using this method Harris [7] was able to prove the impossibility of an "inner raceway control" situation, even with bearings operating with "dry film" lubrication. Moreover, a speed transition point seems to occur in a thrust-loaded angular-contact ball bearing at which a radical shift of the ball speed pitch angle β must occur to achieve load equilibrium in the

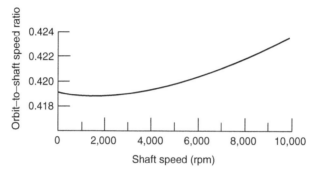

Bearing design data	
Ball diameter	8.731 mm (0.34375 in.)
Pitch diameter	48.54 mm (1.9110 in.)
Free contact angle	24.5°
Inner raceway grove radius/ball diameter	0.52
Outer raceway groove radius/ball diameter	0.52
Thrust load per ball	31.6 N (7.1 1b)

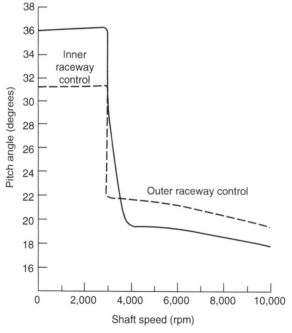

FIGURE 6.6 Orbit/shaft speed ratio vs. shaft speed for a thrust-loaded angular-contact ball bearing. (From Harris, T., *ASME Trans., J. Lubr. Technol.*, 93, 32–38, 1971. With permission.)

bearing. Figure 6.6 and Figure 6.7 from Ref. [7] illustrate the results of this analytical method for a thrust-loaded angular-contact ball bearing.

Additionally, Table 6.2 shows the corresponding locations of rolling lines in the inner and outer contact ellipses for this example.

FIGURE 6.7 Ball speed vector pitch angle vs. shaft speed for a thrust-loaded angular-contact ball bearing. (From Harris, T., *ASME Trans., J. Lubr. Technol.*, 93, 32–38, 1971. With permission.)

TABLE 6.2
Locations of Lines of Zero Sliding in Elliptical Contact Areas of a Thrust-Loaded Angular-Contact Ball Bearing

Shaft	Outer Raceway		Inner Raceway	
	T_1	T_2	T_1	T_2
1000	0.0001	—	−0.00605	0.92123
1500	0.00183	—	−0.00672	0.92376
2000	0.00129	—	−0.00537	0.93140
2500	0.00047	—	−0.00353	0.94272
3000	—	0.02975	0.02995	—
3500	—	−0.00156	—	−0.00190
4000	−0.95339	0.00156	—	0.00052
4500	−0.93237	0.00376	—	0.00064
5000	−0.91449	0.00627	—	0.00077
5500	−0.89730	0.01055	—	−0.00039

Source: From Harris, T., *ASME Trans., J. Lubr. Technol.*, 93, 32–38, 1971.

6.3.2 ROLLER BEARINGS

A similar approach may be applied to roller bearings that have point contact at each raceway. Usually, however, roller bearings are designed to operate in the line contact or modified line contact regime. In the former, the area of roller–raceway contact is basically rectangular, with a "dogbone" effect at the lengthwise limits. This is discussed in Chapter 6 of the first volume of this handbook. The dogbone portion of the contact occupies only a very small area and therefore does not influence friction significantly. In modified line contact (achieved as a result of crowned profile roller or raceway or both), the contact area is approached analytically as elliptical in shape with the lengthwise extremities of the ellipse truncated. In both cases, the major sliding forces acting on the contact are essentially parallel to the direction of rolling and are principally due to the deformation of the surfaces. Thus, the sliding forces acting on the contact surfaces of a loaded roller bearing are usually less complex than those for ball bearings.

Dynamic loading of roller bearings does not generally affect contact angles to any significant extent, and hence the geometry of the contacting surfaces is virtually identical to that occurring under static loading. Because of the relatively slow speeds of operation necessitated when the contact angle differs from zero, gyroscopic moments are negligible. In any event, gyroscopic moments of any magnitude do not substantially alter the normal motion of the rollers. In this analysis, therefore, the sliding on the contact surface of a properly designed roller bearing will be assumed to be a function only of the radius of the deformed surface in a direction transverse to rolling.

To perform the analysis, it is assumed that the contact area between the roller and either raceway is substantially rectangular, and that the normal stress at any distance from the center of the rectangle is adequately defined by the following formula given in Chapter 6 of the first volume of this handbook:

$$\sigma = \frac{2Q}{\pi l b} \left(1 - t^2\right)^{1/2} \tag{6.18}$$

where $t = y/b$ and y is the distance in the rolling direction from the centerline of the contact. Thus, the differential friction force acting at any distance x from the center of the contact rectangle is given by

$$\mathrm{d}F_y = \frac{2\mu Q}{\pi l}\left(1 - t^2\right)^{1/2}\mathrm{d}t\,\mathrm{d}x \tag{6.19}$$

Integrating Equation 6.19 between $t = \pm 1$ yields

$$\mathrm{d}F_y = \frac{\mu Q}{l}\,\mathrm{d}x \tag{6.20}$$

Referring to Figure 6.8, it can be determined that the differential friction moment in the direction of rolling at either raceway is given by

$$\mathrm{d}M_R = \left[\left(R^2 - x^2\right)^{1/2} - \left(R - \frac{D}{2}\right)\right]\mathrm{d}F \tag{6.21}$$

or

$$\mathrm{d}M_R = \frac{2\mu Q}{\pi l}\left(1 - t^2\right)^{1/2}\left[\left(R^2 - x^2\right)^{1/2} - \left(R - \frac{D}{2}\right)\right]\mathrm{d}t\,\mathrm{d}x \tag{6.22}$$

where R is the radius of curvature of the deformed surface. Integrating Equation 6.22 with respect to t between the limits of ± 1 yields

$$\mathrm{d}M_R = \frac{\mu Q}{\pi l}\left[\left(R^2 - x^2\right)^{1/2} - \left(R - \frac{D}{2}\right)\right]\mathrm{d}x \tag{6.23}$$

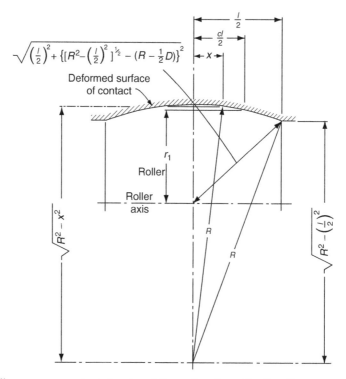

FIGURE 6.8 Roller–raceway contact showing deformed surface of radius R.

Because of the curvature of the deformed surface, pure rolling exists at most at two points $x = \pm cl/2$ on the deformed surface. The radius of rolling measured from the roller axis of rotation is r'; therefore,

$$F_y = \frac{2\mu Q}{l} \left(\int_0^{cl/2} dx - \int_{cl/2}^l dx \right) \tag{6.24}$$

or

$$F_y = \mu Q (2c - 1) \tag{6.25}$$

Also,

$$M_R = \frac{2\mu Q}{l} \left\{ \int_0^{cl/2} \left[(R^2 - x^2)^{1/2} - \left(R - \frac{D}{2} \right) \right] dx - \int_{cl/2}^{l/2} \left[(R^2 - x^2)^{1/2} - \left(R - \frac{D}{2} \right) \right] dx \right\} \tag{6.26}$$

or

$$M_R = \mu Q \left\{ \frac{R^2}{l} \left(2 \sin^{-1} \frac{cl}{2R} - \sin^{-1} \frac{l}{2R} \right) + (1 - 2c) \left(R - \frac{D}{2} \right) \right.$$
$$\left. + cR \left[1 - \left(\frac{cl}{2R} \right)^2 \right]^{1/2} - \frac{R}{2} \left[1 - \left(\frac{2R}{l} \right)^2 \right]^{1/2} \right\} \tag{6.27}$$

Considering the equilibrium of forces acting on the roller at the inner and outer raceway contacts (see Figure 6.9), $F_{yo} = -F_{yi}$; therefore, from Equation 6.25 assuming $\mu_o = \mu_i$

$$c_o + c_i = 1 \tag{6.28}$$

Furthermore, since in uniform rolling motion the sum of the torques at the outer and inner raceway contacts is equal to zero, therefore,

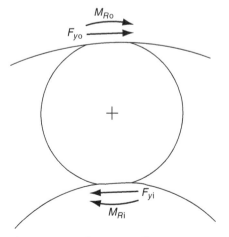

FIGURE 6.9 Friction forces and moments acting on a roller.

$$M_{Ro} \frac{\left(\frac{d_m}{2} + r_o'\right)}{r_o'} + M_{Ri} \frac{\left(\frac{d_m}{2} - r_i'\right)}{r_i'} = 0 \tag{6.29}$$

From Figure 6.8, it can be seen that the roller radius of rolling is

$$r' = \left[R^2 - \left(\frac{cl}{2}\right)^2\right]^{1/2} - \left(R - \frac{D}{2}\right) \tag{6.30}$$

Hence, assuming $\mu_o = \mu_i$, from Equation 6.27, Equation 6.29, and Equation 6.30,

$$
\begin{aligned}
&\left\{ \frac{R_o^2}{l}\left(2\sin^{-1}\frac{c_o l}{2R_o} - \sin^{-1}\frac{l}{2R_o}\right) + (1 - 2c_o)\left(R_o - \frac{D}{2}\right) \right. \\
&\left. \quad + c_o R_o\left[1 - \left(\frac{c_o l}{2R_o}\right)^2\right]^{1/2} - \frac{R_o}{2}\left[1 - \left(\frac{l}{2R_o}\right)^2\right]^{1/2} \right\} \\
&\times \left\{ 1 + \frac{d_m}{2\left\{\left[R_o^2 - \left(\frac{c_o l}{2}\right)^2\right]^{1/2} - \left(R_o - \frac{D}{2}\right)\right\}} \right\} \\
&- \left\{ \frac{R_i^2}{l}\left(2\sin^{-1}\frac{c_i l}{2R_i} - \sin^{-1}\frac{l}{2R_i}\right) + (1 - 2c_i)\left(R_i - \frac{D}{2}\right) \right. \\
&\left. \quad + c_i R_i\left[1 - \left(\frac{c_i l}{2R_i}\right)^2\right]^{1/2} - \frac{R_i}{2}\left[1 - \left(\frac{l}{2R_i}\right)^2\right]^{1/2} \right\} \\
&\times \left\{ 1 + \frac{d_m}{2\left\{\left[R_i^2 - \left(\frac{c_i l}{2}\right)^2\right]^{1/2} - \left(R_i - \frac{D}{2}\right)\right\}} \right\} = 0
\end{aligned}
\tag{6.31}
$$

Equation 6.28 and Equation 6.31 can be solved simultaneously for c_o and c_i. Note that if R_o and R_i, the radii of curvature of the outer and inner contact surfaces, respectively, are infinite, the analysis does not apply. In this case, sliding on the contact surfaces is obviated and only rolling occurs.

Having determined c_o and c_i, one may revert to Equation 6.25 to determine the net sliding forces F_{yo} and F_{yi}. Similarly, M_{Ro} and M_{Ri} may be calculated from Equation 6.27. Figure 6.9 shows the friction forces and moments acting on a roller.

6.4   BEARING OPERATION WITH FLUID-FILM LUBRICATION: EFFECTS OF FRICTION FORCES AND MOMENTS

6.4.1   BALL BEARINGS

6.4.1.1   Calculation of Ball Speeds

As shown in Chapter 5, the surface friction shear stresses $\tau_{y'}$ and $\tau_{x'}$ at a given point (x', y') in the contact surface can be represented by the following equations:

$$\tau_{y'} = c_v \frac{A_c}{A_0} \mu_a \sigma + \left(1 - \frac{A_c}{A_0}\right)\left(\frac{h}{\eta v_{y'}} + \frac{1}{\tau_{\lim}}\right)^{-1} \qquad (5.48)$$

$$\tau_{x'} = c_v \frac{A_c}{A_0} \mu_a \sigma + \left(1 - \frac{A_c}{A_0}\right)\left(\frac{h}{\eta v_{x'}} + \frac{1}{\tau_{\lim}}\right)^{-1} \qquad (5.49)$$

Means were also demonstrated to permit the calculation of $\tau_{y'}$ and $\tau_{x'}$ for a given lubricating fluid and a given condition Λ of rolling contact surface separation. Figure 6.10 shows the force and moment loading of a ball in thrust-loaded oil-lubricated angular-contact ball bearing. The coordinate system is the same as that used in Figure 2.4 to describe ball speeds.

The sliding velocities in the y' (rolling motion) and x' (gyroscopic motion) directions as determined from Chapter 2 are as follows:

$$v_{y'n} = \frac{D}{2}\left\{\frac{\omega_n}{\gamma} + \varphi_n[(c_n\omega_n - \omega_{x'})\cos(\alpha_n + \theta_n) - \omega_{z'}\sin(\alpha_n + \theta_n)]\right\} \qquad (6.32)$$

$$v_{x'n} = \frac{D}{2}\varphi_n\omega_{y'} \qquad (6.33)$$

where

$$\omega_n = c_n(\omega_m - \Omega_n) \qquad (6.34)$$

$$\varphi_n = \left\{\left(\frac{2x_n}{D}\right)^2 + \left[\left(4\zeta_n^2 - \left(\frac{2x_n}{D}\right)^2\right)^{1/2} - \left(4\zeta_n^2 - \left(\frac{2a_n}{D}\right)^2\right)^{1/2} + \left(1 - \left(\frac{2a_n}{D}\right)^2\right)^{1/2}\right]^2\right\}^{1/2}$$

$$(6.35)$$

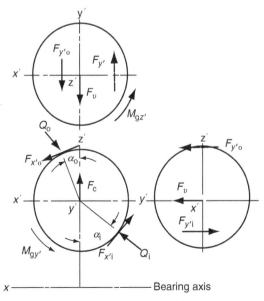

FIGURE 6.10 Forces and moments acting on a ball in an oil-lubricated, thrust-loaded angular-contact ball bearing.

$$\theta_n = \sin^{-1}\left(\frac{x_n}{r_n}\right) \tag{6.36}$$

$$r'_n = \frac{D}{2}\varphi_n \tag{6.37}$$

and

$$\zeta_n = \frac{2f_n}{2f_n + 1} \tag{6.38}$$

In Equation 6.32 and Equation 6.33, $c_o = 1$ and $c_i = -1$.

To calculate the plateau lubricant film thickness h used in the determination of $\tau_{y'}$ and $\tau_{x'}$, the entrainment velocities may be determined from the following equation:

$$u_{y'n} = \frac{D}{4}\left\{\frac{\omega_n}{\gamma} + \varphi_n[(c_n\omega_n + \omega_{x'})\cos(\alpha_n + \theta_n) + \omega_{z'}\sin(\alpha_n + \theta_n)]\right\} \tag{6.39}$$

In the calculation of $\tau_{y'}$ and $\tau_{x'}$, it is important to determine lubricant viscosities at the appropriate temperatures. For calculation accuracy, it is necessary to estimate the lubricant temperature at the entrance to each contact and in the film separating the rolling–sliding components.

Assuming that contact loading is known, the friction forces acting over the contact areas are given by

$$F_{y'n} = a_nb_n\int_{-1}^{1}\int_{-\sqrt{1-q^2}}^{\sqrt{1-q^2}} \tau_{y'n}\,dt\,dq, \quad n = o, i \tag{6.40}$$

$$F_{x'n} = a_nb_n\int_{-1}^{1}\int_{-\sqrt{1-q^2}}^{\sqrt{1-q^2}} \tau_{x'n}\,dt\,dq, \quad n = o, i \tag{6.41}$$

The moments due to the surface friction shear stresses are given by

$$M_{x'n} = \frac{1}{2}Da_nb_n\int_{-1}^{1}\int_{\sqrt{1-q^2}}^{\sqrt{1-q^2}} \tau_{y'n}\varphi_n\cos(\alpha_n + \theta_n)\,dq\,dt, \quad n = o, i \tag{6.42}$$

$$M_{z'n} = \frac{1}{2}Da_nb_n\int_{-1}^{1}\int_{-\sqrt{1-q^2}}^{\sqrt{1-q^2}} \tau_{y'n}\varphi_n\sin(\alpha_n + \theta_n)\,dq\,dt, \quad n = o, i \tag{6.43}$$

$$M_{y'n} = \frac{1}{2}Da_nb_n\int_{-1}^{1}\int_{-\sqrt{1-q^2}}^{\sqrt{1-q^2}} \tau_{x'n}\varphi_n\,dq\,dt, \quad n = o, i \tag{6.44}$$

Using Equation 6.40 through Equation 6.44, the equations for force and moment equilibrium for a bearing ball are

$$Q_o \sin \alpha_o + F_{x'o} \cos \alpha_o - \frac{F_a}{Z} = 0 \tag{6.45}$$

$$\sum_{n=0}^{n=i} c_n (Q_n \cos \alpha_n - F_{x'n} \sin \alpha_n) - F_c = 0, \qquad n = o, i; \quad c_o = 1, c_i = -1 \tag{6.46}$$

$$\sum_{n=0}^{n=i} c_n (Q_n \sin \alpha_n + F_{x'n} \cos \alpha_n) = 0, \quad n = o, i; \quad c_o = 1, c_i = -1 \tag{6.47}$$

$$\sum_{n=0}^{n=i} c_n F_{y'n} + F_v = 0, \quad n = o, i; \quad c_o = 1, c_i = -1 \tag{6.48}$$

$$\sum_{n=0}^{n=i} M_{z'n} = 0 \tag{6.49}$$

$$\sum_{n=0}^{n=i} M_{y'n} - M_{gy'} = 0 \tag{6.50}$$

$$\sum_{n=0}^{n=i} M_{z'n} - M_{gz'} = 0 \tag{6.51}$$

where

$$M_{gy'} = J \omega_m \omega_{y'} \tag{6.52}$$

$$M_{gz'} = J \omega_m \omega_{z'} \tag{6.53}$$

and J is the polar moment of inertia. Viscous drag force F_v in Equation 6.48 is determined from Equation 6.1. Since only a simple thrust load is applied, the cage speed is identical to ball orbital speed ω_m. Another simplification for this example is an assumption of a ball–riding cage that has negligible friction between the cage pockets and balls. The unknown variables in Equation 6.45 through Equation 6.51 are:

- Inner and outer raceway–ball contact deformations δ_i and δ_o
- Ball contact angles α_i and α_o
- Ball speeds $\omega_{x'}$, $\omega_{y'}$, $\omega_{z'}$, and ω_m
- Bearing axial deflection δ_a

Hence, there are seven equations and nine unknown variables. The remaining two equations pertain to the position of the ball center; as obtained from Chapter 3 they are

$$(A_1 - X_1)^2 + (A_2 - X_2)^2 - [(f_i - 0.5)D + \delta_i]^2 = 0 \tag{6.54}$$

$$X_1^2 + X_2^2 - [(f_o - 0.5)D + \delta_o]^2 = 0 \tag{6.55}$$

In Chapter 3 from Equation 3.72 and Equation 3.73, it is shown that the position variables A_1 and A_2 are given by

$$A_1 = BD \sin \alpha_o + \delta_a \qquad (6.56)$$

$$A_2 = BD \cos \alpha_o \qquad (6.57)$$

where $B = f_i + f_o - 1$. Moreover, the position variables X_1, X_2, A_1, and A_2 are related to contact angles α_i and α_o, and contact deformations δ_i and δ_o as follows:

$$\sin \alpha_o = \frac{X_1}{(f_o - 0.5)D + \delta_o} \qquad (6.58)$$

$$\cos \alpha_o = \frac{X_2}{(f_o - 0.5)D + \delta_o} \qquad (6.59)$$

$$\sin \alpha_i = \frac{A_1 - X_1}{(f_i - 0.5)D + \delta_i} \qquad (6.60)$$

$$\cos \alpha_i = \frac{A_2 - X_2}{(f_i - 0.5)D + \delta_i} \qquad (6.61)$$

This system of equations was first solved by Harris [8] using the simplifying assumption of an isothermal Newtonian lubricant, adequately supplied to the ball–raceway contacts. Figure 6.11 and Figure 6.12 show the comparison of the analytical results with the experimental data of Shevchenko and Bolan [9] and Poplawski and Mauriello [10]. The deviations from the solution using the outer raceway control approximation are apparent.

6.4.1.2 Skidding

Resulting from the analyses by Harris [8] as shown in Figure 6.11 and Figure 6.12, investigation of the rolling direction sliding velocity, that is, $v_{y'}$ as a function of location x' along the major axis of the ball–inner raceway and ball–outer raceway contacts, reveals no change in the sliding velocity direction. This means that no points of rolling motion occur over the contacts. This condition of gross sliding is called skidding. An important application with regard to skidding is the mainshaft split inner ring ball bearing in gas turbine engines. This predominantly thrust-loaded angular-contact bearing operates at high speeds, typically in the range exceeding 2 million dn (bearing bore in mm · shaft speed in rpm). Even though the thrust load is high, skidding tends to occur.

Skidding results in surface friction shear stresses of significant magnitudes over the contact areas. If the lubricant film generated by the relative motion of the ball–raceway surfaces is insufficient to completely separate the surfaces, surface damage called smearing will occur. An example of smearing is shown in Figure 6.13. Smearing is defined as a severe type of wear characterized by the metal tightly bonded to the surface in locations to which it has been transferred from remote locations of the same or opposing surfaces. The transferred metal is present in sufficient volume to connect more than one distinct asperity contact. When the number of asperity contacts connected is small, it is called microsmearing. When the number of such contacts is large enough to be seen with the unaided eye, it is called gross smearing or macrosmearing.

If possible, skidding is to be avoided in any bearing application because at the very least it results in increased friction and heat generation even if smearing does not occur. It can occur

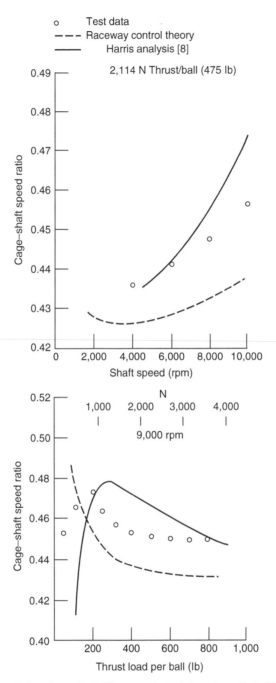

FIGURE 6.11 Experimental data from Ref. [9] vs. analytical data from Ref. [8] for an angular-contact ball bearing having three 28.58-mm (1.125 in.) balls.

in high-speed bearing applications, particularly if the applied load accommodated by each rolling element is relatively small compared with its centrifugal force. The latter causes increased normal loads at the outer raceway contacts compared with the inner raceway contacts. Thus, the balance of friction forces and moments requires higher friction coefficients

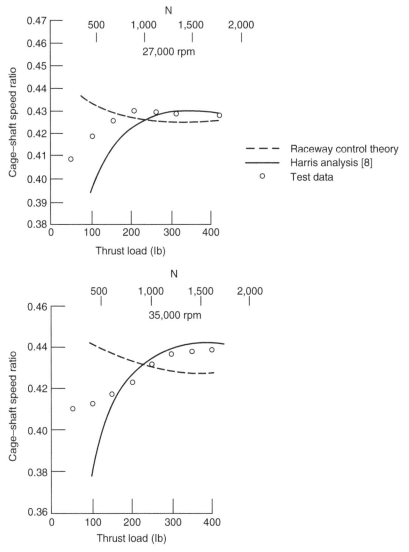

FIGURE 6.12 Experimental data from Ref. [10] vs. analytical data from Ref. [8] for a 35-mm bore–62-mm OD angular-contact ball bearing.

at the inner raceway contacts to compensate for the lower normal contact loads. In Chapter 4, it was shown that the lubricant film thickness generated in an oil-lubricated rolling element–raceway contact depends on the velocities of the surfaces in contact. Considering Newtonian lubrication as a simplified case, the surface friction shear stress is a direct function of the sliding velocity of the surfaces and an inverse function of the lubricant film thickness. Now, considering Equation 5.3, the coefficient is a function of sliding velocity; this is greatest at the inner raceway contacts.

Generally, skidding can be minimized by increasing the applied load on the bearing, thus decreasing the relative magnitude of rolling element centrifugal force to the contact load at the most heavily loaded rolling element. Unfortunately, this remedy tends to reduce bearing fatigue endurance. Another approach is to employ reduced mass rolling elements. These can

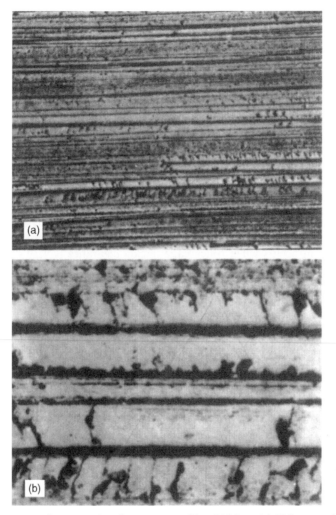

FIGURE 6.13 Raceway surface smearing damage caused by skidding: (a) 100× magnification; (b) 500× magnification.

be manufactured from silicon nitride, a rolling bearing capable ceramic that has a specific gravity 40% that of steel. Hollow rolling elements also might be used; however, bending stresses at the inside diameter also tend to cause earlier fatigue failure.

Skidding is also aggravated by rolling element–lubricant, rolling element–cage, and cage–ring rail friction, each of which tends to retard motion. The most significant of these is the viscous drag of the lubricant on the rolling elements. Therefore, a high-speed bearing operating submerged in lubricant will skid more than the same bearing operating in oil mist-type lubrication. In this case, a compromise is required because, in a high-speed application, a copious supply of fluid lubricant is generally used to carry away the friction heat generated by the bearing.

In general, a compromise between the degree of skidding and bearing endurance must be accepted unless by making the contacting surfaces extremely smooth, the effectiveness of the lubricant film thickness is improved to the point that skidding may occur without surface damage.

6.4.2 Cylindrical Roller Bearings

6.4.2.1 Calculation of Roller Speeds

Roller speeds in oil-lubricated, cylindrical roller bearings can be determined by a consideration of the balance of friction forces and moments on the individual rollers and on the bearing as a unit. Considering the roller–raceway contacts to be divided into laminae as in Chapter 1, the sliding velocity at a selected lamina is given by

$$v_{\lambda nj} = \tfrac{1}{2}\left\{\left[d_m + c_n\left(D_\lambda + \tfrac{2}{3}\delta_{\lambda nj}\right)\right]\omega_{nj} - \left(D_\lambda - \tfrac{1}{3}\delta_{\lambda nj}\right)\omega_{Rj}\right\}$$
$$n = \text{o, i}; \quad c_\text{o} = 1, \quad c_\text{i} = -1; \quad j = 1 \text{ to } Z \tag{6.62}$$

where D_λ is the equivalent roller diameter at lamina λ. It is assumed in Equation 6.62 that $1/3$ of the elastic contact deformation occurs in the roller and $2/3$ in the raceway. Further, to simplify the analysis it is assumed that the roller orbital speed is constrained to equal cage speed. This condition occurs when roller–cage pocket clearance is very small in the circumferential direction. Raceway relative speed ω_{nj} is given by

$$\omega_{nj} = c_n(\omega_m - \Omega_n), \quad n = \text{o, i}; \quad c_\text{o} = 1, \quad c_\text{i} = -1 \tag{6.63}$$

Fluid entrainment velocities are given by Equation 4.54 and Equation 4.55; minimum lubricant film thicknesses are obtained using Equation 4.57. Plateau lubricant film thicknesses are obtained using Equation 4.58. As for ball–raceway contacts, the surface friction shear stress at a point on the contact surface is obtained using Equation 5.48. In this case, normal stress or contact pressure is determined at each lamina λ using Equation 6.50:

$$\sigma_{\lambda nj} = \frac{2q_{\lambda nj}\left(1 - t^2\right)^{1/2}}{\pi b_{nj}} \tag{6.64}$$

where $t = y/b_{nj}$ and $q_{\lambda nj}$ is the load per unit length on lamina λ at roller–raceway contact nj. The friction force acting over a contact is then given by

$$F_{nj} = 2w_n \sum_{\lambda=1}^{\lambda=k} b_{\lambda nj} \int_0^1 \tau_{\lambda nj}\, dt \tag{6.65}$$

where w_n is the lamina thickness.

Figure 6.14 shows the friction and normal forces acting on a roller in a radially loaded cylindrical roller bearing with negligible roller end–ring guide flange friction.

From Figure 6.14, the following force equilibrium equations are obtained:

$$\sum_{n=\text{o}}^{n=\text{i}} c_n Q_{nj} - F_\text{c} = 0, \quad n = \text{o, i}; \quad c_\text{o} = 1, \quad c_\text{i} = -1; \quad j = 1 \text{ to } Z \tag{6.66}$$

$$\sum_{n=\text{o}}^{n=\text{i}} c_n F_{nj} + F_\text{v} - Q_{CGj} = 0 \tag{6.67}$$

where F_c is obtained from Equation 3.38, and in Equation 6.67 F_v is obtained from Equation 6.2. Note that if there is sufficient clearance between the roller and the cage web, then the roller is free to orbit at other than the cage speed. Equation 6.67 then becomes

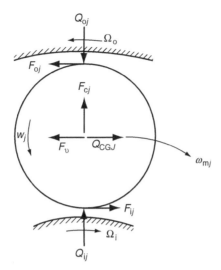

FIGURE 6.14 Forces acting on a roller in a radially loaded cylindrical roller bearing.

$$\sum_{n=0}^{n=i} c_n F_{nj} + F_v - Q_{CGj} = \frac{1}{2} m \frac{dv}{dt} = \frac{1}{2} m d_m \omega_{mj} \frac{d\omega_{Rj}}{d\psi} \qquad (6.68)$$

where m is the mass of the roller.

The moments about the roller axis due to surface friction shear stresses are given by

$$M_{nj} = w_n \sum_{\lambda=1}^{\lambda=k} b_{\lambda nj} D_\lambda \int_0^1 \tau_{\lambda nj}\, dt \qquad (6.69)$$

The summation of moments about the roller axis is

$$\sum_{n=0}^{n=i} M_{nj} - \frac{1}{2} \mu_{CG} D Q_{CGj} = J \omega_m \frac{d\omega_{Rj}}{d\psi} \qquad (6.70)$$

Finally, the equilibrium of radial forces acting on the bearing is expressed by

$$\sum_{j=1}^{j=Z} Q_{ij} \cos \psi_j - F_r = 0 \qquad (6.71)$$

and if the bearing operates at constant speed, the sum of the moments acting on the cage in the circumferential direction must equate to zero, or

$$d_m \sum_{j=1}^{j=Z} Q_{CGj} \pm D_{CR} F_{CL} = 0 \qquad (6.72)$$

where F_{CL} the friction force between the cage rail and the bearing ring land is given by Equation 6.4.

As in Chapter 3, the normal loads Q_{nj} can be written in terms of contact deformations δ_{nj}, and bearing radial deflection can be related to contact deformations and radial clearance.

Accordingly, Equation 6.66, Equation 6.67, and Equation 6.70 through Equation 6.72, a set of $3Z+2$ simultaneous equations, can be solved for δ_r, δ_{ij}, ω_m, ω_{Rj}, and Q_{CGj}. Ref. [1] gives the general solution for all types of roller bearings; that is, for five degrees of freedom in applied bearing loading, freedom for each roller to orbit at a speed other than the cage speed (ω_{mj} instead of ω_m), and a raceway with any shape or roller profile.

6.4.2.2 Skidding

Skidding is a problem in cylindrical roller bearings used to support the mainshaft in aircraft gas turbine engines. These high-speed bearings, used principally for location, arc subjected to very light radial load. Harris [11], using a simpler form of the analysis, considering only isothermal lubrication conditions, and neglecting viscous drag on the rollers, nevertheless managed to demonstrate the adequacy of the analytical method. Figure 6.15, from Ref. [11], compares analytical data against experimental data on cage speed vs. applied load and speed. The analysis showed that skidding, as indicated by the reduction in cage speed compared with kinematic speed, tends to decrease as the load applied is increased. It also appears relatively insensitive to lubricant type.

Some aircraft engine manufacturers assemble their bearings in an oval-shaped or out-of-round outer raceway to achieve the load distribution in Figure 6.16. This selective radial

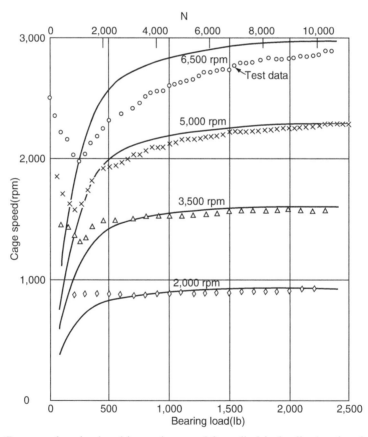

FIGURE 6.15 Cage speed vs. load and inner ring speed for cylindrical roller bearing, lubricant-diester type according to MIL-L-7808 specification. $Z=36$ rollers, $l=20$ mm (0.787 in.), $D=19$ mm (0.551 in.), $d_m=183$ mm (7.204 in.), $P_d=0.0635$ mm (0.0025 in). (From Harris, T., An analytical method to predict skidding in high speed roller bearings, *ASLE Trans.*, 9, 229–241, 1966.)

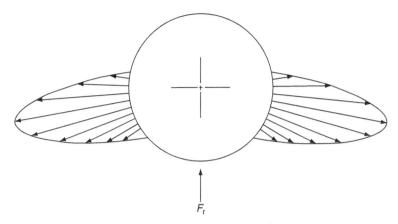

FIGURE 6.16 Distribution of load among the rollers of a bearing having an out-of-round outer ring and subjected to radial load F_r.

preloading of the bearing increases the maximum roller load and doubles the number of rollers so-loaded. Figure 6.17, from Ref. [11], illustrates the effect on skidding of an out-of-round outer raceway. Another method to minimize skidding is to use a few, for example, three equally spaced hollow rollers that provide an interference fit with the raceways under zero applied radial load and static conditions. Figure 6.18, from Ref. [12], illustrates such an assembly, while Figure 6.19 and Figure 6.20 indicate the effectiveness to minimize skidding.

6.5   CAGE MOTIONS AND FORCES

6.5.1   INFLUENCE OF SPEED

With respect to rolling element bearing performance, cage design has become more important as bearing rotational speeds increase. In instrument ball bearings, undesirable torque variations have been traced to cage dynamic instabilities. In the development of solid-lubricated bearings for high-speed, high-temperature gas turbine engines, the cage is a major concern.

A key to successful cage design is a detailed analysis of the forces acting on the cage and the motions it undergoes. Both steady-state and dynamic formulations of varying complexity have been developed.

6.5.2   FORCES ACTING ON THE CAGE

The primary forces acting on the cage are due to the interactions between the rolling element and cage pocket (F_{CP}) and the cage rail and the piloting land (F_{CL}). As Figure 6.21 shows, a roller can contact the cage on either side of the pocket, depending on whether the cage is driving the roller, or *vice versa*. The direction of the cage pocket friction force (F_{CP}) depends on which side of the pocket the contact occurs. For an inner land riding cage, a friction torque (T_{CL}) in the direction of cage rotation develops at the cage–land contact. For an outer land riding cage, a friction torque tending to retard cage rotation develops at the cage–land contact.

A lubricant viscous drag force (f_{DRAG}) develops on the cage surfaces resisting motion of the cage. Centrifugal body forces (shown as F_{CF}) due to cage rotation make the cage expand uniformly outward radially and induce tensile hoop stresses in the cage rails. An unbalanced force (F_{UB}), the magnitude of which depends on how accurately the cage is balanced, acts radially outward.

Hydrodynamic short bearing theory can be used to model the cage–land interaction as indicated in Ref. [13]. The contact between the rolling element and cage pocket can be hydrodynamic, elastohydrodynamic, or elastic in nature, depending on the proximity of the two

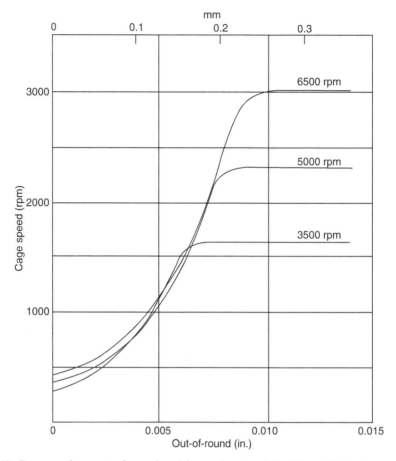

FIGURE 6.17 Cage speed vs. out-of-round and inner ring speed. Lubricant-diester type according to MII-L-7808 specification. $Z = 36$, $i = 1$, $l = 20$ mm (0.787 in.), $D = 14$ mm (0.551 in.), $d_m = 183$ mm (7.204 in.), $P_d = 0.0635$ mm (0.0025 in.), $F_r = 222.5$ N (50 lb). (From Harris, T., An analytical method to predict skidding in high speed roller bearings, *ASLE Trans.*, 9, 229–241, 1966.)

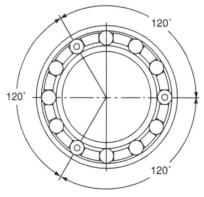

FIGURE 6.18 Cylindrical roller bearing with three preloaded annular rollers. (From Harris, T., and Aaronson, S., An analytical investigation of skidding in a high-speed, cylindrical roller bearing having circumferentially spaced, preloaded hollow rollers, Lub, Eng., 30–34, 1968.)

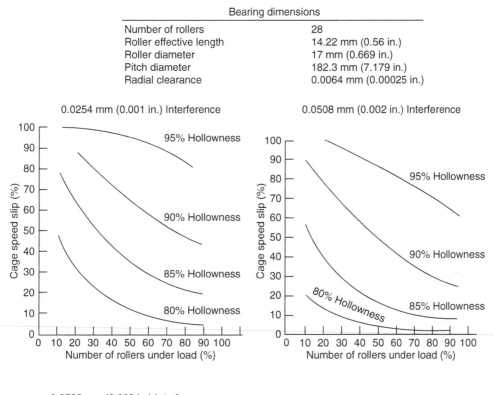

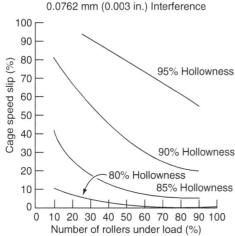

FIGURE 6.19 Skidding in cylindrical roller bearings having spaced preloaded hollow rollers.

bodies and the magnitude of the rolling element forces. In most cases, the rolling element–cage interaction forces are small enough so that hydrodynamic lubrication considerations prevail.

6.5.3 STEADY-STATE CONDITIONS

In section 6.4, it was demonstrated that analytical means exist to predict skidding in ball and roller bearings in any fluid-lubricated application. All the calculations, even for the least complex applications, require the use of a computer. As a spin-off from the skidding analysis, rolling element–cage forces are determined. For an out-of-round outer raceway

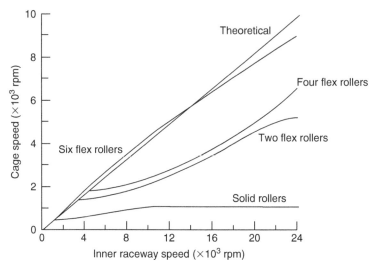

FIGURE 6.20 Cage speed vs. inner raceway speed: 207 roller bearing, $F_r = 0$, $P_d = -0.061$ mm (-0.0024 in.), 90% hollow rollers, lubricant MIL-L-6085A at 0.85 kg/min.

cylindrical roller bearing under radial load, Figure 6.22, from Ref. [14], illustrates cage web loading for steady-state, centric cage rotation.

Whereas the analysis of Ref. [13] considered only centric rotation in the radial plane, Kleckner and Pirvics [15] used three degrees of freedom in the radial plane; that is, the cage rotational speed and two radial displacements locating the cage center in the plane of rotation. The corresponding cage equilibrium equations are

$$\sum_{j=1}^{Z} \left[(F_{\text{CP}j}) \sin \psi_j - (f_{\text{CP}j}) \cos \psi_j \right] - W_y = 0 \tag{6.73}$$

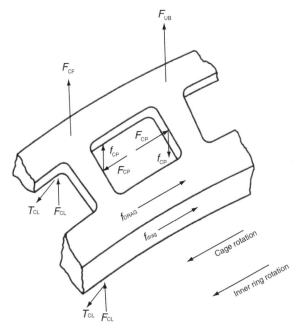

FIGURE 6.21 Cage forces.

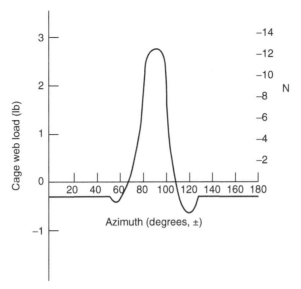

FIGURE 6.22 Cage-to-roller load vs. azimuth for a gas turbine mainshaft cylindrical roller bearing. Thirty 12 mm × 12 mm rollers on a 152.4 mm (6 in.) pitch diameter. Roller i.d./o.d. = 0.6, outer ring out-of-roundness = 0.254 mm (0.01 in.), radial load = 445 N (100 lb), shaft speed = 25,000 rpm. (From Wellons, F., and Harris, T., Bearing design Considerations *Interdisciplinary Approach to the Lubrication of Concentrated Contacts*, NASASP-237, pp. 529–549, 1970.)

$$\sum_{j=1}^{Z} [(-F_{\text{CP}j}) \cos \psi_j - (f_{\text{CP}j}) \sin \psi_j] - W_z = 0 \tag{6.74}$$

$$\frac{1}{2} d_{\text{m}} \sum_{j=1}^{Z} (F_{\text{CP}j}) \pm T_{\text{CL}} = 0 \tag{6.75}$$

where W_y and W_z are the components of F_{CL} in the y and z direction; $F_{\text{CP}j}$ is the cage pocket normal force for the jth rolling element; and $f_{\text{CP}j}$ is the cage pocket friction force for the jth rolling element.

The cage coordinate system is shown in Figure 6.23.

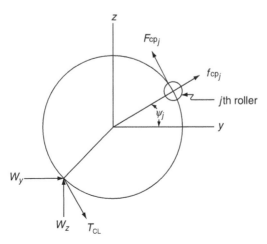

FIGURE 6.23 Cage coordinate system.

Equation 6.73 and Equation 6.74 represent equilibrium of cage forces in the radial plane of motion. The summation of the cage pocket normal forces and friction forces equilibrates the cage–land normal force. Equation 6.75 establishes torque equilibrium for the cage about its axis of rotation. The cage pocket normal forces are assumed to react at the bearing pitch circle. The sign of the cage–land friction torque T_{CL} depends on whether the cage is inner ring land–riding or outer ring land–riding. In the formulations of Ref. [15], each roller is allowed to have different rotational and orbital speeds.

6.5.4 Dynamic Conditions

Rolling element bearing cages are subjected to transient motions and forces due to accelerations caused by contact with rolling elements, rings, and eccentric rotation. In some applications, notably with very high-speed or rapid acceleration, these transient cage effects may be of sufficient magnitude to warrant evaluation. The steady-state analytical approaches discussed do not address the time-dependent behavior of a rolling element bearing cage. Several researchers have developed analytical models for transient cage response [13,16–19]. Because of the complexity of the calculation involved, such performance analyses generally require extensive time on present-day computers.

In general, the cage is treated as a rigid body subjected to a complex system of forces. These forces may include the following:

1. Impact and frictional forces at the cage–rolling element interface
2. Normal and frictional forces at the cage–land surface (if land-guided cage)
3. Cage mass unbalance force
4. Gravitational force
5. Cage inertial forces
6. Others (that is, lubricant drag on the cage and lubricant churning forces)

Forces 1 and 2 are intermittent; for example, the cage may or may not be in contact with a given rolling element or guide flange at a given time, depending on the relative position of the bodies in question. Frictional forces can be modeled as hydrodynamic, elastohydrodynamic lubrication (EHL), or dry friction, depending on the lubricant, contact load, and geometry. Both elastic and inelastic impact models appear in the literature. General equations of motion for the cage may be written. The Euler equations describing cage rotation about its center of mass (in Cartesian coordinates) are as follows:

$$I_x \dot{\omega}_x - (I_y - I_z)\omega_y \omega_z = M_x \tag{6.76}$$

$$I_y \dot{\omega}_y - (I_z - I_x)\omega_z \omega_x = M_y \tag{6.77}$$

$$I_z \dot{\omega}_z - (I_x - I_y)\omega_x \omega_y = M_z \tag{6.78}$$

where I_x, I_y, I_z are the cage principal moments of inertia, and ω_x, ω_y, ω_z are the angular velocities of the cage about the inertial x, y, z axes. The total moment about each axis is denoted by M_z, M_y, and M_z, respectively. The equations of motion for translation of the cage center of mass in the inertial reference frame are

$$m\ddot{r}_x = F_x \tag{6.79}$$

$$m\ddot{r}_y = F_y \tag{6.80}$$

$$m\ddot{r}_z = F_z \tag{6.81}$$

where m is the cage mass, r_x, r_y, r_z describe the position of the cage center of mass, and F_x, F_y, F_z are the net force components acting on the cage.

Once the cage force and moment component are determined, accelerations can be computed. A numerical integration of the equations of motion (with respect to discrete time increments) will yield cage translational velocity, rotational velocity, and displacement vectors. In some approaches [13,17], the cage dynamics model is solved in conjunction with roller and ring equations of motion. Other researchers have devised less cumbersome approaches by limiting the cage to in-plane motion [16] or by considering simplified dynamic models for the rolling elements [18].

Meeks and Ng [18] developed a cage dynamics model for ball bearings, which treats both ball- and ring land-guided cages. This model considers six cage degrees of freedom and inelastic contact between the balls and cage and between the cage and rings. This model was used to perform a cage design optimization study for a solid-lubricated, gas turbine engine bearing [19].

The results of the study indicated that ball–cage pocket forces and wear are significantly affected by the combination of cage–land and ball–pocket clearances. Using the analytical model to identify more suitable clearance values improved experimental cage performance. Figure 6.24 and Figure 6.25 contain typical output data from the cage dynamics analysis.

In Figure 6.24 the cage center of mass motion is plotted vs. time for X and Y (radial plane) direction. The time scale relates to approximately five shaft revolutions at a shaft speed of 40,000 rpm. Figure 6.25 shows the plots of ball–cage pocket normal force for two representative pockets positioned approximately 90° apart.

In addition to the work of Meeks [19], Mauriello et al. [20] succeeded in measuring ball-to-cage loading in a ball bearing subjected to combined radial and thrust loading. They observed impact loading between balls and cage to be a significant factor in high-speed bearing cage design.

6.6   ROLLER SKEWING

Thus far in this section, rollers have been assumed to run "true" in roller bearings. This is an ideal situation. Because of slightly imperfect geometry, there is a tendency for imbalance of friction loading between the roller–inner raceway and roller–outer raceway contacts, creating a tendency for rollers to undergo yaw motions such that each roller's axis of rotation assumes an angle ξ_j with a plane passing through the bearing axis of rotation. ξ_j is called the skewing angle, and the rollers are said to skew.

In a misaligned radial cylindrical roller bearing as shown schematically in Figure 1.6, the rollers are "squeezed" at one end and thereby are forced against the ring guide flange. The sliding contact between each roller end and the guide flange causes a friction force and hence a roller skewing moment. Depending on the clearances between (1) the roller and the guide flange, (2) the roller length and the cage pocket in the direction transverse to rolling motion, and (3) the roller diameter and the cage pocket in the circumferential direction, the roller skewing angle may be limited by one of these constraints. If all these clearances are too great, then, as indicated in Chapter 3, the roller skewing angle will be limited by the outer raceway curvature in the direction of motion. Also, as discussed in Chapter 3, the thrust load applied to radial cylindrical roller bearings with guide flanges on both inner and outer rings results in roller skewing moments that are resisted by one or more of the mechanisms discussed here. Figure 6.26 illustrates cylindrical roller loading that results in both roller tilting angle ζ_j, roller skewing angle ξ_j and the roller–raceway contact normal and surface friction stresses that ensue.

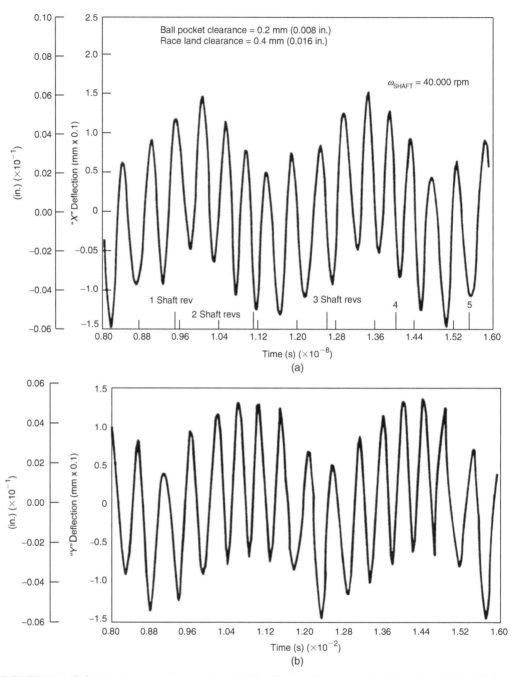

FIGURE 6.24 Calculated cage motion vs. time. (a) Prediction of cage motion X vs. time. (b) Prediction of cage motion, Y vs. time. (From Meeks, C., The dynamics of ball separators in ball bearings—Part II: Results of optimization study, ASLE Paper No. 84-AM-6C-3, May 1984. With permission.)

In tapered roller bearings, even without misalignment, the rollers are forced against the large end flange, and skewing moments occur. These are resisted by either the cage or the outer raceway curvature in the rolling direction. In any case, the roller skewing angles tend to be very small.

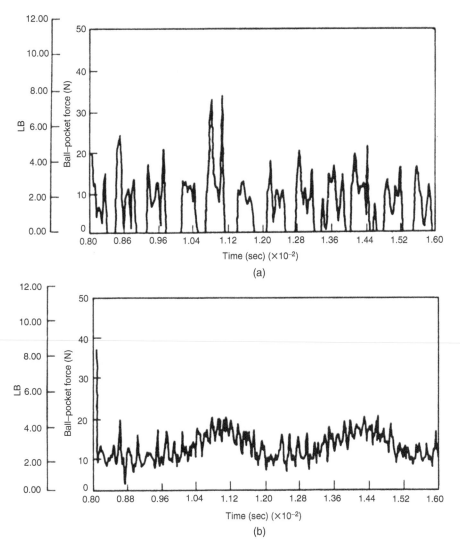

FIGURE 6.25 Calculated ball–pocket force vs. time. (a) Prediction of cage ball–pocket force vs. time (pocket No. 1). (b) Prediction of cage ball–pocket force vs. time (pocket No. 4). (From Meeks, C., ASLE Paper No. 84-AM-6C-3, May 1984. With permission.)

In most cases, roller skewing is detrimental to roller bearing operation because it causes increased friction torque and friction heat generation as well as requiring a cage sufficiently strong to resist the roller skewing moment loading.

6.6.1 ROLLER EQUILIBRIUM SKEWING ANGLE

That rollers skew until skewing moment equilibrium is achieved has implications beyond the determination of roller end–flange load or roller end–cage load. In spherical roller bearings containing rollers with symmetrical profiles, management of roller skewing can minimize friction losses and corresponding friction torque. Early spherical roller bearing designs employing asymmetrical roller profiles, because of their close osculations and primary skewing guidance from cage and flange contacts, exhibit greater friction than current bearings with

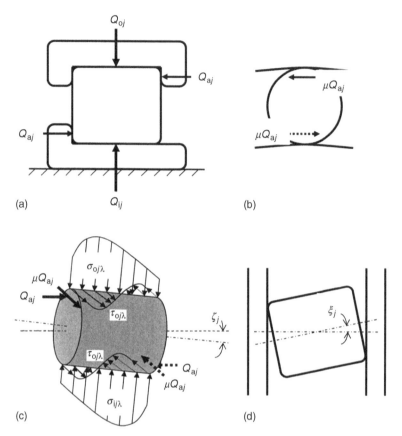

FIGURE 6.26 In a radial cylindrical roller bearing that has crowned rollers and subjected to combined radial and thrust loading, (a) roller–raceway and roller end–guide flange forces, (b) roller end–guide flange friction forces, (c) roller–raceway contact normal and surface friction stresses and roller tilting angle, and (d) roller skewing angle as limited by the roller end–guide flange axial clearance.

symmetrical roller designs. The temperature rise associated with friction limits performance in many applications. Designing the bearings so that skewing equilibrium is provided by raceway guidance alone lowers losses and increases load-carrying capacity. Kellstrom [21,22] investigated skewing equilibrium in spherical roller bearings considering the complex changes in roller force and moment balance caused by roller tilting and skewing in the presence of friction.

Any rolling element that contacts a raceway along a curved contact surface will undergo sliding in the contact. For an unskewed roller there will be at most two points along each contact where the sliding velocity is zero. These zero sliding points form the generatrices of a theoretical "rolling" cone, which represents the contact surface on which pure kinematic rolling would occur for a given roller orientation. At all other points along the contact, sliding is present in the direction of rolling or opposite to it, depending on whether the roller radius is greater or lesser than the radius to the theoretical rolling cone. This situation is illustrated in Figure 6.27.

Friction forces or tractions due to sliding will be oriented to oppose the direction of sliding on the roller. In the absence of tangential roller forces from cage or flange contacts, the roller–raceway traction forces in each contact must sum to zero. Additionally, the sum of the inner

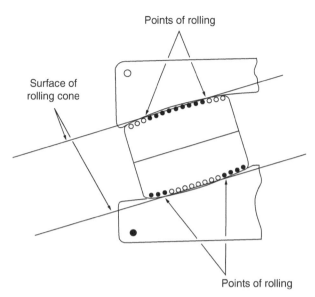

FIGURE 6.27 Spherical roller bearing, symmetrical roller–tangential friction force directions. Motion and force direction: ○ out of page; • into page.

and outer raceway contact skewing moments must equal zero. These two conditions will determine the position of the rolling points along the contacts and thus the theoretical rolling cone. These conditions are met at the equilibrium skewing angle. If the moments tend to restore the roller to the equilibrium skewing angle when it is disturbed, the equilibrium skewing angle is said to be stable.

As a roller skews relative to its contacting raceway, a sliding component is generated in the roller axial direction and traction forces are developed that oppose axial sliding. These traction forces may be beneficial in that, if suitably oriented, they help to carry the axial bearing load, as indicated in Figure 6.28.

Skewing angles that produce axial tractions opposing the applied axial load and reducing the roller contact load required to react with the applied axial load are termed positive (Figure 6.28a). Conversely, skewing angles producing axial tractions that add to the applied axial load are termed negative (Figure 6.28b). For a positive skewing roller, the normal contact is reduced, and an improvement in contact fatigue life achieved.

The axial traction forces acting on the roller also produce a second effect. These forces, acting in different directions on the inner and on outer ring contacts, create a moment about the roller and cause it to tilt. The tilting motion repositions the inner and outer ring contact load distributions with respect to the theoretical points of rolling and distribution of sliding velocity. Detailed evaluations [21,22] of this behavior have shown that skewing in excess of the equilibrium skewing angle generates a net skewing moment opposing the increasing skewing motion. A roller that skews less than the equilibrium skewing angle will generate a net skewing moment tending to increase the skew angle. This set of interactions explains the existence of stable equilibrium skewing angles.

To apply this concept to the design of spherical roller bearings, specific design geometries over a wide range of operating conditions must be evaluated. There are tradeoffs involved between minimizing friction losses and maximizing contact fatigue life. Some designs may exhibit unstable skewing control in certain operating regimes or stable skewing equilibrium and require impractically large skewing angles.

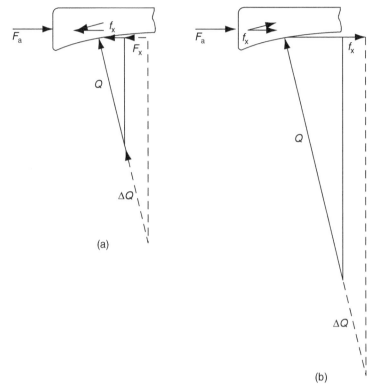

FIGURE 6.28 Forces on outer raceway of axially loaded spherical roller bearing with positive and negative skewing. (a) Positive skewing angle. (b) Negative skewing angle.

6.7 CLOSURE

In the first volume of this handbook, ball and roller speeds were determined using kinematic relationships; these depend on simple rolling motion. While these speed calculations are adequate for many applications, in this chapter, it has been demonstrated that ball and roller speeds are functions of sliding conditions occurring in the rolling element–raceway contacts, sliding conditions between the rolling elements and cage, and between the cage and bearing rings, as well as viscous drag of the lubricant on the orbiting rolling elements. To calculate the rolling element speeds under these conditions, it was shown necessary to create friction force and moment balances about each rolling element and about the bearing as a unit. Solution of the system of equations yields not only the rolling element speeds, but also the cage–rolling element forces and cage–ring land force. This determination enables improved design of cages and bearing internal clearances.

In cylindrical and tapered roller bearings under combined radial, axial, and moment loadings, tendencies toward roller skewing and its effects on speeds and endurance can only be determined by the friction force and moment balance methods introduced in this chapter.

Rolling bearing friction is manifested as temperature rises in the rolling bearing structure and lubricant unless effective heat removal means are employed or naturally occur. In the next chapter, means to estimate bearing internal temperatures will be discussed. It is necessary to note that bearing performance is very sensitive to temperature as (1) alteration of internal dimensions can affect load distribution, (2) lubricant film thickness decreases as temperature-dependent viscosity decreases, (3) friction depends on lubricant film thickness, and (4) in many cases, fatigue endurance is sensitive to the lubricant film thickness and resulting contact surface friction shear stresses.

REFERENCES

1. Harris, T., Rolling element bearing dynamics, *Wear*, 23, 311–337, 1973.
2. Streeter, V., *Fluid Mechanics*, McGraw-Hill, New York, 313–314, 1951.
3. Parker, R., Comparison of predicted and experimental thermal performance of angular-contact ball bearings, NASA Tech. Paper 2275, 1984.
4. Bisson, E. and Anderson, W., *Advanced Bearing Technology*, NASA SP-38, 1964.
5. Rydell, B., New spherical roller thrust bearings, the e design, *Ball Bear. J.*, SKF, 202, 1–7, 1980.
6. Brown, P., et al., Mainshaft high speed cylindrical roller bearings for gas turbine engines, U.S. Navy Contract N00140–76-C-0383, Interim Report FR-8615, April 1977.
7. Harris, T., Ball motion in thrust-loaded, angular-contact ball bearings with coulomb friction, *ASME Trans., J. Lubr. Technol.*, 93, 32–38, 1971.
8. Harris, T., An analytical method to predict skidding in thrust-loaded angular-contact ball bearings, *ASME Trans., J. Lubr. Technol.*, 93, 17–24, 1971.
9. Shevchenko, R. and Bolan, P., Visual study of ball motion in a high speed thrust bearing, SAE Paper No. 37, January 14–18, 1957.
10. Poplawski, J. and Mauriello, J., Skidding in lightly loaded, high speed, ball thrust bearings, ASME Paper 69-LUBS-20, 1969.
11. Harris, T., An analytical method to predict skidding in high speed roller bearings, *ASLE Trans.*, 9, 229–241, 1966.
12. Harris, T. and Aaronson, S., An analytical investigation of skidding in a high-speed, cylindrical roller bearing having circumferentially spaced, preloaded hollow rollers, *Lub. Eng.*, 30–34, January 1968.
13. Walters, C., The dynamics of ball bearings, *ASME Trans., J. Lubr. Technol.*, 93(1), 1–10, January 1971.
14. Wellons, F. and Harris, T., Bearing design considerations, *Interdisciplinary Approach to the Lubrication of Concentrated Contacts*, NASA SP-237, 1970, pp. 529–549.
15. Kleckner, R. and Pirvics, J., High speed cylindrical roller bearing analysis—SKF computer program CYBEAN, Vol. I: analysis, SKF Report AL78P022, NASA Contract NAS3–20068, July 1978.
16. Kannel, J. and Bupara, S., A simplified model of cage motion in angular-contact bearings operating in the EHD lubrication regime, *ASME Trans., J. Lubr. Technol.*, 100, 395–403, July 1078.
17. Gupta, P., Dynamics of rolling element bearings—Part I–IV. cylindrical roller bearing analysis, *ASME Trans., J. Lubr. Technol.*, 101, 293–326, 1979.
18. Meeks, C. and Ng, K., The dynamics of ball separators in ball bearings—Part I: analysis, ASLE Paper No. 84-AM-6C-2, May 1984.
19. Meeks, C., The dynamics of ball separators in ball bearings—Part II: results of optimization study, ASLE Paper No. 84-AM-6C-3, May 1984.
20. Mauriello, J., et al., Rolling element bearing retainer analysis, U.S. Army AMRDL Technical Report 72–45, November 1973.
21. Kellstrom, M. and Blomquist, E., Roller bearings comprising rollers with positive skew angle, U.S. Patent 3,990,753, 1979.
22. Kellstrom, M., Rolling contact guidance of rollers in spherical roller bearings, ASME Paper 79-LUB-23, 1979.

7 Rolling Bearing Temperatures

LIST OF SYMBOLS

Symbol	Description	Units
c	Specific heat	$W \cdot sec/g \cdot {}^\circ C$ (Btu/lb $\cdot$ ${}^\circ F$)
D	Rolling element diameter	mm (in.)
$\mathcal{D}$	Diameter	m (ft)
$\mathcal{E}$	Thermal emissivity	
f	r/D	
f_0	Viscous friction torque coefficient	
f_1	Load friction torque coefficient	
F_a	Applied axial (thrust) load	N (lb)
F_r	Applied radial load	N (lb)
F_β	Equivalent applied load to calculate friction torque	N (lb)
F	Temperature coefficient	$W \cdot sec/{}^\circ C$ (Btu/${}^\circ F$)
g	Acceleration due to gravity	m/sec^2 (in./sec^2)
Gr	Grashof number	
h	Film coefficient of heat transfer	$W/m^2 \cdot {}^\circ C$ (Btu/hr $\cdot$ ft^2 $\cdot$ ${}^\circ F$)
H	Heat flow rate, friction heat generation rate	W (Btu/hr)
J	Conversion factor, 10^3 N $\cdot$ mm $= 1$ W $\cdot$ sec	
k	Thermal conductivity	$W/m \cdot {}^\circ C$ (Btu/hr $\cdot$ ft $\cdot$ ${}^\circ F$)
$\mathcal{L}$	Length of heat conduction path	m (ft)
M	Friction torque	N $\cdot$ mm (in. $\cdot$ lb)
n	Rotational speed	rpm
Pr	Prandtl number	
q	Error function	
Re	Reynolds number	
$\mathcal{R}$	Radius	m (ft)
s	Surface roughness	μm (μin.)
S	Area normal to heat flow	m^2 (ft^2)
T	Temperature	${}^\circ C$ (${}^\circ F$)
u_s	Fluid velocity	m/sec (ft/sec)
v	Velocity	m/sec (ft/sec)
w	Weight flow rate	g/sec (lb/sec)
$\mathcal{W}$	Width	m (ft)
x	Distance in x direction	m (ft)
z	Number of rolling elements	
ε	Error	
η	Absolute viscosity	cp (lb $\cdot$ sec/in.^{2})
ν	Fluid kinematic viscosity	m^2/sec (ft^2/sec)
σ	Rolling element–raceway contact normal stress	MPa (psi)

τ	Surface friction shear stress	MPa (psi)
ω	Rotational velocity	rad/sec
Ω	Rotational velocity	rad/sec

Subscripts

a	Air or ambient condition
BRC	Ball–raceway contact
c	Heat conduction
CRL	Contact between the cage rail and ring land
CPR	Contact between the cage pocket and rolling element
f	Friction
fdrag	Viscous drag on the rolling elements
i	Inner raceway
j	Rolling element position
n	raceway
o	Oil or outer raceway
r	Heat radiation
REF	Roller end–flange contact
RRC	Roller–raceway contact
tot	Bearing total friction heat generation
v	Heat convection
x	x direction, transverse to rolling direction
y	y direction, rolling direction
1	Temperature node 1
2	Temperature node 2, and so on

7.1 GENERAL

The overall temperature level at which a rolling bearing operates depends on many variables among which are:

- Applied load
- Operating speeds
- Lubricant type and its rheological properties
- Bearing mounting arrangement and housing design
- Operational environment

In the steady-state operation of a rolling bearing, the friction heat generated must be dissipated. Therefore, the steady-state temperature level of one bearing system compared with that of another using identical sizes and number of bearings is a measure of that system's efficiency of heat dissipation.

If the rate of heat dissipation is less than the rate of heat generation, then an unsteady state exists and the system temperatures will rise, most likely until lubricant deterioration occurs, ultimately resulting in bearing failure. The temperature at which this occurs depends greatly on the type of lubricant and the bearing materials. The discussion in this chapter is limited to the steady-state thermal operation of rolling bearings, since this is the principal concern of bearing users.

Most ball and roller bearing applications perform at relatively cool temperature levels and, therefore, do not require any special consideration regarding thermal adequacy. This is due to either one of the following conditions:

- The bearing friction heat generation rate is low because of light load and relatively slow operating speed.
- The bearing heat dissipation rate is sufficient because the bearing assembly is located in a moving air stream or there is adequate heat conduction through adjacent metal.

Some applications experience adverse environmental conditions such that external heat removal means is required. A rapid determination of the bearing cooling requirements may then suffice to establish the cooling capability that must be applied to the lubricating fluid. In applications where it is not obvious whether external cooling means is required, it may be economically advantageous to analytically determine the thermal conditions of bearing operation.

7.2 FRICTION HEAT GENERATION

7.2.1 BALL BEARINGS

Rolling bearing friction represents a power loss manifested in the form of heat generation. The friction heat generated must be effectively removed from the bearing or an unsatisfactory temperature condition will obtain in the bearing. In a ball–raceway contact, the friction heat generation rate is given by

$$H_{nyj} = \frac{1}{J} \int \tau_{nyj} v_{nyj} \, dA_{nj} = \frac{a_{nj}b_{nj}}{J} \int_{-1}^{+1} \int_{-\sqrt{1-q^2}}^{+\sqrt{1-q^2}} \tau_{nyj} v_{nyj} \, dt \, dq, \quad n = i, o; \; j = 1 - Z \quad (7.1)$$

where J is a constant converting N · m/sec to watts. In Equation 7.1, the surface friction shear stress τ_{ny} may be obtained directly from Equation 5.49 or from Equation 5.5 recognizing that $\tau_{nyj} = \mu_{nyj}\sigma$. The values of sliding velocity v_{yj} may be obtained from Equation 2.9 and Equation 2.20. Similarly,

$$H_{nxj} = \frac{1}{J} \int \tau_{nxj} v_{nxj} \, dA_{nj} = \frac{a_{nj}b_{nj}}{J} \int_{-1}^{+1} \int_{-\sqrt{1-t^2}}^{+\sqrt{1-t^2}} \tau_{nyj} v_{nxj} \, dq \, dt, \quad n = i, o; \; j = 1 - Z \quad (7.2)$$

where τ_{nxj} may be obtained directly from Equation 5.50 and v_{xj} may be obtained from Equation 2.10 and Equation 2.21. For an entire bearing, the friction heat generated in the ball–raceway contacts is

$$H_{BRC} = \sum_{n=i}^{n=o} \sum_{j=1}^{j=Z} \left(H_{ynj} + H_{xnj} \right) \quad (7.3)$$

For an oil-lubricated bearing, in addition to the friction heat generated in the ball–raceway contacts, friction heat is generated due to the balls passing through the lubricant in the bearing free space. Using Equation 6.1 to define the viscous drag force F_v, the friction heat generation rate thereby effected is given by

$$H_{fdrag} = \frac{d_m \omega_m F_v Z}{2J} \tag{7.4}$$

where d_m is the bearing pitch diameter, ω_m is the ball orbital speed, and Z is the number of balls.

Finally, friction heat is generated due to sliding between the cage and the inner ring land for an inner ring piloted cage; due to sliding between the cage and the outer ring land for an outer ring piloted cage; and between the balls and the cage pockets for any cage design execution. These heat generation rates generally tend to be small; however, they may be calculated using the ball and cage speed equations of Chapter 2 together with estimations of cage rail–ring land loading and cage–ball loading. These may be determined using a complete friction force and moment balance according to Chapter 6.

The total friction heat generation rate is obtained by summation of the component heat generation rates

$$H_{tot} = H_{BRC} + H_{fdrag} + H_{CRL} + H_{CPB} \tag{7.5}$$

It is noted that H_{tot} does not include the friction heat generation rate due to the contact between integral seals and the bearing ring surface. This heat component will most likely be greater than H_{tot} as defined in Equation 7.5.

Bearing friction torque about the shaft can be derived from H_{tot} using the following equation:

$$M = 10^3 \times \frac{H_{tot}}{\Omega_n} \tag{7.6}$$

where H_{tot} is in watts, friction torque M is in N $\cdot$ mm, and ring speed Ω_n is in rad/sec. For ring speed in rpm,

$$M = 9.551 \times 10^3 \times \frac{H_{tot}}{n_n} \tag{7.7}$$

7.2.2 Roller Bearings

To find the roller–raceway contact friction heat generation rate, as introduced in Chapter 1, each contact area of effective length l_{eff} is divided into k laminae, each lamina having thickness w_n and width $2b_{nj\lambda}$, subscript λ referring to the specific lamina. Hence,

$$H_{nj} = \frac{2b_{nj}w_n}{3kJ} \sum_{\lambda=1}^{\lambda=k} S_k \tau_{nj\lambda} v_{nj\lambda} \tag{7.8}$$

In Equation 7.8, S_k is the Simpson's rule coefficient, $\tau_{nj\lambda}$ is the average surface friction shear stress over the lamina area $2b_{nj\lambda}w_n$, and $v_{nj\lambda}$ is the sliding velocity at the lamina surface. For cylindrical roller bearings, the sliding velocity over the lamina may be obtained from Equation 6.62. The friction heat generation rate for all roller–raceway contacts is then

$$H_{RRC} = \sum_{n=i}^{n=o} \sum_{j=1}^{j=Z} H_{nj} \tag{7.9}$$

Using Equation 6.2 to define the viscous drag force F_v, the friction heat generation rate thereby effected is given by Equation 7.4.

As with ball bearings, friction heat is generated due to sliding between cage and the inner ring land for an inner ring piloted cage, due to sliding between cage and the outer ring land for an outer ring piloted cage, and between the rollers and the cage pockets for any cage design execution. These heat generation rates generally tend to be small; however, they may be calculated using the roller and cage speed equations of Chapter 2 together with estimations of cage rail–ring land loading and cage–roller loading. These may be determined using a complete friction force and moment balance according to Chapter 6.

In addition to the above-mentioned sources of friction heat generation, in cylindrical roller bearings that are misaligned or otherwise subjected to combined radial and thrust loadings, significant friction heat generation can occur between the roller ends and inner and outer ring roller guide flanges. To estimate the heat generation rates, it is first necessary to calculate the roller end–flange loads Q_{aj} using the analytical methods indicated in Chapter 1 and Chapter 3. Then, using the methods indicated in Chapter 6, the cage speed ω_m and roller speeds ω_{Rj} need to be estimated. Knowing the ring speeds, it is possible to estimate an average sliding velocity between the roller ends and ring flange. Finally, depending on the lubrication method, a coefficient of sliding friction for the roller end–flange contacts needs to be assumed or calculated. Generally, for oil-lubricated bearings, a value of $0.03 \le \mu \le 0.07$ should be obtained. The friction heat generation rate for a roller end–flange contact is then

$$H_{\text{REF}nj} = \frac{\mu Q_{aj} v_{\text{REF}nj}}{J} \tag{7.10}$$

and

$$H_{\text{REF}} = \sum_{n=\text{i}}^{n=\text{o}} \sum_{j=1}^{j=Z} H_{\text{REF}nj} \tag{7.11}$$

Each roller in a tapered roller bearing experiences contact between the roller end and the large end flange as indicated in Chapter 5 of the first volume of this handbook. In this case,

$$H_{\text{REF}j} = \frac{\mu Q_{fj} v_{\text{REF}j}}{J} \tag{7.12}$$

and

$$H_{\text{REF}} = \sum_{j=1}^{j=Z} H_{\text{REF}j} \tag{7.13}$$

Equation 7.12 and Equation 7.13 may be used to calculate H_{REF} for spherical roller bearings with asymmetrical contour rollers.

For roller bearings the total friction heat generation rate, exclusive of seals, is

$$H_{\text{tot}} = H_{\text{RRC}} + H_{\text{fdrag}} + H_{\text{CRL}} + H_{\text{CPR}} + H_{\text{REF}} \tag{7.14}$$

See Example 7.1 and Example 7.2.

Methods for calculating friction torque and heat generation rates may also be found in bearing catalogs; for example, Ref [8].

7.3 HEAT TRANSFER

7.3.1 MODES OF HEAT TRANSFER

There exist three fundamental modes for the transfer of heat between masses with different temperature levels. These are the conduction of heat within solid structures, the convection of heat from solid structures to fluids in motion (or apparently at rest), and the radiation of heat between masses separated by space. Although other modes exist, such as radiation to gases and conduction within fluids, their effects are minor for most bearing applications and may usually be neglected.

7.3.2 HEAT CONDUCTION

Heat conduction, which is the simplest form of heat transfer, may be described for the purpose of this discussion as a linear function of the difference in temperature level within a solid structure, that is,

$$H_c = \frac{kS}{\mathcal{L}}(T_1 - T_2) \tag{7.15}$$

The quantity S in Equation 7.15 is the area normal to the flow of heat between two points and $\mathcal{L}$ is the distance between the same two points. The thermal conductivity k is a function of the material and temperature levels; however, the latter variation is generally minor for structural solids and will be neglected here. For heat conduction in a radial direction within a cylindrical structure such as a bearing inner or outer ring, the following equation is useful:

$$H_c = \frac{2\pi k \mho (T_i - T_o)}{\ln(\Re_o/\Re_i)} \tag{7.16}$$

In Equation 7.16, $\mho$ is the width of the annular structure and $\Re_o$ and $\Re_i$ are the inner and outer radii defining the limits of the structure through which heat flow occurs. If $\Re_i = 0$, an arithmetic mean area is used and the equation assumes the form of Equation 7.15.

7.3.3 HEAT CONVECTION

Heat convection is the most difficult form of heat transfer to estimate quantitatively. It occurs within the bearing housing as heat is transferred to the lubricant from the bearing and from the lubricant to other structures within the housing as well as to the inside walls of the housing. It also occurs between the outside of the housing and the environmental fluid— generally air, but possibly oil, water, another gas, or a working fluid medium.

Heat convection from a surface may generally be described as follows:

$$H_v = h_v S(T_1 - T_2) \tag{7.17}$$

where h_v, the film coefficient of heat transfer, is a function of surface and fluid temperatures, fluid thermal conductivity, fluid velocity adjacent to the surface, surface dimensions and attitude, fluid viscosity, and density. It can be seen that many of these properties are temperature dependent. Therefore, heat convection is not a linear function of temperature unless fluid properties can be considered reasonably stable over a finite temperature range.

Heat convection within the housing is most difficult to describe, and a rough approximation will be used for the heat transfer film coefficient. As oil is used as a lubricant and the viscosity is high, laminar flow is assumed. Eckert [2] states for a plate in a laminar flow field:

$$h_v = 0.0332k \, \mathrm{Pr}^{1/3} \left(\frac{u_s}{\nu_o x} \right)^{1/2} \tag{7.18}$$

The use of Equation 7.18 taking u_s equal to bearing cage surface velocity and x equal to bearing pitch diameter seems to yield workable values for h_v, considering heat transfer from the bearing to the oil that contacts the bearing. For heat transfer from the housing inside surface to the oil, taking u_s equal to one third cage velocity and x equal to housing diameter yields adequate results. In Equation 7.18, ν_o represents kinematic viscosity and Pr the Prandtl number of the oil.

If cooling coils are submerged in the oil sump, it is best that they be aligned parallel to the shaft so that a laminar cross-flow is obtained. In this case, Eckert [2] shows that for a cylinder in cross-flow, the outside heat transfer film coefficient may be approximated by

$$h_v = 0.06 \frac{k_o}{\mathcal{D}} \left(\frac{u_s \mathcal{D}}{\nu_o} \right)^{1/2} \tag{7.19}$$

where $\mathcal{D}$ is the outside diameter of the tube and k_o is the thermal conductivity of the oil. It is recommended that u_s be taken as approximately one fourth of the bearing inner ring surface velocity.

These approximations for film coefficient are necessarily crude. If greater accuracy is required, Ref. [2] indicates more refined methods for obtaining the film coefficient. In lieu of a more elegant analysis, the values yielded by Equation 7.18 and Equation 7.19, and Equation 7.20 and Equation 7.21 that follow, should suffice for general engineering purposes.

In quiescent air, heat transfer by convection from the housing external surface may be approximated by using an outside film coefficient in accordance with Equation 7.20 (see Jakob and Hawkins [3]):

$$h_v = 2.3 \times 10^{-5} (T - T_a)^{0.25} \tag{7.20}$$

For forced flow of air of velocity u_s over the housing, Ref. [2] yields:

$$h_v = 0.03 \frac{k_a}{\mathcal{D}} \left(\frac{u_s \mathcal{D}_h}{\nu_a} \right)^{0.57} \tag{7.21}$$

where $\mathcal{D}_h$ is the approximate housing diameter. Palmgren [4] gives the following formula to approximate the external area of a bearing housing or pillow block:

$$S = \pi \mathcal{D}_h \left(\mathcal{W}_h + \tfrac{1}{2}\mathcal{D}_h \right) \tag{7.22}$$

where $\mathcal{D}_h$ is the maximum diameter of the pillow block and $\mathcal{W}_h$ is the width.

The calculations of lubricant film thickness as specified in Chapter 4 depend on the viscosity of the lubricant entering the rolling/sliding contact, while the calculations of traction over the contact as specified in Chapter 5 depend on the viscosity of the lubricant in the contact. Since lubricant viscosity is a function of temperature, a detailed performance analysis of ball and roller bearings entails the estimation of temperatures of lubricants both entering, and residing in, the individual contacts. To do this requires the estimation of heat dissipation rates from the rotating components and rings. The coefficient of convection heat transfer for a rotating sphere (ball) is provided by Kreith [5] as follows:

$$\frac{h_v D}{k} = 0.33 \text{Re}_D^{0.5} \text{ Pr}^{0.4} \tag{7.23}$$

where Re_D, the Reynolds number for a rotating ball, is given by

$$\text{Re}_D = \frac{\omega D^2}{\nu} \tag{7.24}$$

In Equation 7.24, D is the diameter of the ball, ω is the ball speed about its own axis, and ν is the lubricant kinematic viscosity. Equation 7.23 is valid for $0.7 < \text{Pr} < 217$ and $\text{Gr}_D < 0.1 \cdot \text{Re}_D^2$. The Grashof number is given by

$$\text{Gr} = \frac{Bg(T_s - T_\infty)D^2}{\nu^2} \tag{7.25}$$

where B is the thermal coefficient of fluid volume expansion, g the acceleration due to gravity, T_s the temperature at the ball surface, and T_∞ is the fluid stream temperature. The Prandtl number is given by

$$\text{Pr} = \frac{\eta g c}{k} \tag{7.26}$$

where c is the specific heat of the fluid.

For a rotating cylindrical ring or roller,

$$\frac{h_v D}{k} = 0.19(\text{Re}_D^2 + \text{Gr}_D) \tag{7.27}$$

In Equation 7.27, D is the outside diameter of the ring or roller. Equation 7.27 is valid for $\text{Re}_D < 4 \times 10^5$.

7.3.4 Heat Radiation

The remaining mode of heat transfer to be considered is the radiation from the housing external surface to the surrounding structures. For a small structure in a large enclosure, Ref. [3] gives

$$H_r = 5.73 \,\varepsilon S \left[\left(\frac{T}{100} \right)^4 - \left(\frac{T_a}{100} \right)^4 \right] \tag{7.28}$$

where the temperature is in degrees Kelvin (absolute). Equation 7.28, nonlinear being in temperatures, is sometimes written in the following form:

$$H_r = h_r S(T - T_a) \tag{7.29}$$

where

$$h_r = 5.73 \times 10^{-8} \varepsilon (T + T_a)(T^2 + T_a^2) \tag{7.30}$$

Equation 7.29 and Equation 7.30 are useful for hand calculation in which problem T and T_a are not significantly different. On assuming a temperature T for the surface, the pseudofilm coefficient of radiation h_r may be calculated. Of course, if the final calculated value of T is significantly different from that assumed, then the entire calculation must be repeated. Actually, the same consideration is true for calculation of h_v for the oil film. Since k_o and ν_o are dependent on temperature, the assumed temperature must be reasonably close to the final calculated temperature. How close is dictated by the actual variation of those properties with oil temperature.

7.4 ANALYSIS OF HEAT FLOW

7.4.1 SYSTEMS OF EQUATIONS

Because of the discontinuities of the structures that comprise a rolling bearing assembly, classical methods of heat transfer analysis cannot be applied to obtain a solution describing the system temperatures. By classical methods we mean the description of the system in terms of differential equations and the analytical solution of these equations. Instead, methods of finite difference as demonstrated by Dusinberre [6] must be applied to obtain a mathematical solution.

For finite difference methods applied to steady-state heat transfer, various points or nodes are selected throughout the system to be analyzed. At each of these points, the temperature is determined. In steady-state heat transfer, heat influx to any point equals heat efflux; there-fore, the sum of all heat flowing toward a temperature node is equal to zero. Figure 7.1 is a heat flow diagram at a temperature node, demonstrating that the nodal temperature is

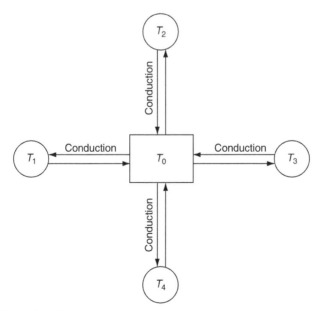

FIGURE 7.1 Two-dimensional temperature node system.

affected by the temperatures of each of the four indicated surrounding nodes. (Although the system depicted in Figure 7.1 shows only four surrounding nodes, this is purely by choice of grid and the number of nodes may be greater or smaller.) Since the sum of the heat flows is zero,

$$H_{1-0} + H_{2-0} + H_{3-0} + H_{4-0} = 0 \tag{7.31}$$

For this example, it is assumed that heat flow occurs only by conduction and that the grid is nonsymmetrical, making all areas S and lengths of flow path different. Furthermore, the material is assumed nonisotropic so that thermal conductivity is different for all flow paths. Substitution of Equation 7.15 into Equation 7.31 therefore yields

$$\frac{k_1 S_1}{\mathscr{L}_1}(T_1 - T_0) + \frac{k_2 S_2}{\mathscr{L}_2}(T_2 - T_0) + \frac{k_3 S_3}{\mathscr{L}_3}(T_3 - T_0) + \frac{k_4 S_4}{\mathscr{L}_4}(T_4 - T_0) = 0 \tag{7.32}$$

By rearranging terms, one obtains

$$\frac{k_1 S_1}{\mathscr{L}_1} T_1 + \frac{k_2 S_2}{\mathscr{L}_2} T_2 + \frac{k_3 S_3}{\mathscr{L}_3} T_3 + \frac{k_4 S_4}{\mathscr{L}_4} T_4 - \sum_{i=1}^{i=4} \frac{k_i S_i}{\mathscr{L}_i} T_0 = 0 \tag{7.33}$$

or

$$F_1 T_1 + F_2 T_2 + F_3 T_3 + F_4 T_4 - \sum_{i=1}^{i=4} F_i T_0 = 0 \tag{7.34}$$

Dividing by ΣF_i yields

$$\frac{F_1}{\Sigma F_i} T_1 + \frac{F_2}{\Sigma F_i} T_2 + \frac{F_3}{\Sigma F_i} T_3 + \frac{F_4}{\Sigma F_i} T_4 - T_0 = 0 \tag{7.35}$$

More concisely, Equation 7.35 may be written

$$\phi_i T_i = 0 \tag{7.36}$$

where ϕ_i are influence coefficients of temperature equal to $F_i/\Sigma F_i$. If the material were isotropic and a symmetrical grid was chosen, then $\phi_0 = 1$ and the other $\phi_i = 0.25$.

In the example, only heat conduction was illustrated. If, however, heat flow between points 4 and 0 was by convection, then according to Equation 7.17, $F_4 = h_{v4} S_4$. For a multi-nodal system, a series of equations similar to Equation 7.35 may be written. If the equations are linear in temperature T, they may be solved by classical methods for the solution of simultaneous linear equations or by numerical methods (see Ref. [7]).

The system may include heat generation and be further complicated, however, by non-linear terms caused by heat radiation and free convection. Consider the example schematic-ally illustrated in Figure 7.2. In that illustration, heat is generated at point 0, dissipated by free convection and radiation between points 1 and 0 and dissipated by conduction between points 2 and 0. Thus,

$$H_{f0} + H_{1-0,v} + H_{1-0,r} + H_{2-0} = 0 \tag{7.37}$$

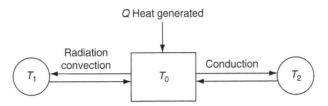

FIGURE 7.2 Convective, radiation, and conductive heat transfer system.

The use of Equation 7.15, Equation 7.17, Equation 7.20, and Equation 7.28 gives

$$H_{f0} + 2.3 \times 10^{-5} S_1 (T_1 - T_0)^{1.25} + 5.73 \times 10^{-8} \varepsilon S_1 (T_1^4 - T_0^4) + \frac{K_2 S_2}{L_2} (T_2 - T_0) = 0 \quad (7.38)$$

or

$$H_{f0} + F_{1v}(T_1 - T_0)^{1.25} + F_{1r}(T_1^4 - T_0^4) + F_2(T_2 - T_0) = 0 \quad (7.39)$$

7.4.2 SOLUTION OF EQUATIONS

A system of nonlinear equations similar to Equation 7.39 is difficult to solve by direct numerical methods of iteration or relaxation. Therefore, the Newton–Raphson method [7] is recommended for solution.

The Newton–Raphson method states that for a series of nonlinear functions q_i of variables T_j

$$q_i + \sum \frac{\partial q_i}{\partial T_j} \varepsilon_j = 0 \quad (7.40)$$

Equation 7.40 represents a system of simultaneous linear equations that may be solved for ε_j (error on T_j).

Then, the new estimate of T_j is

$$T_j' = T_j(0) + \varepsilon_j \quad (7.41)$$

and new values q_i may be determined. The process is continued until the functions q_i are virtually zero. With a system of nonlinear equations similar to Equation 7.39, such equations must be linearized according to Equation 7.40. Thus, let Equation 7.39 be rewritten as follows:

$$H_{f0} = F_{1v}(T_1 - T_0)^{1.25} + F_{1r}(T_1^4 - T_0^4) + F_2(T_2 - T_0) = q_0 \quad (7.42)$$

Now,

$$\frac{\partial q_0}{\partial T_0} = -1.25 F_{1v}(T_1 - T_0)^{0.25} - 4F_{1r}T_0^3 + F_2$$

$$\frac{\partial q_0}{\partial T_1} = 1.25 F_{1v}(T_1 - T_0)^{0.25} + 4F_{1r}T_1^3 \quad (7.43)$$

$$\frac{\partial q_0}{\partial T_2} = F_2$$

Substitution of Equation 7.42 and Equation 7.43 into Equation 7.40 yields one equation in variables ε_0, ε_1, and ε_2.

The system of nonlinear equations is solved for T_0, T_1, and T_2 when the root mean square (rms) error is sufficiently small, for example, less than $0.1°$.

7.4.3 TEMPERATURE NODE SYSTEM

A simple system of temperature nodes that could be used to determine the temperatures in an oil-lubricated, spherical roller bearing pillow block assembly is illustrated in Figure 7.3. In this illustration, the dimensions of a 23072 double-row bearing are shown together with pertinent dimensions of the pillow block. This illustration has been designed to be as simple as possible such that all equations and methods of solution may be demonstrated. To do this, the following conditions have been assumed:

1. Ten temperature nodes are sufficient to describe the system shown in Figure 7.3. Node A is ambient temperature; nine temperatures need to be determined.
2. The inside of the housing is coated with oil and may be described by a single temperature.
3. The inner ring raceway may be described by a single temperature.

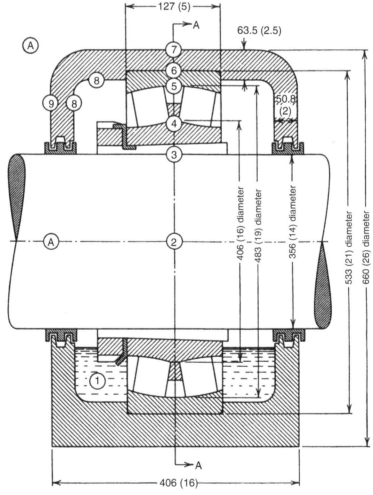

FIGURE 7.3 Simple temperature node system selected for analysis of a spherical roller bearing pillow block assembly.

4. The outer ring raceway may be described by a single temperature.
5. The housing is symmetrical about the shaft centerline and vertical section A–A. Thus, heat transfer in the circumferential direction does not have to be considered.
6. The sump oil may be considered at a single temperature.
7. The shaft ends at the axial extremities of the pillow block are at ambient temperature.

Considering the temperature node system of Figure 7.3, the heat transfer system with pertinent equations is given in Table 7.1. The heat flow areas and lengths of heat flow paths are obtained from the dimensions of Figure 7.3 considering the location of each temperature node. Based on Figure 7.3 and Table 7.1, a set of nine simultaneous nonlinear equations with unknown variables $T_1 - T_9$ can be developed. This system is nonlinear because of free convection from the pillow block external surface to ambient air and radiation from the external surface to structures at ambient temperature; the Newton–Raphson method may be used to obtain a solution.

See Example 7.3.

The system chosen for evaluation was necessarily simplified for the purpose of illustration. A more realistic system would consider variation of bearing temperature in a circumferential direction also. This would entail many more temperature nodes and corresponding heat transfer equations. In this case, viscous friction torque may be constant with respect to angular location; however, friction torque due to load varies as the individual rolling element load on the stationary ring. The latter, however, may be considered invariant with respect to angular location on the rotating ring. A three-dimensional analysis such as that indicated by load friction torque variation on the stationary ring should, however, show little variation in temperature around the bearing ring circumferences so that a two-dimensional system should suffice for most engineering applications. Of course, if temperatures of structures surrounding or abutting the housing vary significantly, then a three-dimensional study is required.

It is not intended that the results of this method of analysis will be of extreme accuracy, but only that accuracy will be sufficient to determine the approximate thermal level of operation. Then, corrective measures may be taken if excessive steady-state operating temperatures are indicated. In the event cooling of the assembly is required, the same methods may be used to evaluate the adequacy of the cooling system.

Generally, the more temperatures selected, that is, the finer the heat transfer grid, the more accurate will be the analysis.

7.5   HIGH TEMPERATURE CONSIDERATIONS

7.5.1   Special Lubricants and Seals

Having established the operating temperatures in a rolling bearing assembly while using a conventional mineral oil lubricant and lubrication system, and having estimated that the bearing or lubricant temperatures are excessive, it then becomes necessary to redesign the system to either reduce the operating temperatures or make the assembly compatible with the temperature level. Of the two alternatives, the former is safest when considering prolonged duration of operation of the assembly. When shorter lubricant or bearing life is acceptable, it may be expeditious and even economical to simply accommodate the increased temperature level by using special lubricants in the bearing operations or a bearing manufactured from a high-temperature capacity steel. The latter approach is effective when space and weight limitations preclude the use of external cooling systems. It is further necessitated in applications in which the bearing is not the prime source of heat generation.

TABLE 7.1
Heat Transfer System Matrix and Heat Transfer Equations for Figure 7.3

Node	A	1	2	3	4	5	6	7	8	9
1	—	—	—	Convection (7.17) (7.7)	Convection (7.17) (7.7)	Convection (7.17) (7.7)	—	—	Convection (7.17) (7.7)	—
2	Conduction (7.15)	—	—	Conduction (7.16)	—	—	—	—	—	—
3	—	Convection (7.17) (7.7)	Conduction (7.16)	—	Conduction (7.16)	—	—	—	—	—
4	—	Convection (7.17) (7.7)	—	Conduction (7.16)	Heat generation (7.14)	—	—	—	—	—
5	—	Convection (7.17) (7.7)	—	—	—	Heat generation (7.14)	Conduction (7.16)	—	—	—
6	—	—	—	—	—	Conduction (7.16)	—	Conduction (7.16)	Conduction (7.15)	—
7	Convection (7.17) (7.20) Radiation (7.28)	—	—	—	—	—	Conduction (7.16)	—	Conduction (7.16)	—
8	—	Convection (7.17) (7.7)	—	—	—	—	—	Conduction (7.15)	Conduction (7.16)	Conduction (7.15)
9	Convection (7.17) (7.20) Radiation (7.28)	—	—	—	—	—	—	Conduction (7.15)	Conduction (7.15)	Conduction (7.15)

7.5.2 HEAT REMOVAL

For situations in which the bearing is the prime source of heat generation and in which the ambient conditions do not permit an adequate rate of heat removal, placing the bearing housing in a moving air stream may be sufficient to reduce operating temperatures. This may be accomplished by using a fan of sufficient air moving capacity.

Additional heat removal capacity may be effected by designing a housing with fins to increase the effective area for heat transfer.

See Example 7.4 and Example 7.5.

When the bearing is not the prime source of heat generation, cooling of the housing in the foregoing manner will generally not suffice to maintain the bearing and lubricant cool. In this case, it is generally necessary to cool the lubricant and permit the lubricant to cool the bearing. The most effective way of accomplishing this is to pass the oil through an external heat exchanger and direct jets of cooled oil on the bearing. To save space when a supply of moving coolant is readily available, it may be possible to place the heat exchanger coils directly in the sump of the bearing housing. The cooled lubricant is then circulated by bearing rotation. The latter method is not quite as efficient thermally as jet cooling although bearing friction torque and heat generation may be less by not resorting to jet lubrication and the attendant churning of excess oil.

See Example 7.6.

Several researchers have applied these methods to effectively predict temperatures in rolling bearing applications. Initially, Harris [9,10] applied the method to relatively slow-speed, spherical roller bearings. Subsequently, these methods have been successfully applied to both high-speed ball and roller bearings [11–13]. Good agreement with experimentally measured temperatures was reported [15] using the steady-state temperature calculation operation mode of SHABERTH [14], a computer program to analyze the thermo-mechanical performance of shaft-rolling bearing systems. Figure 7.4 shows a nodal network model and the associated heat flow paths for a 35-mm bore ball bearing. Figure 7.5 shows the agreement achieved between calculated and experimentally measured temperatures. It must be pointed out, however, that construction of a thermal model that mathematically simulates a bearing accurately often requires a considerable amount of effort and heat transfer expertise.

7.6 HEAT TRANSFER IN A ROLLING–SLIDING CONTACT

Accurate calculation of lubricant film thickness and traction in a rolling contact depends on the determination of lubricant viscosity at the appropriate temperatures. For lubricant film thickness, this means calculation of the lubricant temperature entering the contact. For traction, this means calculation of the lubricant temperature for its duration in the contact. In Ref. [14], the heat transfer system illustrated in Figure 7.6 was used.

Designating subscript k to represent the raceway and j the rolling element location, the following heat flow equations describe the system:

$$H_{c,2kj-1kj} + H_{v,\text{tout}-1kj} = 0 \qquad (7.44)$$

Since the lubricant is essentially a solid slug during its time in the contact, heat transfer from the film to the rolling body surfaces is by conduction. Then, assuming that the minute slug exists at an average temperature T_3

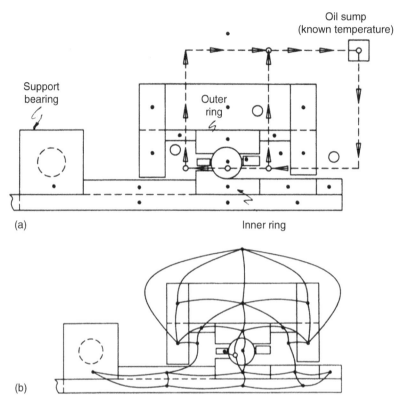

FIGURE 7.4 Bearing system nodal network and heat flow paths for steady-state thermal analysis. (a) Metal, air, and lubricant temperature nodes: (•) metal or air node; (○) lubricant node; (- →) lubricant flow path. (b) Conduction and convection heat flow paths (From Parker, R., NASA Technical Paper 2275, February 1984.)

$$H_{c,1kj-2kj} + H_{c,3kj-2kj} = 0 \qquad (7.45)$$

The lubricant slug is transported through the contact; it enters at temperature T_3^0 and exits at T_3'. Therefore, for heat transfer totally within the slug

$$H_{c,2kj-3kj} + H_{c,4kj-3kj} + H_{gen,j} + wc\left(T_{3kj}' - T_{3kj}^0\right) = 0 \qquad (7.46)$$

$$H_{c,3kj-4kj} + H_{c,5kj-4kj} = 0 \qquad (7.47)$$

$$H_{c,4kj-5kj} + H_{v,tout-5kj} = 0 \qquad (7.48)$$

Finally, the lubricant acts as a heat sink carrying heat away from the contact

$$H_{v,6-1kj} + H_{v,6-5kj} - wc(T_6 - T_{1.in}) = 0 \qquad (7.49)$$

In high-speed bearing frictional performance analyses such as those indicated in Chapter 6, the rolling–sliding contact heat transfer analyses are performed thousands of times to achieve consistent solutions. The analyses are begun by assuming a set of system temperatures. Lubricant viscosities are then determined at these temperatures, and frictional heat generation rates are calculated. These are subsequently used to recalculate temperatures and temperature-dependent parameters. The process is repeated until the calculated temperatures

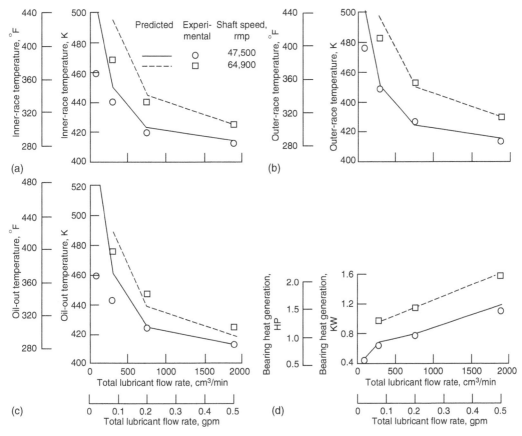

FIGURE 7.5 Comparison of predicted and experimental temperatures using SHABERTH. (a) Inner raceway temperature. (b) Outer raceway temperature. (c) Oil-out temperature. (d) Bearing heat generation (From Parker, R., NASA Technical Paper 2275, February 1984.).

substantially match the assumed temperatures. This method while producing more accurate calculations for bearing heat generations and friction torques requires rather sophisticated computer programs for its execution see Ref. [1,16]. For slow-speed bearing applications in which the bearing rings are rigidly supported, the simpler calculation methods for bearing heat generations provided in Chapter 10 of the first volume of this handbook will generally suffice.

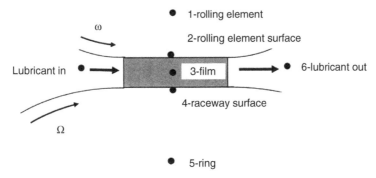

FIGURE 7.6 Rolling element–lubricant–raceway–ring temperature node system.

7.7 CLOSURE

The temperature level at which a rolling bearing operates dictates the type and amount of lubricant required as well as the materials from the bearing components that may be fabricated. In some applications, the environment in which the bearing operates establishes the temperature level whereas in other applications the bearing is the prime source of heat. In either case, depending on the bearing materials and the endurance required of the bearing, it may be necessary to cool the bearing using the lubricant as a coolant.

General rules cannot be formulated to determine the temperature level for a given bearing operating under a given load at a given speed. The environment in which the bearing operates is generally different for each specialized application. Using the friction torque formulas of Chapter 10 of the first volume of this handbook or Chapter 6 in the second volume to establish the rate of bearing heat generation in conjunction with the heat transfer methods presented in this chapter, however, it is possible to estimate the bearing system temperatures with an adequate degree of accuracy.

REFERENCES

1. Harris, T., Establishment of a new rolling bearing fatigue life calculation model, Final Report U.S. Navy Contract N00421-97-C-1069, February 23, 2002.
2. Eckert, E., *Introduction to the Transfer of Heat and Mass*, McGraw-Hill, New York, 1950.
3. Jakob, M. and Hawkins, G., *Elements of Heat Transfer and Insulation*, 2nd Ed., Wiley, New York, 1950.
4. Palmgren, A., *Ball and Roller Bearing Engineering*, 3rd Ed., Burbank, Philadelphia, 1959.
5. Kreith, F., Convection heat transfer in rotating systems, *Adv. Heat Transfer*, 5, 129–251, 1968.
6. Dusinberre, G., *Numerical Methods in Heat Transfer*, McGraw-Hill, New York, 1949.
7. Korn, G. and Korn, T., *Mathematical Handbook for Scientists and Engineers*, McGraw-Hill, New York, 1961.
8. SKF, *General Catalog 4000 US*, 2nd Ed., 49, 1997.
9. Harris, T., Prediction of temperature in a rolling bearing assembly, *Lubr. Eng.*, 145–150, April 1964.
10. Harris, T., How to predict temperature increases in rolling bearings, *Prod. Eng.*, 89–98, December 9, 1963.
11. Pirvics, J. and Kleckner, R., Prediction of ball and roller bearing thermal and kinematic performance by computer analysis, *Adv. Power Transmission Technol.*, NASA Conference Publication 2210, 185–201, 1982.
12. Coe, H., Predicted and experimental performance of large-bore high speed ball and roller bearings, *Adv. Power Transmission Technol.*, NASA Conference Publication 2210, 203–220, 1982.
13. Kleckner, R. and Dyba, G., High speed spherical roller bearing analysis and comparison with experimental performance, *Adv. Power Transmission Technol.*, NASA Conference Publication 2210, 239–252, 1982.
14. Crecelius, W., User's manual for SKF computer program SHABERTH, steady state and transient thermal analysis of a shaft bearing system including ball, cylindrical, and tapered roller bearings, SKF Report AL77P015, submitted to U.S. Army Ballistic Research Laboratory, February 1978.
15. Parker, R., Comparison of predicted and experimental thermal performance of angular-contact ball bearings, NASA Technical Paper 2275, February 1984.
16. Harris, T. and Barnsby, R. Tribological performance prediction of aircraft gas turbine mainshaft ball bearings, *Tribol. Trans.*, 41(1), 60–68, 1998.

8 Application Load and Life Factors

LIST OF SYMBOLS

Symbol	Description	Units
a	Semimajor axis of projected contact ellipse	mm (in.)
A_c	Fatigue life reduction factor for clearance	
A_c/A_o	Contact area fraction in asperity–asperity contact	
A_{steel}	Fatigue life factor for steel	
A_1	Reliability–life factor	
A_2	Material–life factor	
A_3	Lubrication–life factor	
A_4	Contamination–life factor	
A_{ISO}	Life modification factor based on ISO systems approach of life calculation	
A_{SL}	Stress–life factor	
b	Semiminor axis of projected contact ellipse	mm (in.)
c	Simpson's rule coefficients	
C	Bearing basic dynamic capacity	N (lb)
C_L	Particulate contamination parameter	
C_{L1}	Parameter used to calculate C_L	
C_{L2}	Constant used to calculate C_L	
C_{L3}	Constant used to calculate C_{L1}	
d_m	Bearing pitch diameter	mm (in.)
d_r	Raceway diameter	mm (in.)
D	Ball or roller diameter	mm (in.)
e	Weibull slope	
F_r	Applied radial load	N (lb)
F_a	Applied axial load	N (lb)
F_e	Equivalent applied load	N (lb)
F_{lim}	Fatigue limit load	N (lb)
FR	Filter rating	μm
h^0	Minimum lubricant film thickness	μm (μin.)
I	Life integral	
K_C	Stress concentration factor due to particulate contamination	
K_L	Stress concentration factor associated with lubrication effectiveness	
L	Fatigue life	
L_{10}	Fatigue life that 90% of a group of bearings will endure	revolutions $\times 10^6$
L_{50}	Fatigue life that 50% of a group of bearings will endure	revolutions $\times 10^6$
N	Number of stress cycles	

n	Rotational speed	rpm
q	Load on a roller–raceway contact lamina	N (lb)
q_c	Basic dynamic capacity for a roller–raceway contact lamina	N (lb)
Q	Ball or roller load	N (lb)
Q_c	Basic dynamic capacity of a raceway contact	N (lb)
r	z/b	
R	Oil bath and grease contamination parameter	
$\mathcal{S}$	Probability of survival	
SF	Composite rms surface roughness of mating surfaces	μm (μin.)
T	Temperature	°C (°F)
u	Number of stress cycles per revolution	
$\mathcal{V}$	Volume under stress	mm^3 (in.^{3})
w	Width of a roller–raceway contact lamina	mm (in.)
z_0	Depth of maximum orthogonal shear stress	mm (in.)
Z	Number of rolling elements per row	
α	Contact angle	rad,°
β_x	Filter effectiveness ratio for particles of size x μm	
γ	$D\cos\alpha/d_m$	
δ_a	Bearing axial deflection	mm (in.)
δ_r	Bearing radial deflection	mm (in.)
Λ	h^0/SF	
ν	Kinematic viscosity	mm^2/sec (in.^{2}/sec)
ν_1	Kinematic viscosity for adequate lubrication	mm^2/sec (in.^{2}/sec)
σ_{VM}	von Mises stress	MPa (psi)
τ_0	Maximum orthogonal subsurface shear stress	MPa (psi)
ϕ	Oscillation angle	rad,°
ψ	Rolling element azimuth angle	rad,°

Subscripts

B	Ball
i	Inner ring or raceway
j	Rolling element azimuth location
k	Roller–raceway contact lamina location
m	Raceway
n	Probability of failure
o	Outer ring or raceway
R	Roller
RE	Equivalent rotating bearing or rolling element
μ	Rotating raceway
ν	Nonrotating raceway

8.1 GENERAL

The Lundberg–Palmgren theory and the standard load and fatigue life calculations that resulted [1–5] are only the first step toward determining the bearing fatigue lives in applications. Use of the standard methods should be limited to those applications in which the internal geometries and rolling component materials of the bearings employed conform to the standard specifications, and the operating conditions are bounded as follows:

- The bearing outer ring is mounted and properly supported in a rigid housing.
- The bearing inner ring is properly mounted on a nonflexible shaft.
- The bearing is operated at a steady speed under invariant loading.
- Operational speed is sufficiently slow such that rolling element centrifugal and gyro-scopic loadings are insignificant.
- Bearing loading can be adequately defined by a single radial load, a single axial load, or a combination of these.
- Bearing loading does not cause significant permanent deformations or material trans-formations.
- For bearings under radial loading, mounted internal clearance is essentially nil.
- For angular-contact ball bearings, nominal contact angle is constant.
- For roller bearings, uniform loading is maintained at each roller–raceway contact.
- The bearing is adequately lubricated.

Many applications can be considered to be included within these conditions.

In many applications, these simple conditions are exceeded. For example, many applications do not operate at a steady speed or load, rather, they operate under a load–speed cycle. Furthermore, the bearing may support, as indicated in Chapter 1, combined radial, axial, and moment loadings under which the distribution of internal loading is significantly different from the standard limitations. Bearings may operate at speeds that cause substantial rolling element inertial loading and variation in contact angles between inner and outer raceway contacts. These conditions may be addressed by applying the Lundberg–Palmgren equations in detail using computer programs to perform the complex calculations.

After the development of the Lundberg–Palmgren theory, the ability of a lubricant to separate rolling elements from raceways, as discussed in Chapter 4, was established. This condition has been shown to have, probably, the most profound effect on extending bearing fatigue life compared with any other condition. Improvements in modern bearing steel manufacturing methods have provided steels of very high cleanliness and homogeneity, as compared with the basic air-melt AISI 52100 steel used in the development of the Lundberg–Palmgren theory and the standards. With the advent of substantially extended life, increased reliability in bearing life prediction can be considered.

Finally, as the improvements in bearing manufacture and lubrication were applied, it became apparent that, similar to other steel structures subjected to cyclic loading, bearing raceways and rolling elements also exhibit an endurance limit in fatigue. This means that in a given application, a ball or roller bearing does not have to fail in fatigue, provided that applied loading and conditions of operation are such that the bearing material fatigue limit stress is not exceeded.

All of these conditions will be addressed in this chapter.

8.2 EFFECT OF BEARING INTERNAL LOAD DISTRIBUTION ON FATIGUE LIFE

8.2.1 Ball Bearing Life

8.2.1.1 Raceway Life

When the distribution of load among the balls is different from that resulting from the applied loading conditions specified in the load rating standards, it is necessary to revert to the Lundberg–Palmgren load–life relationships as given in Chapter 11 in the first volume of this handbook for individual ball–raceway contacts. For example, for a contact on a rotating raceway

$$L_{\mu j} = \left(\frac{Q_{c\mu j}}{Q_{\mu j}} \right)^3 \tag{8.1}$$

where $Q_{c\mu j}$ is the basic dynamic capacity of the contact of ball j on the rotating raceway, and $Q_{\mu j}$ is the load acting on the contact. It is to be noted that the capacity may be different from point to point around the raceway because the contact angle may vary with the azimuth angle. For a nonrotating raceway contact,

$$L_{vj} = \left(\frac{Q_{cvj}}{Q_{vj}} \right)^3 \tag{8.2}$$

It is also to be noted that the ball–raceway load may differ between raceways due to ball inertial loading. From Equation 8.1 and Equation 8.2, it can be determined that the life of a bearing that has a complement of Z balls is given by

$$L = \left(\sum_{j=1}^{j=Z} L_{\mu j}^{-e} + \sum_{j=1}^{j=Z} L_{vj}^{-e} \right)^{-1/e} \tag{8.3}$$

where exponent e is the slope of the Weibull distribution. It is further to be noted that the life calculated according to Equation 8.3 does not include ball lives.

8.2.1.2 Ball Life

Notwithstanding the fact that the Lundberg–Palmgren equations are based on bearing fatigue failure dependent only on raceway fatigue failure, there is ample evidence that balls, as well as raceways, can succumb to fatigue failure. Assuming that in rolling bearings subjected to reasonable levels of loading, the balls contact the raceways over defined tracks, starting with Equation 11.41 of the first volume of this handbook, an equation for basic dynamic capacity of the ball portion of a ball–raceway contact can be developed. In that equation, it is observed that for a ball, track diameter at the rotating raceway contact $d_\mu = D \cos \alpha_{\mu j}$; also, $d_v = D \cos \alpha_{vj}$. Furthermore, for a ball track, the number of stress cycles per ball revolution $u = 2$. Making these substitutions, the basic dynamic capacity for the ball in a ball–raceway contact is given by

$$Q_{Bnj} = 77.9 \left(\frac{2f_n}{2f_n - 1} \right)^{0.41} \left(1 + c_n \gamma_{nj} \right)^{1.69} \frac{D^{1.8}}{\left(\cos \alpha_{nj} \right)^{0.3}}, \quad n = \mu; \nu \tag{8.4}$$

where $c_n = +1$ for a ball–outer raceway contact; $c_n = -1$ for a ball–inner raceway contact. Using Equation 8.4, the equation for bearing life becomes

$$L = \left(\sum_{j=1}^{j=Z} L_{\mu j}^{-e} + \sum_{j=1}^{j=Z} L_{vj}^{-e} + \sum_{j=1}^{j=Z} L_{B\mu j}^{-e} + \sum_{j=1}^{j=Z} L_{Bvj}^{-e} \right)^{-1/e} \tag{8.5}$$

In using Equation 8.5, it must be recognized that bearing life is defined in revolutions of the rotating ring. For example, for a simple rolling motion, the number of ball revolutions per inner ring revolution, as determined from Equation 10.14 of the first volume of this handbook, is

$$\frac{n_B}{n_i} = \frac{d_m}{2D}\left(1 - \gamma_{nj}^2\right), \qquad n = \mu, \nu \qquad (8.6)$$

Therefore, the ball lives indicated in Equation 8.5 must first be divided by the ratio of Equation 8.6. In cases where ball speeds are calculated considering frictional effects, ball speeds are calculated according to the methods in Chapter 2, and the ratio of Equation 8.6 may be replaced by the calculated speed ratio.

Also, in using Equation 8.5, it must be recognized that the Weibull slope for ball failures may be somewhat different from that for raceway failures. For example, in a fatigue failure investigation of vacuum-induction-melted, vacuum-arc-remelted (VIMVAR) M50 steel balls, the data of Harris [6] indicated an average Weibull slope of 3.33. In such a case, an average value of e may be used in Equation 8.5.

8.2.2   ROLLER BEARING LIFE

8.2.2.1   Raceway Life

In Chapter 1, it was shown that to determine the distribution of load among the rollers for nonstandard applied loading, the roller–raceway contacts may be divided into a number of laminae. Hence, for a roller–raceway contact of length l, if the contact is divided into m laminae, each of width w, $l = mw$ and

$$Q_{\mu j} = w \sum_{k=1}^{k=m} q_{\mu kj} \qquad (8.7)$$

Therefore, referring to Equation 8.1 and Equation 8.2 for ball bearings and considering, as indicated in Chapter 11 of the first volume of this handbook, a fourth power load–life relationship for line contact, the following equations may be written for the fatigue lives of roller–raceway contact laminae:

$$L_{\mu jk} = \left(\frac{q_{c\mu j}}{q_{\mu jk}}\right)^4 \qquad (8.8)$$

$$L_{\nu jk} = \left(\frac{q_{c\nu j}}{q_{\nu jk}}\right)^{9/2} \qquad (8.9)$$

Accordingly, roller bearing fatigue life may be calculated using

$$L = \left(\sum_{j=1}^{j=Z}\sum_{k=1}^{k=m} L_{\mu jk}^{-e} + \sum_{j=1}^{j=Z}\sum_{k=1}^{k=m} L_{\nu jk}^{-e}\right)^{-1/e} \qquad (8.10)$$

8.2.2.2   Roller Life

Similar to Equation 8.4, the basic dynamic capacity for a roller track at a roller–raceway contact lamina is given by

$$q_{cnjk} = 464(1 + c_n\gamma_n)^{1.324} w^{7/9} \frac{D^{29/27}}{(\cos\alpha_n)^{2/9}} \qquad (8.11)$$

Roller bearing fatigue life, including the lives of the rollers, may be calculated using

$$L = \left(\sum_{j=1}^{j=Z} \sum_{k=1}^{k=m} L_{\mu jk}^{-e} + \sum_{j=1}^{j=Z} \sum_{k=1}^{k=m} L_{\nu jk}^{-e} + \sum_{j=1}^{j=Z} \sum_{k=1}^{k=m} L_{R\mu jk}^{-e} + \sum_{j=1}^{j=Z} \sum_{k=1}^{k=m} L_{R\nu jk}^{-e} \right)^{-1/e} \quad (8.12)$$

As for balls, roller life must be reduced by the speed ratio for use in the above equation.

8.2.3 CLEARANCE

The fatigue life of a rolling bearing is strongly dependent on the maximum rolling element load Q_{max}; if Q_{max} is significantly increased, fatigue life is significantly decreased. Any parameter that affects Q_{max}, therefore, affects bearing fatigue life. One such parameter is radial (diametral) clearance. In Chapter 7 of the first volume of this handbook, the effect of clearance on load distribution in radial bearings was examined. Figure 8.1 illustrates the variation of load distribution among the rolling elements for some conditions of radial clearance as defined by the projection of the bearing load zone on a diameter.

The effect of clearance on bearing fatigue life may be expressed in terms of the standard rating life; for example, $L_{10c} = A_c L_{10}$. Figure 8.2, from Ref. [7], which gives the L_{10} life reduction factor A_c as a function of the extent of rolling element loading, was developed by using the load distribution data of Chapter 7 of the first volume of this handbook in accordance with the contact life and bearing life (Equation 8.1 through Equation 8.3 and Equation 8.8 through Equation 8.10). As shown in Figure 8.1, an increase in Q_{max} for rigidly supported bearings is accompanied by a decrease in the numbers of rolling elements loaded.

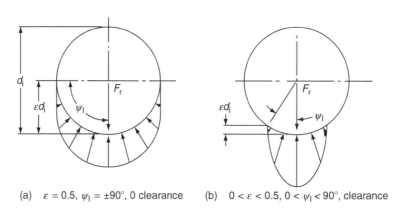

(a) $\varepsilon = 0.5$, $\psi_l = \pm 90°$, 0 clearance   (b) $0 < \varepsilon < 0.5$, $0 < \psi_l < 90°$, clearance

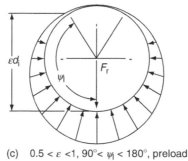

(c) $0.5 < \varepsilon < 1$, $90° < \psi_l < 180°$, preload

FIGURE 8.1 Rolling element load distribution for different radial clearance conditions.

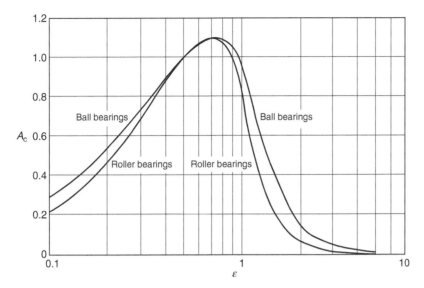

FIGURE 8.2 Fatigue life reduction factor A_c based on diametral clearance.

This decrease in load zone, however, has less effect on the mean effective rolling element load than does the increase in Q_{max}.

See Example 8.1.

8.2.4 FLEXIBLY SUPPORTED BEARINGS

If one or both rings of a rolling bearing bend under the applied loads such as in a planet gear application [8,9] or other aircraft bearing applications in which ring and housing cross-sections are optimized for aircraft weight reduction, then load distribution may be considerably different from that of a rigid ring bearing. Depending on the flexibility of the ring and bearing clearance, it may be possible for a flexible ring to yield superior endurance characteristics when compared with a rigid ring bearing. Figure 8.3 from Jones and Harris [8] shows the variation of bearing fatigue life with outer ring section and clearance for a planet gear bearing as shown in Figure 1.22 and Figure 1.23. The load distribution obtained is illustrated in Figure 1.31.

When the bearing rings are flexibly supported, it may be possible to alter bearing design and obtain increased fatigue life. Harris and Broschard [9] applied clearance selectively at the planet gear bearing maximum load positions by making the bearing inner ring elliptical. Figure 8.4 demonstrates the variation of fatigue life with diametral clearance and out-of-round. Out-of-round is the difference between the major and minor axes of the elliptical ring. A further reference [10] also demonstrates that rolling bearing ring dimensions can be optimized to maximize fatigue life.

8.2.5 HIGH-SPEED OPERATION

Operation at high speeds, as shown in Chapter 2, affects the bearing load distribution due to the increased magnitude of rolling element centrifugal forces and gyroscopic moments. The standard methods of fatigue life calculation [3–5] do not account for these inertial forces and moments and subsequent effects such as changes in ball bearing contact angles. Hence, the deviation in fatigue life from that calculated according to the standard method can be

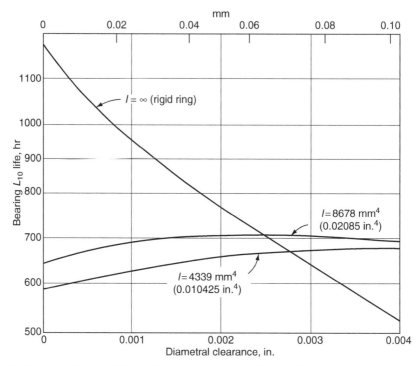

FIGURE 8.3 Planet gear bearing life vs. diametral clearance and outer ring cross-section moment of inertia.

considerable. In Chapter 2, methods were developed to calculate load distribution in high-speed ball and roller bearings. Methods for using these load distributions in the estimation of fatigue life have been given in this chapter. Figure 8.5 demonstrates the variation of life with load and speed for the 218 angular-contact ball bearing of Figure 3.12 through Figure 3.14.

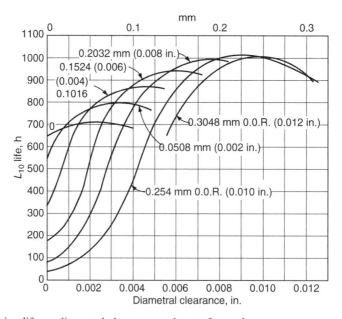

FIGURE 8.4 Bearing life vs. diametral clearance and out-of-round.

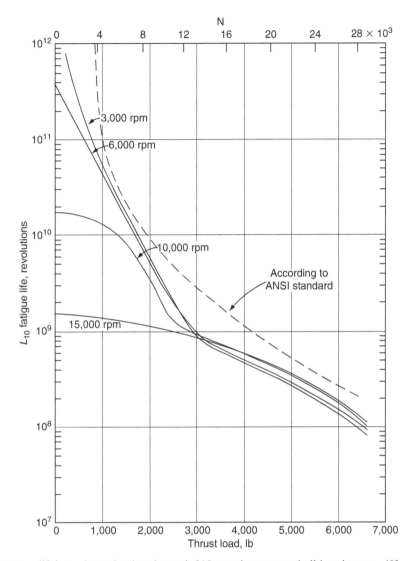

FIGURE 8.5 L_{10} life* vs. thrust load and speed; 218 angular-contact ball bearing, $\alpha = 40°$.

Note that the data shown in Figure 8.5 do not consider the effect of skidding, which results in a reduction in ball orbital speed, and hence reduced ball centrifugal and gyroscopic loadings. This, in turn, tends to result in an increase in fatigue life; however, depending on the thickness of the lubricant films separating the balls from the raceways, sliding in the ball–raceway contacts, with its potential deleterious effect on fatigue endurance, may more than eliminate the beneficial effect of reduced inertial loading.

Figure 8.6 compares the fatigue life of the 218 angular-contact ball bearing operating at a high speed with light-weight silicon nitride balls to that of the bearing that has steel balls. Whereas the silicon nitride balls operate with reduced inertial loading, the elastic modulus of hot isostatically pressed (HIP) silicon nitride is approximately 50% greater than that of steel. This results in reduced contact area between the steel raceways and ceramic balls; therefore, Hertz stresses are increased causing a reduction in fatigue life. Thus, the beneficial effect of

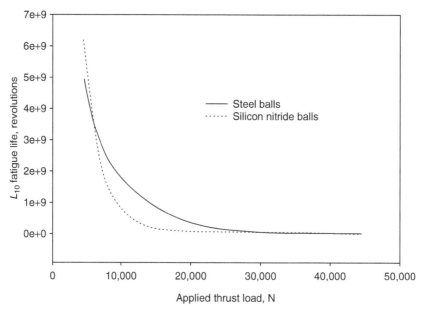

FIGURE 8.6 Life vs. thrust load for a 218 angular-contact ball bearing operating at approximately 1.50 million dn. (Bearing bore in millimeter times shaft speed in rpm.)

light-weight balls is counteracted. By decreasing the radii of the raceway grooves somewhat, the Hertz stresses may be decreased. This, however, causes an increase in frictional stresses and higher operating temperatures that may have to be accommodated by cooling the lubricant or bearing. Optimum bearing design may be achieved for a given application by parametric study using a bearing performance analysis computer program. It can be seen from Figure 8.6 that there is little difference in the fatigue life performance of the bearing under relatively heavy loading.

Figure 8.7 shows life vs. speed for the 209 cylindrical roller bearing of Figure 3.19. Skidding effects are not included in this illustration.

8.2.6 MISALIGNMENT

Misalignment in nonaligning rolling bearings distorts the internal load distribution, and thus alters fatigue life. In Chapter 1, methods were described to determine the misalignment angle in ball and roller bearings as a function of the applied moment. In ball bearings, the load distribution from ball to ball is altered by misalignment; in roller bearings, however, the distribution of the roller load per unit length becomes nonuniform as shown in Figure 1.8. The variable load per unit length is given by Equation 1.36.

The analysis of roller bearing lives indicated in Chapter 11 of the first volume of this handbook pertained only to bearings that have a uniform distribution of load per unit length along the roller length at each roller–raceway contact. As indicated in Chapter 1, roller–raceway loading varies not only from contact to contact, but also from lamina to lamina along a contact. The methods defined in Chapter 1 allow the determination of the load per unit length q_{njk} at each roller–raceway lamina contact, where $n = 1$ (outer raceway) or 2 (inner raceway), $j = 1, \ldots, Z$, and $k = 1, \ldots, m$.

It should be apparent that misalignment can quickly lead to edge loading in the roller–raceway contacts; edge loading of even small magnitude can rapidly diminish fatigue life. In

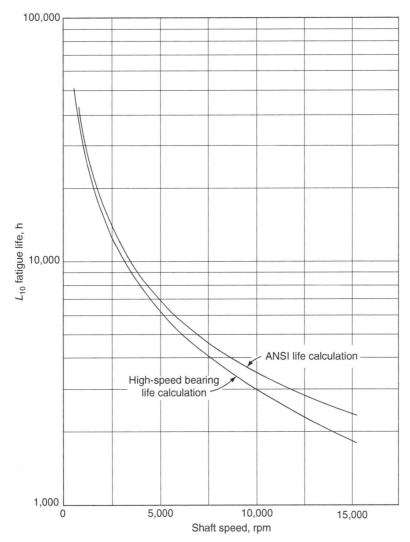

FIGURE 8.7 Life vs. speed; 209 cylindrical roller bearing with zero mounted clearance supporting 44,500 N (10,000 lb) radial load.

Chapter 6 of the first volume of this handbook, references were cited indicating that the magnitude of edge stressing can be calculated for any roller–raceway contact profile. Alternatively, the methods defined in section 1.6 allow calculation of the contact stresses, including edge stresses, for any roller-raceway crowning, load and misalignment combination. Figure 8.8, from Ref. [11], shows the effect of misalignment on the life of a 309 cylindrical roller bearing as a function of roller crowning and applied load. Table 8.1 indicates, based on experience data in manufacturers' catalogs, maximum acceptable misalignments for the various rolling bearing types.

8.3   EFFECT OF LUBRICATION ON FATIGUE LIFE

In Chapter 4, it was indicated that if a rolling bearing is adequately designed and lubricated, the rolling surfaces can be completely separated by a lubricant film. Endurance testing of rolling bearings as shown by Tallian et al. [12] and Skurka [13] has demonstrated the

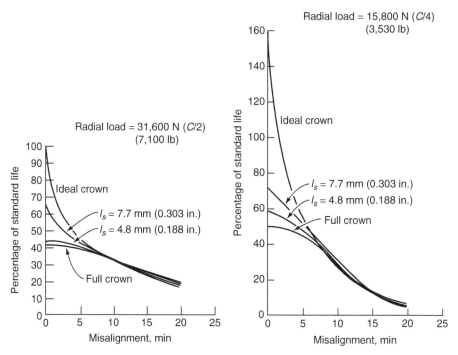

FIGURE 8.8 Life vs. misalignment for a 309 cylindrical roller bearing as a function of crowning and applied load. (From Harris, T., The Effect of misalignment on the fatigue life of cylindrical roller bearings having crowned rolling members, *ASME Trans., J. Lubr. Technol.*, 294–300, April 1969.)

considerable effect of lubricant film thickness on bearing fatigue life. In Chapter 4, methods for estimating this lubricant film thickness were given. It was also demonstrated that lubricant film thickness is sensitive to bearing operating speed and lubricant viscous properties. Moreover, the film thickness is virtually insensitive to load.

The test results reported in Refs. [12,13] showed that at high operational speeds a considerable improvement in fatigue life occurs. Moreover, a similar effect can be achieved by using a sufficiently viscous lubricant at slower speeds. The effectiveness of the lubricant film thickness generated depends on its magnitude relative to the surface topographies of the contacting rolling elements and raceways. For example, a bearing with very smooth raceway and rolling element surfaces requires less of a lubricant film than does a bearing with relatively rough surfaces (see Figure 8.9).

TABLE 8.1
Estimated Maximum Allowable Rolling Bearing Misalignment Angle[a]

Bearing Type	Minutes	Radians
Cylindrical roller bearing	3–4	0.001
Tapered roller bearing	3–4	0.001
Spherical roller bearing	30	0.0087
Deep-groove ball bearing	12–16	0.0035–0.0047

[a]Based on acceptable reduction in fatigue life.

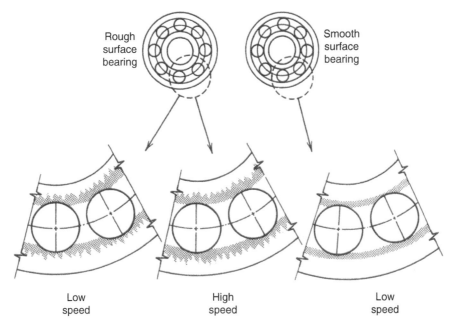

Rough surface bearing

Smooth surface bearing

Low speed

High speed

Low speed

FIGURE 8.9 Illustration of the effect of surface roughness on the lubricant film thickness required to prevent metal-to-metal contact.

The relationship of lubricant film thickness to surface roughness has been signified in rolling bearing literature by Λ, which utilizes the simple root mean square (rms) value of the roughnesses of the surfaces of the contacting bodies. Tallian [14] among many other researchers introduced the use of asperity slopes as well as peak heights of asperities. Chapter 5, which covers microcontact phenomena, provides additional means to evaluate the effect of a "rough" surface on contact, and hence bearing lubrication and performance. Using Λ, Harris [15] indicated the effect of lubrication on bearing fatigue life, as in Figure 8.10. According to Ref. [15], if $\Lambda \geq 4$, fatigue life can be expected to exceed standard L_{10} estimates by at least 100%. Conversely, if $\Lambda < 1$, the bearing may not attain calculated L_{10} estimates because of surface distress such as smearing that can lead to rapid fatigue failure of the rolling surfaces. Figure 8.10 shows the various operating regions just described. In Figure 8.10, the ordinate "percent film" is a measure of the time during which the "contacting" surfaces are fully separated by an oil film.

Tallian [14] showed a more definitive estimate of rolling bearing fatigue life vs. Λ as did Skurka [13]. Bamberger et al. [16] show the combination of the foregoing in Figure 8.11, recommending the use of the mean curve. Experimental data indicate that for $\Lambda > 4$, the L/L_{10} ratios are substantially greater than those given in Figure 8.11 for accurately manufactured bearings lubricated by minimally contaminated oil.

Using a microtransducer to measure the pressure distribution in the direction of rolling in an oil-lubricated line contact, Schouten [17] showed that edge stress in a line contact is substantially reduced if an adequate lubricant film separates the contacting bodies. In this situation, the lubricant film tends to permit an increase in fatigue life by reducing the magnitude of normal stress at the end(s) of a heavily loaded contact.

The mean curve in Figure 8.11 is frequently used to estimate the effect of lubrication on bearing fatigue life.

See Example 8.2 and Example 8.3.

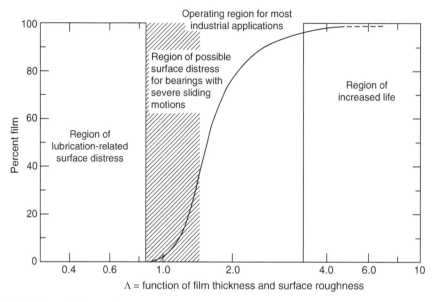

FIGURE 8.10 Percent film vs. Λ.

Unfortunately, if gross sliding occurs, the reduction in fatigue life can be much more severe than that predicted in Figure 8.11. In Chapter 11 of the first volume of this handbook it was shown that fatigue life is a strong function of normal stresses acting on the contacts between mating rolling surfaces. Surface friction shear stresses augment the subsurface stresses effected by the normal contact stresses. In fact, from Lundberg–Palmgren theory it can be shown that for point contacts $L \propto \tau_0^{-9.3}$. Hence, small increases in stress cause large decreases in life. Thus, lubricant film parameter Λ may only be regarded as a qualitative

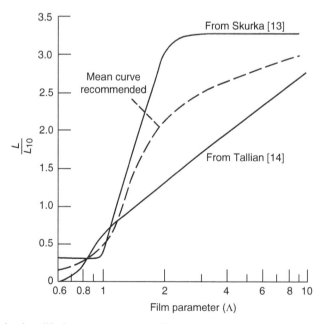

FIGURE 8.11 Lubrication–life factor vs. lubricant film parameter Λ. (From Bamberger, E., et al., *Life Adjustment Factors for Ball and Roller Bearings*, AMSE Engineering Design Guide, 1971. With permission.)

TABLE 8.2
A_{chem} **vs. Steel Type**

Steel Type	A_{chem}
AISI 52100	3
M50	2
M50NiL	4

measure of lubrication effectiveness. How to include the surface friction shear stresses in the prediction of bearing fatigue life will be discussed later in this chapter.

8.4 EFFECT OF MATERIAL AND MATERIAL PROCESSING ON FATIGUE LIFE

In Chapter 11 of the first volume of this handbook, the effect on fatigue endurance of the basic steel used in modern bearing manufacture was included in the b_m or f_{cm} factors in the calculation of basic load rating C. This standard steel is assumed to be carbon vacuum degassed (CVD) 52100, through-hardened at least to Rockwell C 58. Many roller bearings, particularly tapered roller bearings manufactured in the United States, are however fabricated from carburized (case-hardened) steel. Since the load and life rating methods for such bearings are assumed to be included in the standards [35], it has been historically assumed that the endurance performances of the CVD 52100 through-hardened steel and the basic carburizing steels are equivalent.

To attain high-temperature, long-life performance, VIMVAR M50 tool steel was developed for aircraft gas turbine mainshaft bearing applications. This VIMVAR steel provides excellent fatigue endurance characteristics for bearing rings and rolling elements. Because of the necessity to operate modern gas turbine mainshaft bearings at ultrahigh speed, for example, at 3 million dn, a carburizing version of this steel, VIMVAR M50NiL, was developed. In this case, it is intended that the "softer" core will arrest any fatigue cracks that emanate in the hardened case and thus prevent through-cracking of bearing rings.

A number of specialty steels have been developed to provide superior corrosion resistance while not sacrificing fatigue endurance properties; for example, Cronidur 30. Additionally, ceramic materials, for example, HIP silicon nitride, are now used in the manufacture of balls and rollers.

STLE [18] has attempted to codify the effect of some of these materials on rolling bearing fatigue life. Moreover, STLE [18] has also separated the effects of heat treatment and metalworking. A material–life factor A_{steel} has been recommended such that

$$L'_{10} = A_{steel} \left(\frac{C}{F} \right)^p \tag{8.13}$$

TABLE 8.3
$A_{heattreat}$ **vs. Heat Treatment**

Heat Treatment	$A_{heattreat}$
Air-melt	1
Carbon vacuum degassed (CVD)	1.5
Vacuum arc remelted (VAR)	3
Double VAR	4.5
Vacuum induction melted, vacuum arc remelted (VIMVAR)	6

TABLE 8.4
$A_{process}$ vs. Metalworking Process

Metalworking Process	$A_{process}$
Deep-groove ball bearing raceways	1.2
Angular-contact ball bearing raceways	1
Angular-contact ball bearing raceways—forged rings	1.2
Cylindrical roller bearings	1

where $A_{steel} = A_{chem} \times A_{heattreat} \times A_{process}$. The data in Table 8.2 through Table 8.4 were obtained from Ref. [18].

From the tabular data, it can be determined that an angular-contact bearing with forged rings manufactured from VIMVAR M50NiL steel would be given an $A_{steel} = 28.8$.

No value has been universally established to date for HIP silicon nitride. Endurance testing of single balls in ball/v-ring endurance test has, however, yielded high multiples of the endurance for steel balls tested under the same loading conditions. To date, owing to relative weakness in tensile strength in bending tests and extremely low coefficient of thermal expansion, silicon nitride has been principally used for balls and rollers in high-precision, high-speed applications; for example, machine tool spindle bearings.

8.5 EFFECT OF CONTAMINATION ON FATIGUE LIFE

Excessive contamination in the lubricant will severely shorten bearing fatigue life. The standards [3–5] and manufacturers' catalogs contain warning statements about this. Contaminants may be either particulate or liquid, usually water. Even small amounts of contaminants have significant limiting effects on bearing fatigue life.

Particulate contaminants such as gear wear metal particles, alumina, silica, and so on will cause dents in the raceway and rolling element surfaces, which disrupt the lubricant films that tend to separate the rolling body surfaces. This tends to locally increase the frictional shear stresses produced in the rolling–sliding contacts. Furthermore, the raised material on the shoulder of the dent tends to cause stress concentrations. Ville and Nélias [19], using a two-disk rolling–sliding test rig, demonstrated the stress concentration phenomenon. They further showed that combined rolling–sliding motion is a more severe condition with regard to generation of surface distress and fatigue than rolling alone. Both the film disruption and dent shoulder stress-increasing effects accelerate the onset of rolling contact fatigue and component failure. Figure 8.12 from a study of the effects of surface topography on fatigue failure by Webster et al. [20] indicates the relative risk of failure effected by the shoulders of dents.

Hamer et al. [21] and Sayles et al. [22] pointed out that even relatively soft particles can generate significant denting, assuming bearing speeds and loads are sufficiently high. They further indicate that the particle diameter to lubricant film thickness ratio appears to be a critical parameter with regard to denting. In Figure 8.13 through Figure 8.15, Nélias and Ville [23] characterized the types of dents generated by hard and soft particles.

Using the same rolling–sliding disk endurance test rig of Ref. [19], Nélias and Ville [23] showed that fatigue microspalling commences on the surface ahead of the dent in the friction direction; see Figure 8.16. Xu et al. [24] also noted that spalling due to dents can initiate at either the leading or trailing edge depending on the direction of surface traction.

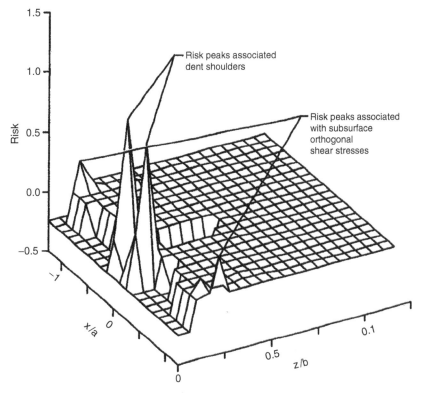

Note: z/b = 0 represents surface

FIGURE 8.12 Plot showing relative risk of fatigue failure throughout raceway subsurface including effect of dent shoulders. (From Webster, M., Ioannides, E., and Sayles, R., *Proc. 12th Leeds–Lyon Symp. Tribol.*, 207–226, 1986. With permission.)

Nélias and Ville [23] also demonstrated the dent location using transient elastohydrodynamic lubrication (EHL) analysis; see Figure 8.17. Xu et al. [24] in an analytical and experimental study presented similar results to those of Nélias and Ville [23]. They also showed that the location of spall initiation depends on the EHL and dent condition, and that spalls can initiate at either the leading or trailing edge of the dent depending on the direction of surface traction; see Figure 8.18.

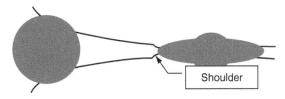

FIGURE 8.13 Dent generated by a ductile metallic particle; for example M50 steel. (From Nélias, D. and Ville, F., *ASME Trans.*, *J. Tribol.*, 122, 1, 55–64, 2000. With permission.)

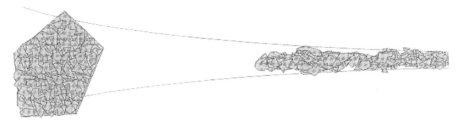

FIGURE 8.14 Dent generated by hard brittle material; for example Arizona road dust. (From Nélias, D. and Ville, F., *ASME Trans.*, *J. Tribol.*, 122, 1, 55–64, 2000. With permission.)

The experimental data of Sayles and MacPherson [25] demonstrated the effect of different levels of particulate contamination on bearing fatigue life by endurance testing cylindrical roller bearings with varying degrees of absolute lubricant filtration; for example, from 40 μm

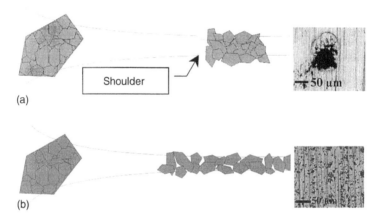

FIGURE 8.15 (a) Coarse dent generated by ceramic material at slow speed; for example boron carbide or silicon carbide at 2.51 m/sec (98.8 in./sec). (b) Fine dents generated by ceramic material at high speed; for example boron carbide or silicon carbide at 20 m/sec (787.4 in./sec). (From Nélias, D. and Ville, F., *ASME Trans.*, *J. Tribol.*, 122, 1, 55–64, 2000. With permission.)

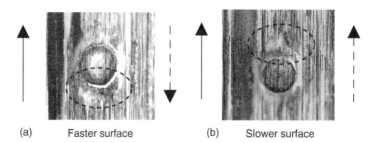

FIGURE 8.16 Surface distress (in dotted ellipses) associated with dent in rolling–sliding motion, endurance tested 52100 steel components. Solid arrows signify rolling direction; dashed arrows signify friction direction. (From Nélias, D. and Ville, F., *ASME Trans.*, *J. Tribol.*, 122, 1, 55–64, 2000. With permission.)

 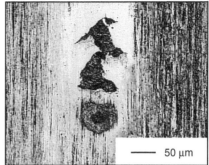

FIGURE 8.17 For the slower surface in Figure 8.16, formation of microspalls ahead of the dent in the sliding direction on the surface of a 52100 steel component after 60×10^6 stress cycles at 3500 MPa $(5.08 \times 10^5 \, \text{psi})$. Rolling speed is 40 m/sec (1575 in./sec); slide–roll ratio $= +0.015$. Solid arrow signifies rolling direction; dashed arrow signifies friction direction. (From Nélias, D. and Ville, F., *ASME Trans.*, *J. Tribol.*, 122, 1, 55–64, 2000. With permission.)

(0.0016 in.) down to 1 μm (0.00004 in.). Particulate matter was deemed typical of that generated in gearboxes. Figure 8.19 is a photograph of dents incurred under the Sayles–MacPherson [25] operating conditions with 40 μm (0.0016 in.) filtration. The dents are approximately 10–30 μm (0.0004–0.0012 in.) long and about 2 μm (0.00008 in.) deep. Comparing this depth with the thickness of a good lubricant film $(\Lambda > 1.5)$, it can be determined that the film can easily collapse in the dent. Evaluation of the Sayles–MacPherson [25] operating conditions according to the methods discussed in Chapter 1, Chapter 3, and Chapter 4, indicates Λ values from approximately 0.45 at 40 μm (0.0016 in.) filtration to nearly 1 using magnetic filtration.

Figure 8.20 from Ref. [25] shows L_{50} life vs. filter rating. According to Figure 8.20, significant improvement in life is achieved with a finer lubricant filtration level; however, little improvement in life is achieved for a filtration level less than 3 μm. Thus, there appears to be a limit to fine filter effectiveness. Sayles–MacPherson [25] data were confirmed by Tanaka et al. [26], who, by using sealed ball bearings in an automotive gearbox, managed to increase fatigue life several fold, compared with that of open (no seals or shields) bearings in the same application. Considering the lubricant film conditions of the test program, the data of Figure 8.20 have been curve-fitted to the following equation for contamination–life factor:

$$A_{\text{contam}} = \left[0.4162 + 3.366 \frac{\ln\left(FR/h^0\right)^2}{FR/h^0}\right]^2 \tag{8.14}$$

where FR is the filter rating.

Based on test results using 3- and 49-μm filtration, Needelman and Zaretsky [27] recommend the following equation for the reduction of fatigue life due to particulate contamination:

$$A_{\text{contam}} = 1.8(FR)^{-0.25} \tag{8.15}$$

It is apparent that equations for fatigue life reduction due to particulate contamination must be applied with care since they depend on the type of particles as well as the size and on the bearing lubrication conditions.

The presence of water in the lubricant is thought to effect hydrogen embrittlement of the surface steel, creating stress concentrations and shortening fatigue life. Figure 8.21, from Ref. [28], illustrates the life reduction effect.

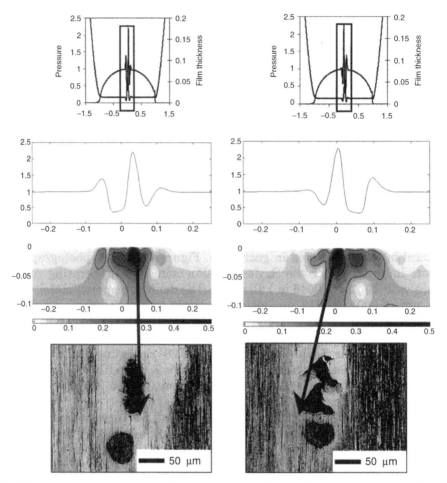

FIGURE 8.18 Comparison of results of numerical simulations and tests for two opposite slide–roll ratios. The upper row shows pressure distribution and film thickness over the line contact, the middle rows show zoom views of the film thickness around the dent and lines of constant maximum shear stress in the metal, and the lower row shows dent area micrographs. (From Nélias, D. and Ville, F., *ASME Trans.*, *J. Tribol.*, 122, 1, 55–64, 2000. With permission.)

Table 8.5 from Ref. [28] for ISO 220 circulating oils indicates that the effect of water in the lubricant also varies with the composition of the lubricant.

It appears that adding 0.5% water to lubricant A caused a life reduction by a factor of 3, which is consistent with the data in Figure 8.21. The results for the remaining lubricant variants, however, demonstrate a wide variation in bearing life, indicating a significant endurance dependency on the lubricant composition as well as on the amount of contained moisture. Because of this, life reduction equations need to be based on the combination of lubricant type, specific composition, and amount of contained moisture.

8.6   COMBINING FATIGUE LIFE FACTORS

It may be observed that nonstandard loading conditions can be accommodated in the estimation of bearing fatigue life by determining the bearing internal load distribution and

FIGURE 8.19 Denting caused by particulate contamination. (From Sayles, R. and MacPherson, P., Influence of wear debris on rolling contact fatigue, *ASTM Special Technical Publication 771*, J. Hoo, Ed., 255–274, 1982. With permission.)

applying the contact life equations presented at the beginning of this chapter. User-friendly computer programs to perform the calculations using the equations and methods presented in Chapter 1 through Chapter 4 are readily available for operation on personal computers. To apply the effects of increased reliability, nonstandard materials, lubrication, and

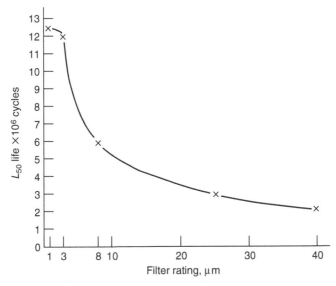

FIGURE 8.20 Bearing fatigue life vs. degree of lubricant filtration. (From Sayles, R. and MacPherson, P., Influence of wear debris on rolling contact fatigue, *ASTM Special Technical Publication 771*, J. Hoo, Ed., 255–274, 1982. With permission.)

TABLE 8.5
Bearing Fatigue Life for 0.5% Water Concentration in Various Lubricants

Lubricant	L_{10}	L_{50}
A (no water)	59.2	171.4
A	20.8	61.2
B	66.7	195.7
C	33.4	77
D	54.5	195
E	20.8	61.2
F	23.9	168
G	32.1	143
H	66.8	410
I	47.4	122

contamination, the simple approach of cascading the life factors has been most frequently taken, and is recommended in Ref. [18] and various bearing manufacturers' catalogs. This approach uses the following equation:

$$L_{na} = A_1 A_2 A_3 A_4 \left(\frac{C}{F}\right)^p \tag{8.16}$$

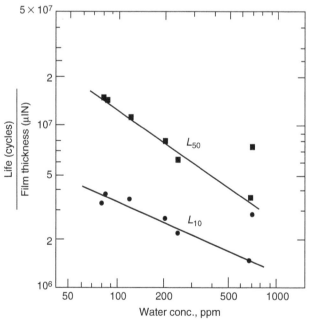

FIGURE 8.21 Effect of water contamination on rolling bearing life. (From Barnsby, R., et al., Life ratings for modern rolling bearings, ASME Paper 98-TRIB-57, presented at the ASME/STLE Tribology Conference, Toronto, October 26, 1998. With permission.)

In the above equation:

- A_1 is the reliability–life factor as determined from Table 11.25 of the first volume of this handbook.
- A_2 is the material–life factor as determined from Table 8.2 through Table 8.4 or similar empirical data.
- A_3 is the lubrication–life factor determined using Figure 8.11 or similar empirical data.
- A_4 is the contamination–life factor using Equation 8.14, Equation 8.15, or similar empirically derived data.
- L_{na} is the adjusted fatigue life at reliability n.

This simple calculation approach has been used since the 1960s when the first improvements in bearing steels and understanding of the role of lubricant films in bearing fatigue endurance occurred. It does not however recognize the interdependency of the various life factors. Therefore, it must be used judiciously. For example, the ANSI standards [3,4] state "It may not be assumed that the use of a special material, process, or design will overcome a deficiency in lubrication. Values of A_2 greater than 1 should therefore normally not be applied if A_3 is less than 1 because of such deficiency." The contamination–life factor is strongly dependent on the thickness of the lubricant film compared with the size of foreign particulate matter; in large bearings it is far less significant than in small bearings.

8.7   LIMITATIONS OF THE LUNDBERG–PALMGREN THEORY

The Lundberg and Palmgren fatigue life theory and accompanying formulas were a significant development in rolling bearing technology; however, it was not possible to correlate the fatigue lives of bearing surfaces in rolling contact so calculated with fatigue lives of other engineering structures. Nor was it possible to correlate rolling contact fatigue in bearings to fatigue of elemental surfaces in rolling contact.

A major consideration in the analysis of fatigue lives of mechanical engineering structures subjected to cyclically applied tension, bending, and torsion is the existence of an endurance limit. This is a cyclically applied stress level that the structure can endure without succumbing to fatigue failure. In other words, if the equivalent stresses cyclically applied to a mechanical structure are everywhere less than the endurance limit, then the structure will survive indefinitely without the possibility of fatigue damage. Conversely, according to the Lundberg–Palmgren theory and the standard methods of rolling bearing fatigue life prediction derived therefrom, irrespective of the magnitude of the applied load, rolling bearing fatigue life is finite in any application. Innumerable modern rolling bearing applications, however, have defied this limitation. Endurance data for bearings of standard design, accurately manufactured from high-quality steel—having minimal impurities and homogeneous chemical and metallurgical structures [28]—have demonstrated that infinite fatigue life is a practical consideration in some rolling bearing applications. Since the Lundberg and Palmgren formulas did not address the concept of infinite fatigue life and did not relate to structural fatigue, an improvement in these formulas beyond the application of empirical life adjustment factors was required.

As explained in Chapter 11 of the first volume of this handbook and illustrated in Figure 8.22, the Lundberg and Palmgren theory considers that a cyclically applied concentrated load results in a Hertz stress on the raceway contact surface, which in turn causes a cyclic subsurface orthogonal shear stress. A sufficiently large magnitude of the latter stress leads to the initiation of a fatigue crack at a point below the raceway surface where its location coincides with a weak point in the material. The weak points are assumed to be randomly distributed throughout the material. The subsurface crack propagates toward the surface resulting eventually in a spall (pit). According to Lundberg and Palmgren, the large-magnitude shear stress is the range of the

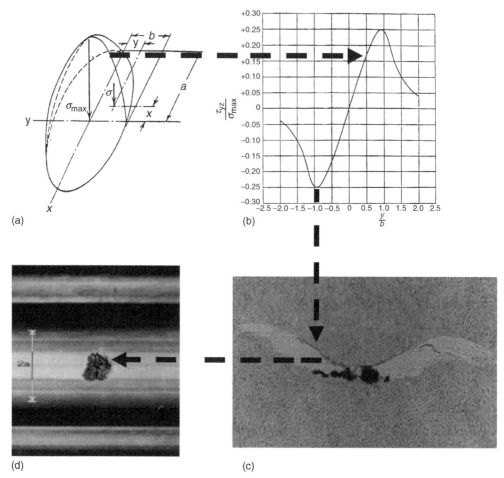

FIGURE 8.22 Basis of Lundberg–Palmgren theory: (a) cyclic Hertz stress on raceway contact surface leads to (b) cyclic subsurface orthogonal shear stress, which leads to (c) a subsurface crack at material weak point, which leads to (d) spall on raceway surface.

maximum orthogonal shear stress, that is, $2\tau_0$; this occurs at depth $z_0 \approx 0.5b$ below the raceway surface for both point and line contact.

In Chapter 4 it was shown that oil-lubricated contact pressure distributions, that is, EHL pressure distributions, are different from the pure Hertzian pressure distribution illustrated in Figure 8.22. Moreover, if the surfaces are not ideal, that is, not smooth but rather having perturbations or roughness peaks on the smooth surfaces, then concepts of micro-EHL as discussed in Chapter 5 obtain. Additionally, in their analysis Lundberg and Palmgren did not include the effect of surface friction shear stresses; these can substantially alter the subsurface stresses as demonstrated in Figure 8.23. In Figure 8.23, the subsurface stress determined is from the distortion energy failure theory of von Mises; a similar situation would occur considering subsurface shear stresses.

There are various opinions concerning which subsurface stress effects rolling contact fatigue. The depths below the surface at which maximum orthogonal shear stress and maximum von Mises stress occur are somewhat different; the latter occurs at a depth approximately 50% deeper than the former. Whichever stress is considered most detrimental, the effect of surface shear stress is to bring the maximum subsurface stress toward the surface. When the ratio $\tau/\sigma \geq 0.30$ (approximately), then the maximum stress occurs on the

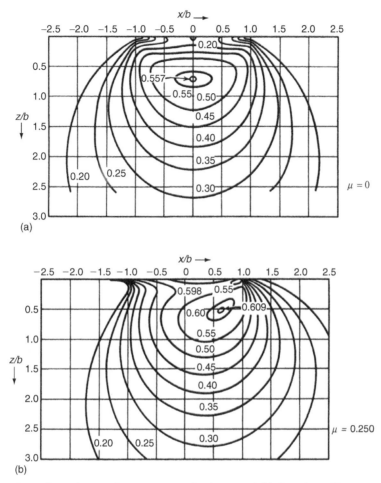

FIGURE 8.23 Lines of equal von Mises stress/σ_{max} in the material below the rolling contact surface for (a) pure rolling with no surface friction stress (coefficient of friction $\mu = 0$) and (b) rolling with surface friction stress (coefficient of friction $\mu = 0.25$).

surface. Figure 8.23b indicates a tendency toward this condition, showing a secondary peak occurring in the upper-right portion of the contact. In general, shear stresses of this magnitude do not occur over the entire concentrated contact area in an effective EHL contact. Such stresses could occur in micro-EHL contacts existing within the overall contact area. When the maximum subsurface stress approaches the surface, the potential for surface-initiated fatigue occurs. Tallian [29] considers competing modes of failure, that is, surface-initiated and subsurface-initiated. Rigorous mathematical analysis requires the consideration of failure at any point in the material from the surface into the subsurface, consistent with the stresses applied to the contact surface, both normal and tangential.

The basic equation stated by Lundberg and Palmgren is

$$\ln \frac{1}{\mathcal{S}} \propto \frac{N^e \tau_0^c \mathcal{V}}{z_0^h} \tag{8.17}$$

In Equation 8.17, τ_0 is the maximum orthogonal shear stress, z_0 is the depth at which it occurs, $\mathcal{V}$ is the volume of stressed material, and $\mathcal{S}$ is the probability of survival of the stressed volume. Actually, Lundberg and Palmgren state that the volume under stress is proportional to the volume of the cylindrical ring defined by

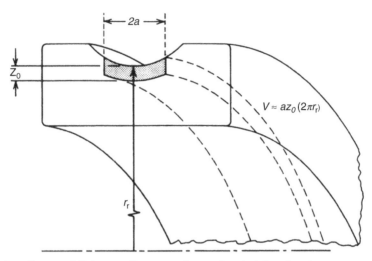

FIGURE 8.24 Lundberg and Palmgren theory—volume of material under stress.

$$V = 2a\,z_0(2\pi r_r) \tag{8.18}$$

where r_r is the raceway radius; thus, Lundberg and Palmgren did not actually define an effective stress volume; see Figure 8.24. The Lundberg and Palmgren proportionality is only valid when simple Hertz is applied to a smooth surface.

The Lundberg and Palmgren theory also does not account for the bearing operating temperatures and their effects on material properties, also not accounting for the effect of temperature on lubrication and hence on surface shear stresses. Furthermore, the theory does not consider the rate at which energy is absorbed by the materials in rolling contact. Bearing speeds are used simply to convert predicted fatigue lives in millions of revolutions to time values. Nor are hoop stresses induced by ring fitting on shafts or in housings or by high-speed centrifugal loading accommodated. Finally, the development of microstructural alterations and residual stresses below the contact surfaces, induced by rolling contact, as indicated by Voskamp [30] must be considered.

8.8 IOANNIDES–HARRIS THEORY

Considering the Lundberg–Palmgren theory limitations, Ioannides and Harris [31] developed the basic equation

$$\ln\left(\frac{1}{\Delta \tilde{s}_i}\right) = F(N, T_i - T_{\text{limit}})\Delta V_i \tag{8.19}$$

In this formula a fatigue crack is presumed incapable of getting initiated until the stress criterion T_i exceeds a threshold value of the criterion T_{limit} at a given elemental volume ΔV_i. It is evident that the crack threshold criterion T_{limit} corresponds to an endurance limit. To be consistent with the Lundberg–Palmgren theory, the stress criterion would be the orthogonal shear stress amplitude $2\tau_0$; however, another criterion, such as the von Mises or maximum shear stress may be used. In Equation 8.19, in lieu of the stress volume used by Lundberg and Palmgren, that is, $2\pi a z_0 d$, in which d is the raceway diameter, only the incremental volume over which $T_i > T_{\text{limit}}$ is considered at risk; see Figure 8.25.

Therefore, the probability of survival in Equation 8.19 is a differential value; that is, $\Delta \tilde{s}_i$. The probability of component survival is determined according to the product law of

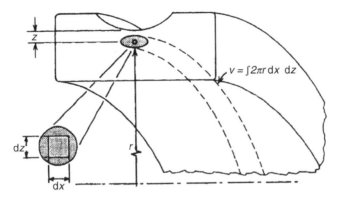

FIGURE 8.25 Risk volume in fatigue theory of Equation 8.19.

probability; subsequently, Equation 8.20, which corresponds to the Lundberg–Palmgren relationship Equation 8.17, is obtained:

$$\ln\left(\frac{1}{\mathcal{S}}\right) \approx \overline{A} N^e \int_{V_R} \frac{(T - T_{\text{limit}})^c}{z'^h} \, dV \tag{8.20}$$

where A is a constant pertaining to the overall material and z' is a stress-weighted average depth to the volume at risk to fatigue. When $T_{\text{limit}} = 0$, Equation 8.20 reduces to Equation 8.17 if it is assumed that $T = \tau_0$.

Harris and McCool [32] applied the Ioannides–Harris theory using octahedral shear stress as the fatigue-initiating stress to 62 different applications involving deep-groove and angular-contact ball bearings and cylindrical roller bearings manufactured from CVD 52100, M50, M50NiL, and 8620 carburizing steels. A value of $\tau_{\text{oct,limit}}$ was determined for each material. Using these values, the L_{10} life for each application was calculated and compared against the measured bearing fatigue life. Also, the L_{10} life calculated according to the Lundberg–Palmgren theory (standard method) was calculated and compared with the measured bearing life. It was thereby determined by statistical analysis that the bearing fatigue lives calculated using the Ioannides–Harris theory were closer to the measured lives than were the lives calculated using the standard method as modified by the life factors discussed above. Subsequently, Harris [33] demonstrated the application of the Ioannides–Harris theory in the prediction of fatigue lives of balls endurance tested in ball/v-ring rigs.

To accurately calculate bearing fatigue lives using the Ioannides–Harris theory requires:

• Selection of a fatigue-initiating stress criterion
• Determination and application of all residual, applied, and induced stresses acting on the material of the rolling element–raceway contacts
• Development and application of a stress–life factor

This was accomplished in the Harris and McCool [32] investigation using the analytical methods defined in this text combined in ball and roller bearing performance analysis computer programs TH-BBAN* and TH-RBAN.* Moreover, it should be apparent that the concept of a stress–life factor fulfills the requirement for the interdependency of the

---

*FORTRAN/VISUAL BASIC computer programs developed by T.A. Harris for operation on personal computers.

various fatigue life-influencing factors cited previously. As an alternative to the life formula indicated by Equation 8.16, resulting from the work initiated by Ioannides and Harris [31], ISO [5] established the bearing life equation format below:

$$L_{nM} = A_1 A_{ISO} L_{10} \qquad (8.21)$$

where L_{nM} is the basic rating life modified for a reliability $(100 - n)\%$, and A_{ISO} is the integrated life factor, including all of the effects considered in the multiplicative life factors A_1 to A_4 and other effects if required. In other words, $A_{ISO} = f(A_1, A_2, A_3, A_4, A_m)$.

ISO [5] states that the reliability–life factor A_1 can be calculated using

$$A_1 = 0.95 \left(\frac{\ln \frac{100}{s}}{\ln \frac{100}{90}} \right)^{1/e} + 0.05 \qquad (8.22)$$

where $\tilde{s}$ is the probability of survival in percent. This equation gives the same values of A_1 as Table 11.25 of the first volume of this handbook when Weibull slope $e = 1.5$. ISO [5] provides the means to establish the magnitude of A_{ISO}. This will be discussed later in this chapter.

8.9 THE STRESS–LIFE FACTOR

8.9.1 Life Equation

In 1995, the Tribology Division of ASME International established a technical committee to investigate life ratings for modern rolling bearings. The result of this effort was Ref. [34], in which the following equation for the calculation of bearing fatigue life was established:

$$L_n = A_1 A_{SL} \left(\frac{C}{F_e} \right)^p \qquad (8.23)$$

In the above equation, C is the bearing basic load rating as given in bearing catalogs, F_e is the equivalent applied load, and A_{SL} is the stress–life factor. As in the ISO Equation 8.21, A_1 is the reliability–life factor; it is not stress-dependent. A_{SL} is calculated considering all the life-influencing stresses acting on rolling element–raceway contacts including normal stresses, frictional shear stresses, material residual stresses due to heat treatment and manufacturing methods, and fatigue limit stress. In Equation 8.23, exponent p is 3 for ball bearings and 10/3 for roller bearings.

Considering nonstandard loading in which life is calculated for each contact, for point contacts

$$L_{mj} = A_1 A_{SLmj} \left(\frac{Q_{cmj}}{Q_{mj}} \right)^p \qquad (8.24)$$

In the above equation, subscript m refers to the raceway contact, exponent $p = 3$ for the rotating raceway, and $p = 10/3$ for the stationary raceway. Equation 8.24 further recognizes that the stress–life factor A_{SLmj} is a function of the raceway and contact azimuth location.

For line contacts,

$$L_{mjk} = A_1 A_{SLmjk} \left(\frac{q_{crj}}{q_{mjk}} \right)^p \qquad (8.25)$$

In the above equation, subscript m refers to the raceway contact, k refers to the lamina, exponent $p = 4$ for the rotating raceway, and $p = 9/2$ for the stationary raceway.

8.9.2 FATIGUE-INITIATING STRESS

In Ref. [34], the von Mises stress is considered the appropriate failure-initiating stress criterion. The von Mises stress defined according to Equation 8.26 is a scalar quantity associated with the commonly used Mises–Hencky distortion energy theory of fatigue failure:

$$\sigma_{VM} = \frac{1}{\sqrt{2}} \left[(\sigma_x - \sigma_y)^2 + (\sigma_y - \sigma_z)^2 + (\sigma_z - \sigma_x)^2 + 6\left(\tau_{xy}^2 + \tau_{yz}^2 + \tau_{zx}^2\right) \right]^{1/2} \tag{8.26}$$

See Ref. [35] or other machine design texts.

It is of interest to note that the octahedral shear stress, a vector quantity, also identified in Ref. [35] as a failure-initiating stress criterion is directly proportional in magnitude to von Mises stress; for example,

$$\tau_{oct} = \frac{\sqrt{2}}{3} \sigma_{VM} \tag{8.27}$$

8.9.3 SUBSURFACE STRESSES DUE TO NORMAL STRESSES ACTING ON THE CONTACT SURFACES

Applied loading in all applications, that is, involving both standard and nonstandard loading, is distributed over the rolling elements. The rolling element loads that are applied perpendicular to the contact areas result in pressure-type (normal to the contact surface) stresses. In Chapter 6 of the first volume of this handbook, assuming "dry" contact, equations to define the magnitudes of these Hertz stresses were provided. In Chapter 4, it was shown that, under the influence of EHL, the normal stress distribution over the contact may be somewhat altered from the Hertzian distribution. Nevertheless, in most rolling bearing applications it is satisfactory to assume the Hertzian stress distribution. On the other hand, if the rolling contact surfaces are not completely separated by a lubricant film, asperities of these surfaces will come into contact, increasing the contact stresses above the Hertz stress values. A stress concentration factor may be applied to the Hertz stress to account for this phenomenon. Equation 8.28 defines the stress concentration factor in terms of the ratio A_c/A_0, the portion of the contact area over which Coulomb friction occurs:

$$K_{Ln,mj} = \frac{Q_{mj,c}}{Q_{mj}} \left(\frac{A_c}{A_0}\right)^{-1} + \frac{1}{1 - \left(\frac{A_c}{A_0}\right)} \left(1 - \frac{Q_{mj,c}}{Q_{mj}}\right) \tag{8.28}$$

where $Q_{mj,c}$ is the load carried by the asperities and Q_{mj} is the total contact normal load at raceway m, azimuth location j. Subscript L refers to the stress concentration caused by an incomplete lubricant film; subscript n means that $K_{Ln,mj}$ is applied to the normal or Hertz stress. A_c/A_0 is determined using the method of Greenwood and Williamson (see Ref. [21] of Chapter 5). At any point (x, y, z) under the contact surface, the stresses resulting from the Hertzian loading may be determined using the methods of Thomas and Hoersch [36].

In applying the latter, the contact surface normal stress σ' at a point (x, y) is given by

$$\sigma' = \frac{3K_{Ln,mj}Q_{mj}}{2\pi a_{mj}b_{mj}} \left[1 - \left(\frac{x}{a_{mj}}\right)^2 - \left(\frac{y}{b_{mj}}\right)^2 \right]^{1/2} \qquad (8.29)$$

8.9.4 SUBSURFACE STRESSES DUE TO FRICTIONAL SHEAR STRESSES ACTING ON THE CONTACT SURFACES

In most rolling bearing contacts, as discussed in Chapter 2, some degree of sliding occurs. In angular-contact ball bearings, spherical roller bearings, and thrust cylindrical roller bearings, a substantial amount of sliding occurs. These sliding motions, occurring in relatively heavily loaded rolling element–raceway contacts, result in significant frictional shear stresses. The magnitude of the friction shear stress at any point (x, y) on the contact surface depends on the local contact pressure, the local sliding velocity, the lubricant rheological properties, and the topographies of contact surfaces.

Depending on the degree of contact surface separation by the lubricant film, sliding in conjunction with the basic rolling motion may produce surface distress that can result in microspalls; these can lead to macrospalls. Nélias et al. [37], conducting endurance tests using a rolling–sliding disk rig, demonstrated that smooth surfaces on 52100 and M50 steel test components, irrespective of the occurrence of sliding, experienced no surface distress. The tests were conducted at 1500–3500 MPa ($2.18–5.08 \times 10^5$ psi) under lubricant film parameter Λ ranging approximately from 0.6 to 1.3. This indicates the need for finely finished rolling element and raceway surfaces, especially in the presence of marginal lubrication. Nélias et al. [37] noted that, in the absence of sliding, microspall progression occurs both in the direction of sliding, and transverse to that direction. This is shown in Figure 8.26 taken from Ref. [37]. In their test rig, the drive disk turns faster than the follower disk, and the friction direction over the contact for the follower disk is in the rolling direction. The friction direction over the contact of the driver disk is, however, in the direction opposite to rolling. Figure 8.27 shows that the microcracks are dependent on the friction direction. It can be seen that the typical arrowhead shape is oriented in the friction direction while crack propagation is in the direction opposite to friction. Nélias et al. [37] further noted that the driven surfaces were prone to greater damage than the driver surfaces.

Another observation of Nélias et al. [37] was that the size and volume of the spalled material increased with the magnitude of normal (Hertz) stress (see Figure 8.28). This situation indicates that sliding damage is more severe under a heavy load than under a lighter load, a condition that must be of concern in heavily loaded angular-contact ball bearings and spherical roller bearings with marginal lubrication.

At any point (x, y, z) under the contact surface, the stresses resulting from the surface shear stresses may be determined using the methods of Ahmadi et al. [38].

8.9.5 STRESS CONCENTRATION ASSOCIATED WITH SURFACE FRICTION SHEAR STRESS

To employ the methods of Ahmadi et al. [38], it is necessary to define the value of surface friction shear stress τ at each point (x, y) on the contact surface. For a contact that incurs sliding in both the rolling and transverse to rolling directions, Equation 5.49 and Equation 5.50 can be used to define the surface friction shear stresses τ_y and τ_x when the lubricant film is insufficient to completely separate the rolling contact surfaces:

$$\tau_d = c_v \frac{A_c}{A_0} \mu_a \sigma + \left(1 - \frac{A_c}{A_0}\right) \left(\frac{h}{\eta v_d} + \frac{1}{\tau_{lim}}\right)^{-1} \qquad d = y, x \quad (5.49, 5.50)$$

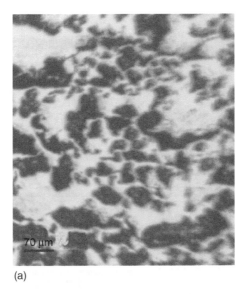

(a)

(b)

FIGURE 8.26 Surfaces of M50 steel endurance test components operated at 3500 MPa (5.08×10^5 psi) under (a) simple rolling and (b) rolling and sliding. (From Nélias, D., et al., *ASME Trans., J. Tribol.,* 120, 184–190, April 1998. With permission.)

In applying Equation 5.49 and Equation 5.50, the normal stress σ is replaced by σ' defined by Equation 8.29; viscosity is also calculated considering σ'.

Alternatively, considering only the fluid friction portion of the surface friction shear stress, the following stress concentration factor may be applied to the latter stress:

$$K_{\text{Lf},mj} = 1 + \frac{\mu_c Q_{mj,c}}{F_{\text{fm}j}} \tag{8.30}$$

where μ_c is the coefficient of friction associated with asperity–asperity interaction. A value $\mu_c = 0.1$ may be used for the lubricated contact.

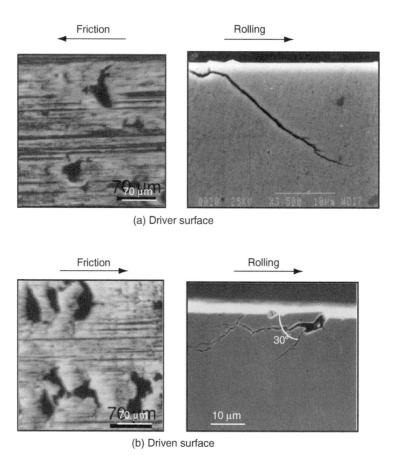

(a) Driver surface

(b) Driven surface

FIGURE 8.27 Microcrack orientation with respect to rolling and friction directions for M50 steel specimens tested at 3500 MPa (5.08×10^5 psi). (From Nélias, D., et al., *ASME Trans., J. Tribol.*, 120, 184–190, April 1998. With permission.)

8.9.6 STRESSES DUE TO PARTICULATE CONTAMINANTS

To determine the surface stresses associated with dents, the methods developed by Ville and Nélias [19,23] or Ai and Cheng [39] may be applied. This requires a definition of the contaminants involved in the application. Also, if the topography of the dented surface can be defined, the methods of Webster et al. [20] may be applied. These methods, while effective for laboratory investigations, typically consume many minutes and even hours of computer time for the stress analysis of a single contact. The analysis of rolling bearing fatigue endurance involves the iterative solution of many thousands of contacts. To include the effect of particulate contamination in the prediction of bearing fatigue life in an engineering application, approximations are necessary regarding the types of particles, their concentration in the lubricant, and their effects on subsurface stresses. In essence, a stress concentration factor based on these parameters would be applied to the contact stress in the determination of subsurface stresses.

The contamination level in an oil-lubricated application may be measured by counting particles in the oil. This information may be used to establish a contamination parameter C_L. In bearing applications, the lubricant contains particles of widely varying sizes and properties.

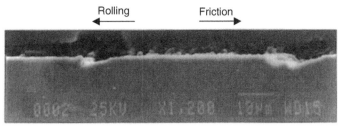

(a) 1500 MPa (2.18 10^5 psi), 5–10 μm spall size

(b) 2500 MPa (3.63 10^5 psi), 20 μm spall size

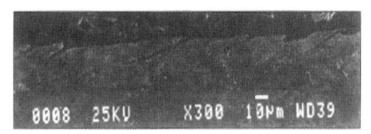

(c) 3500 MPa (5.08 10^5 psi), 40 μm spall size

FIGURE 8.28 Increase in microcrack size (length, depth) with normal stress for M50 steel endurance tested under rolling and sliding conditions. (From Nélias, D., et al., *ASME Trans., J. Tribol.*, 120, 184–190, April 1998. With permission.)

For oil-bath-type lubrication, Ioannides et al. [40] recommend the use of the international cleanliness code for hydraulic fluids, ISO Standard 4406 [41] to codify these. The cleanliness levels are indicated in Table 8.6.

In using Table 8.6, the following guidelines apply:

- If the filter has been validated to withstand system-operating conditions, the lowest level (cleanest) on the left-hand side of the row associated with the specific filter rating should be used.
- If the filter has not been validated for withstanding the operating conditions of the specific system, the highest (most contaminated) level on the right-hand side of the row associated with the specific filter rating should be used.
- If low contaminant ingress is expected, such as with a system having an air-vent filter operating in a clean ambient environment, one level can be subtracted; that is, a move of one level to the left is acceptable.
- If high contaminant ingress is expected, such as for mobile equipment with open reservoirs, one level should be added; that is, a move of one level to the right is required.

TABLE 8.6
ISO 4406 Fluid Cleanliness Levels

Filter Rating $\beta(x_c)^a$	Cleanliness Levels[b]			
2.5	13/10/7	14/11/9	15/12/10	16/13/11
5	15/12/10	16/13/12	16/14/12	17/15/12
7	17/14/12	18/15/12	18/16/13	19/16/14
12	18/16/13	19/16/14	20/17/14	21/18/15
22	20/17/14	21/18/15	22/19/16	23/20/17
35	22/19/16	23/20/17	24/21/18	25/22/18

[a] In $\beta(x_c) \geq 1000$, x is the particle size in μm, and $1000 =$ the filtration ratio. Filtration ratio means that for a given particle size x, the number of particles upstream of the filter is $1000 \times$ the number of downstream particles.
[b] Code $X/Y/Z$; for example 13/10/7, refers to cleanliness levels whereby X is the number of particles of size ≥ 4 μm; Y is the number of particles of size ≥ 6 μm; and Z is the number of particles of size ≥ 14 μm.

- Example 1: For a full-flow 5-μm filter validated for high contaminant ingress operating conditions, it is appropriate to start at 15/12/10 and move one level to the right; that is, 16/13/11.
- Example 2: For a full-flow 12-μm filter validated for moderate contaminant ingress operating conditions, it is appropriate to move to the highest level; that is, 21/18/15.
- If operation is with two full-flow filters of the same rating in series, two levels should be subtracted.

For use with the computer program supplied with Ref. [34] for calculation of bearing fatigue life, Table 8.7 provides some simplified guidelines for the cleanliness level.

The earlier classifications do not account for the hardness of the particles. It has been established, however, that in a wide scope of rolling bearing applications, there exists a similar distribution of hard and soft particles, which produces a generally similar fatigue life-reducing effect. Even if Table 8.6 indicates the number of particles >4 μm, >6 μm, and >14 μm, this does not mean that just a few contaminant particles of such minute size affect the fatigue lives of rolling bearings. The standardized figures are only a statistical measure for the existence of critical particles.

Ioannides et al. [40] state that for circulating oil lubrication, the filtering efficiency of the system can be used in lieu of ISO 4406 [41] to define contaminant size. This may be defined by the filtering capacity as specified by ISO 4372 [42].

TABLE 8.7
ASME Guidelines for Cleanliness Classification vs. Contamination Level

Cleanliness Classification	ISO 4406 Cleanliness Level	Filter Rating $\beta(x_c)$ (μm)
Utmost cleanliness	14/11/8	2.5–5
Improved cleanliness	16/13/10	5
Normal cleanliness	18/15/12	7
Moderate contamination	20/17/14	12–22
Heavy contamination	22/19/16	35 or coarser

TABLE 8.8
Lubricant Contamination Factor Calculation Constants

Type of Lubrication	Contamination Level	C_{L2}	C_{L3}	Restriction
Circulating oil	ISO –/13/10	0.5663	0.0864	
	ISO –/15/12	0.9987	0.0432	
	ISO –/17/14	1.6329	0.0288	
	ISO –/19/16	2.3362	0.0216	
Bath oil	ISO –/13/10	0.6796	0.0864	
	ISO –/15/12	1.141	0.0288	
	ISO –/17/14	1.670	0.0133	
	ISO –/19/16	2.5164	0.00864	
	ISO –/21/18	3.8974	0.00411	
Grease	High cleanliness	0.6796	0.0864	
	Normal cleanliness	1.141	0.0432	
	Slight-to-typical contamination	1.887	0.0177	$d_m < 500\,mm$
		1.677	0.0177	$d_m \geq 500\,mm$
	Severe contamination	2.662	0.0115	
	Very severe contamination	4.06	0.00617	

Depending on the size of the rolling contact areas in a bearing, sensitivity to particulate contamination varies. Ball bearings tend to be more vulnerable than roller bearings; contaminant particles are more harmful in small bearings than in bearings with large rolling elements. Considering the foregoing and using empirically determined data, Ioannides et al. [40] linked the contamination parameter C_L to bearing size, lubrication system, and lubrication effectiveness. Further considering that solid contaminants found in bearings are mainly hard metallic particles resulting from wear of the mechanical system, they developed Figure CD8.1 through Figure CD8.14, which are charts of C_L vs. lubricant effectiveness parameter κ and bearing pitch diameter d_m for various ISO Standard 4406 cleanliness levels. For circulating oil-lubrication systems, filtration levels according to ISO 4572 are also indicated.

The values of C_L may also be obtained using the base equation for the curves provided in Figure CD8.1 through Figure CD8.14. This base equation may be obtained from the appendix of ISO Standard 281 [5].

$$C_L = C_{L1}\left(1 - \frac{C_{L2}}{d_m^{1/3}}\right) \tag{8.31}$$

where

$$C_{L1} = C_{L3}\kappa^{0.68}d_m^{0.55} \qquad C_{L1} \leq 1 \tag{8.32}$$

Values of the constants C_{L2} and C_{L3} may be obtained from Table 8.8 for the various ISO contamination levels. For oil-lubricated bearings, Table 8.9 gives the range of contamination levels corresponding to the basic level given in Table 8.8. For circulating oil-lubricated bearings, Table 8.9 also provides the $\beta(x_c)$ level corresponding to the basic contamination level.

In Equation 8.32 and in Figure CD8.1 through Figure CD8.14, κ is defined as ν/ν_1, where ν is the kinematic viscosity of the lubricant at the operating temperature and ν_1 is the kinematic viscosity required for adequate separation of the contacts. According to ISO 281

TABLE 8.9
Contamination Ranges and $\beta(x_c)$ for Data of Table 8.8

Type of Lubrication	Basic ISO Contamination Level	ISO Contamination Range	$x(c)$	$\beta(x_c)$
Circulating oil	−/13/10	−/13/10, −/12/10, −/13/11, −/14/11	6	200
	−/15/12	−/15/12, −/16/12, −/15/13, −/16/13	12	200
	−/17/14	−/17/14, −/18/14, −/18/15, −/19/15	25	75
	−/19/16	−/19/16, −/20/17, −/21/18, −/22/18	40	75
Bath oil	−/13/10	−/13/10, −/12/10, −/11/9, −/12/9	—	—
	/15/12	−/15/12, −/14/12, −/16/12, −/16/13	—	—
	−/17/14	−/17/14, −/18/14, −/18/15, −/19/15	—	—
	−/19/16	−/19/16, −/18/16, −/20/17, −/21/17	—	—
	−/21/18	−/21/18, −/21/19, −/22/19, −/23/19	—	—

[5] $6.7^{1.3}$, $\kappa \approx \Lambda^{1.12}$. According to ISO 281 [5], the reference viscosity ν_1 may be estimated using Equation 8.33 and Equation 8.34. Alternatively, the chart of Figure CD8.15 may be used to estimate ν_1:

$$\nu_1 = 45,000 \ n^{-0.83} d_m^{-0.5} \quad n < 1,000 \text{ rpm} \tag{8.33}$$

$$\nu_1 = 4,500 \ n^{-0.5} d_m^{-0.5} \quad n \geq 1,000 \text{ rpm} \tag{8.34}$$

For circulating oil, in Figure CD8.1 through Figure CD8.4, as indicated in footnote a of Table 8.6, the parameter β_x is defined in Ref. [40] as

$$\beta_x = \frac{N_{pu} > x}{N_{pd} > x} \tag{8.35}$$

where N_{pu} is the number of particles upstream of size greater than x μm, and N_{pd} is the number of particles downstream of size greater than x μm. Thus, $\beta_6 = 200$ means that for every 200 particles >6 μm upstream of the filter, only 1 particle >6 μm passes through the filter. Although this is a useful method for comparing filter performance, it is not infallible since contaminant particles may have different shapes according to the application.

The C_L values obtained using Figures CD8.1 through Figure CD8.9 are for oil lubricants without additives. When the calculated $\beta_x < 1$, a high-quality lubricant with tested and approved additives may be expected to promote a favorable smoothing of the raceway surfaces during running in. Thereby, β_x may improve and reach a value of 1.

When contamination is not measured or known in detail, the contamination parameter C_L may be estimated using Table 8.10 provided in Refs. [5,43].

For use in determination of rolling contact fatigue life, the contamination parameter C_L needs to be converted to the form of a stress concentration factor to be applied to the contact stress; for example, $\sigma'(x,y) = K_c \sigma(x,y)$. Also, the stress concentration factor may be applied to the surface shear stress as well; for example,

$$\tau'(x,y) = K_c \tau(x,y)$$

TABLE 8.10
Contamination Parameter Levels

Bearing Operation Condition	C_L	
	$d_m < 100\,mm$	$d_m \geq 100\,mm$
Extreme cleanliness	1	1
Particle size of the order of lubricant film thickness		
High cleanliness	0.8–0.6	0.9–0.8
Oil filtered through extremely fine filter; conditions typical of bearings greased for life and sealed		
Normal cleanliness	0.6–0.5	0.8–0.6
Oil filtered through fine filter; conditions typical of bearings greased for life and shielded		
Slight contamination	0.5–0.3	0.6–0.4
A small amount of contaminant in lubricant		
Typical contamination	0.3–0.1	0.4–0.2
Conditions typical of bearings without integral seals; coarse filtering; wear particles and ingress from surroundings		
Severe contamination[a]	0.1–0	0.1–0
Bearing environment heavily contaminated and bearing arrangement with inadequate sealing		
Very severe contamination[a]	0	0

[a]In the cases of severe and very severe contamination, failure may be caused by wear, and the useful life of the bearing may be far less than the calculated rating life.

Barnsby et al. [34], as derived from Ioannides et al. [40], give the following equations for point and line contacts:

$$K_{C,point} = 1 + (1 - C_L^{1/3})\frac{\sigma_{VM,lim}}{\sigma_{VM,max}} \tag{8.36}$$

$$K_{C,line} = 1 + (1 - C_L^{1/4})\frac{\sigma_{VM,lim}}{\sigma_{VM,max}} \tag{8.37}$$

where $\sigma_{VM,max}$ is the maximum value of the von Mises stress occurring below the contact surface, and $\sigma_{VM,lim}$ is the fatigue limit of the von Mises stress for the rolling component material. Values for the fatigue limit stress will be discussed later in this chapter.

Nélias [44] illustrates in Figure 8.29 that for a dented or rough surface the magnitude of the maximum shear stress is strongly influenced by sliding on the surface. Nélias [44] further postulates that failure of rough or dented surfaces may commence near the surface; however, coalescence of microcracks may proceed inward in the direction toward the location of the maximum subsurface stresses due to the average contact loading. Thus, the subsurface failure might be initiated by the surface condition. This competition of subsurface, failure-initiating stresses is illustrated in Figure 8.30. Because most modern ball and roller bearings have relatively smooth raceway and rolling element surfaces, roughness is more indicative of dents in contaminated applications. Thus, competition for initiation of subsurface fatigue failure would tend to occur more in applications with contamination. When calculations for subsurface von Mises stresses (or other assumed failure-initiating stresses) indicate maximum values approaching the surface, it may be presumed that surface pitting will most likely occur first; however, not to the exclusion of subsurface fatigue failure depending on the amount of operational cycles accumulated.

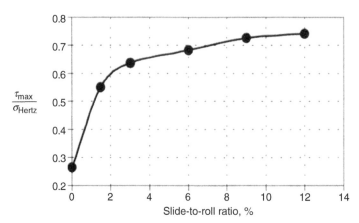

FIGURE 8.29 Maximum shear stress/maximum Hertz stress vs. slide–roll ratio in the vicinity of a dent 1.5 μm deep by 40 μm wide; the dent with a shoulder 0.5 μm. (From Nélias, D., *Contribution a L'etude des Roulements*, Dossier d'Habilitation a Diriger des Recherches, Laboratoire de Mécanique des Contacts, UMR-CNRS-INSA de Lyon No. 5514, December 16, 1999. With permission.)

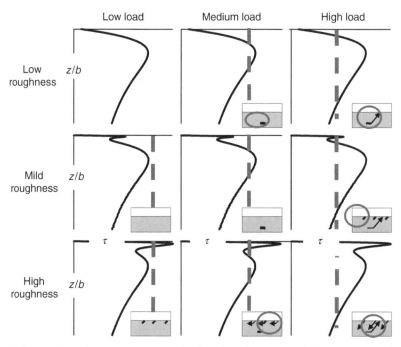

FIGURE 8.30 Competition between surface and subsurface crack growth for various loads and surface roughnesses. Each graph represents shear stress vs. nondimensionalized depth z/b. The dashed line represents the fatigue limit stress below which crack initiation (straight lines in inserts) does not occur and propagation direction (arrow-tip lines in inserts). (From Nélias, D., *Contribution a L'etude des Roulements*, Dossier d'Habilitation a Diriger des Recherches, Laboratoire de Mécanique des Contacts, UMR-CNRS-INSA de Lyon No. 5514, December 16, 1999. With permission.)

8.9.7 COMBINATION OF STRESS CONCENTRATION FACTORS DUE TO LUBRICATION AND CONTAMINATION

The stress concentration factors K_L and K_C occur due to imperfections in the contact surfaces. These stress concentrations do not act independently; rather, their combined value is given by

$$K_{LC,mj} = K_{L,mj} + K_{C,mj} - 1 \qquad (8.38)$$

It can be seen that for very smooth rolling contact surfaces without dents, $K_{LC,mj} = 1$, and for all surfaces with no contaminants present, $K_{LC,mj} = K_{L,mj}$.

8.9.8 EFFECT OF LUBRICANT ADDITIVES ON BEARING FATIGUE LIFE

Thus far, only the effect of the base stock lubricant has been considered with regard to fatigue life. However, a base stock lubricant is supplied to a rolling bearing rarely only. In fact, more often than not, with the exception of bearings that are delivered with integral seals and greased for life, the bearing must survive with the lubricant required to maximize performance of the overall mechanism; for example, a gear-box. Such lubricants typically contain additives to achieve one or more of the following properties: (1) antiwear, (2) antiscuffing or extreme pressure (EP) resistance, (3) antioxidation, (4) antifoaming, (5) rust/corrosion inhibition, (6) control of deposit formations on surface through detergents, (7) demulsification to aid in separation of water, and (8) control of sludge formation through dispersants. Some of these additives tend to influence fatigue endurance significantly; however, it has not been possible to specify these effects through the use of contact stress concentration factors. Rather in Ref. [34] the effects of these additives on life have been specified as ranges on L_{10} lives, as in Table 8.11.

8.9.9 HOOP STRESSES

To prevent rotation of the bearing inner ring about the shaft, and hence prevent fretting corrosion of the bearing bore surface, the bearing inner ring is usually press-fitted to the shaft. The amount of diametral interference, and therefore the required pressure between the ring

TABLE 8.11
Estimated Bearing Life Ranges for Common Lubricant Classes

Lubricant Class	Fatigue Life Range	Average Fatigue Life
Industrial Lubricants		
Hydraulic oils	0.6–1.0 L_{10}	0.8 L_{10}
Rolling bearing oils with no antiwear additive	0.8–1.4 L_{10}	1.1 L_{10}
Rolling bearing oils with antiwear additive	0.6–1.0 L_{10}	0.8 L_{10}
Turbine oils	0.6–1.0 L_{10}	0.8 L_{10}
Circulating oils with no antiwear additive	0.8–1.4 L_{10}	1.1 L_{10}
Circulating oils with antiwear additive	0.6–1.0 L_{10}	0.8 L_{10}
Synthetic antiwear oils	0.8–1.7 L_{10}	1.2 L_{10}
Gear oils	0.4–1.3 L_{10}	0.8 L_{10}
Automotive and Aviation Lubricants		
Gear lubricants	0.3–0.7 L_{10}	0.5 L_{10}
Automatic transmission fluids	0.6–1.0 L_{10}	0.8 L_{10}
Aviation turbine oils	0.8–1.7 L_{10}	1.2 L_{10}

bore and the shaft outside diameter, depends primarily on the amount of applied loading and secondarily on the shaft speed. The greater the applied load and shaft speed, the greater must be the interference to prevent ring rotation. For recommendation of the magnitude of the interference fit required for a given application as dictated only by the magnitude of applied loading, ANSI/ABMA Standard No. 7 [45] may be consulted for radial ball, cylindrical roller, and spherical roller bearings. For tapered roller bearings, ANSI/ABMA Standards No. 19.1 [46] and No. 19.2 [47] may be consulted. Because the ring and shaft dimensions, and materials are defined, standard strength of materials calculations, for example, Timoshenko [48], may be used to determine the radial stresses. The interference fit causes the ring to stretch resulting in tensile hoop stress.

Similarly, for outer ring rotation such as in wheel bearing applications, the outer ring may be press-fitted into the housing. In this case, compressive hoop stress and radial stress will be induced.

Ring rotation, particularly at a high speed, gives rise to radial centrifugal stress, which in turn causes the ring to stretch with attendant hoop stresses resisting the ring expansion. Outer ring rotation results in tensile hoop stresses that tend to counteract the compressive hoop stresses caused by press-fitting of the outer ring in the housing. Timoshenko [48] details the method to calculate the tensile hoop and radial stresses associated with ring rotation.

Each of the stresses due to press-fitting or ring rotation is superimposed on the subsurface stress field caused by contact surface stresses.

8.9.10 RESIDUAL STRESSES

8.9.10.1 Sources of Residual Stresses

Residual stress is that stress which remains in a material when all externally applied forces are removed. Residual stresses arise in an object from any process that produces a nonuniform change in shape or volume. These stresses may be induced mechanically, thermally, chemically, or by a combination of these processes [49]. An example of such a process is as follows:

> If a relatively thin sheet of malleable material such as copper is repeatedly struck with a hammer, the thickness of the sheet is reduced, and the length and width are correspondingly increased; that is, the volume remains constant. If the same number of equally intensive hammer blows were uniformly delivered to the surface of a copper block several centimeters thick, the depth of penetration of plastic deformation would be relatively shallow with respect to the block thickness. The deformed surface layer would be restrained from lateral expansion by the bulk of subsurface material, which experienced less deformation. Consequently, the heavily deformed surface material would be like an elastically compressed spring, prevented from expanding to its unloaded dimensions by its association with elastically extended subsurface material. The resulting residual stress profile is one in which the surface region is in residual compression and the subsurface region is in a balancing residual tension. This example is a literal description of the shot-peening process, wherein a surface is bombarded with pellets of steel or glass. A highly desirable compressive residual stress pattern is established for components that experience high, cyclic tensile stresses at the surface during service. The magnitude of tensile stress experienced by the component during service is functionally reduced by the amount of residual compressive stress, thereby providing significantly increased fatigue lives for parts such as shafts and springs.

The shot-peening example illustrates the essential characteristics of a surface in which residual stress has been induced:

1. Nonuniformity of plastic deformation; the surface material is encouraged to expand laterally.

2. Subsurface material, which experiences less plastic deformation, is elastically strained in tension as it restrains expansion of the surface material, thereby inducing compressive residual stress in the surface region.
3. The resulting state of residual stress is a reflection of the elastic components of strain in the surface and subsurface regions, which are in equilibrium, providing a balanced tensile–compressive system.

Heat treatment used for hardening rolling bearing components can exert very significant influence over the state of residual stress. Depending on the steel composition, austenitizing temperature, quenching severity, component geometry, section thickness, and so forth, heat treatment can provide either residual compressive stress or residual tensile stress in the surface of the hardened component [49–50]. Temperature gradients are established from the surface to the center of a part during quenching after heating. The differential thermal contraction associated with these gradients provides for nonuniform plastic deformation, giving rise to residual stresses. Additionally, volumetric changes associated with the phase transformation occurring during heat treatment of steel occur at different times during quenching at the part surface and interior due to the thermal gradients established. These sequential volumetric changes, combined with differential thermal contractions, are responsible for the residual stress state in a hardened steel component. The sequence and relative magnitudes of these contributing factors determine the stress magnitude and whether the surface is in residual compression or tension.

Grinding of a hardened steel component to finished dimensions also affects the residual surface stress. Generally, if effects of abusive grinding practices that generate heat and produce microstructural alterations are neglected, it is found that the residual stress effects associated with grinding are confined to material within 50 μm (0.002 in.) of the surface. Good grinding practice, as applied to bearing rings, produces residual compressive stress in a shallow surface layer. Grinding also involves some plastic deformation of the surface, producing residual compression as described above.

The residual stress state in a finished bearing component is therefore a function of heat treatment and grinding. If properly ground, the residual stress in a through-hardened bearing component will be 0 to slightly compressive. The subsurface residual stress conditions will be determined by the prior heat treatment. In a surface-hardened component, the surface and subsurface residual stresses will be compressive; in the core of the material, the residual stresses will be tensile. The depth of the case must, therefore, be sufficient with regard to bearing fatigue endurance. This depth has historically been set at approximately four times the depth of the maximum subsurface orthogonal shear stress; see Chapter 6 of the first volume of this handbook.

8.9.10.2 Alterations of Residual Stress Due to Rolling Contact

As a result of cyclic stressing during rolling contact, the bearing steel experiences changes in the microstructure. Associated with these alterations are changes in residual stress and retained austenite; this has been reported in Refs. [30,51–55]. The forms of the changes in circumferential direction residual stress and retained austenite profiles are illustrated in Figure 8.31. Indications are that significant changes in residual stress and retained austenite precede any observable alterations in microstructure. See Figure 11.4 in the first volume of this handbook.

The residual stress data of Figure 8.31 show peak values at increasing depths corresponding to increasing numbers of stress cycles. A similar form is indicated for decomposition of retained austenite, with peak effect depths slightly less than those for residual stress. The data of Figure 8.31 for the high maximum-contact stress indicate more rapid rates of change for both residual stress and retained austenite content.

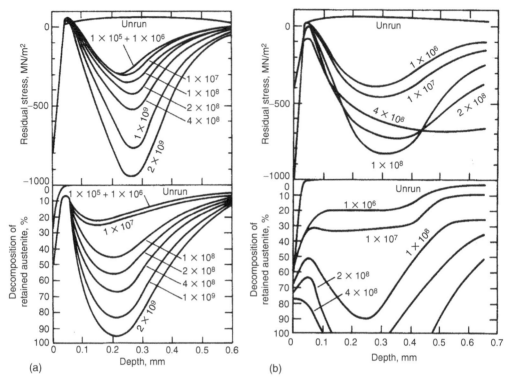

FIGURE 8.31 Residual stress and percent retained austenite decomposition vs. depth below raceway surface for various numbers of inner ring revolutions of a 309 deep-groove ball bearing; the bearing ring was manufactured from 52100 steel through-hardened to R_c 64. (a) Maximum contact stress: 3280 MPa (475 kpsi); depth of maximum orthogonal shear stress 0.19 mm (0.0075 in.); depth of maximum shear stress 0.30 mm (0.0118 in.) (b) Maximum contact stress: 3720 MPa (539 kpsi); depth of maximum orthogonal shear stress 0.21 mm (0.0083 in.); depth of maximum shear stress 0.33 mm (0.0130 in.).

Harris [56] found compressive surface stresses in the range of 600 MPa (87,000 psi) for both M50 and 52100 balls that had not been run. Beneath the surface, in the zone of maximum subsurface applied stress, the compressive stress level reduced to values in the range of 70 MPa (10,000 psi). When the balls were operated under normal bearing Hertz stresses, for example, maximum 2,700 MPa (400,000 psi), these compressive stresses seemed to disappear, most likely, as a result of retained austenite transformation. The slight differences in the depths at which the peak values occur in residual stress and retained austenite decomposition imply correlation with the maximum shear stress and the maximum orthogonal shear stress, respectively. The work of Muro and Tsushima [53] supports the correlation of peak residual stress values with the maximum shear stress. There appears to be no direct relationship between retained austenite decomposition and the generation of residual compressive stress, nor, according to Voskamp et al. [54], any indication of which, if either, of these processes triggers microstructural alterations.

8.9.10.3 Work Hardening

It has also been observed that running-in bearing raceways under heavy loading for a short period of time before normal operation tends to work harden the near-surface regions. This introduces a slight compressive residual stress into the material, increasing its resistance to fatigue. Excessive amounts of compressive stress tend to reduce resistance to fatigue.

8.9.11 Life Integral

The stresses discussed in this section each contribute to the overall subsurface stress distribution. Using superposition and the assumption of von Mises stress as the fatigue failure-initiating criterion, the stress tensor may be calculated for every subsurface point (x, y, z). The basic equation of the Ioannides–Harris theory, that is Equation 8.19, may be restated as follows:

$$\ln\frac{1}{\Delta S_i} \propto \frac{N^e\left(\sigma_{VM,i} - \sigma_{VM,\lim}\right)^c \Delta V_i}{z_i^h} \tag{8.39}$$

The above equation refers to the survival of volume element ΔV_i for N stress cycles with probability ΔS_i. The probability that the entire stressed volume will survive N stress cycles may be determined using the product law of probability; that is, $S = \Delta S_1 \times \Delta S_2 \times \cdots \times \Delta S_n$. Therefore,

$$\ln\frac{1}{S} = \sum_{i=1}^{i=n} \ln\frac{1}{\Delta S_i} \propto N^e \pi d_r \sum_{i=1}^{1=n} \left[\frac{\left(\sigma_{VM,i} - \sigma_{VM,\lim}\right)^c A_i}{z_i^h}\right] \tag{8.40}$$

where A_i is the radial plane cross-sectional area $\Delta x \times \Delta z$ of the volume element on which the effective stress acts, and d_r is the raceway diameter. Letting $q = x/a$ and $r = z/b$, where a and b are the semimajor and semiminor axes, respectively, of the contact ellipse (see Figure 8.22), then $\Delta x = a\Delta q$ and $\Delta z = b\Delta r$. Numerical integration may be performed using Simpson's rule, letting $\Delta q = \Delta r = 1/n$, where n is the number of segments into which the major axis is divided. With the indicated substitutions, Equation 8.39 becomes

$$\ln\frac{1}{S} = \frac{N^e \pi ab^{1-h} d_r}{9n^2} \sum_{j=1}^{j=n} c_j \sum_{k=1}^{k=n} c_k \left[\frac{\left(\sigma_{VM,jk} - \sigma_{VM,\lim}\right)^c}{r_k^h}\right] \tag{8.41}$$

where c_j and c_k are Simpson's rule coefficients. The number of stress cycles survived is $N = uL$, where u is the number of stress cycles per revolution and L is the life in revolutions. Therefore,

$$L \propto u\left\{\frac{\pi ab^{1-h} d_r}{9n^2} \sum_{j=1}^{j=n} c_j \sum_{k=1}^{k=n} c_k \left[\frac{\left(\sigma_{VM,jk} - \sigma_{VM,\lim}\right)^c}{r_k^h}\right]\right\}^{1/e} \tag{8.42}$$

The above equation may be used to find the stress–life factor A_{SL} by (1) evaluating the equation for the stress conditions assumed by Lundberg and Palmgren, (2) evaluating the equation for the actual bearing stress conditions occurring in the application, and (3) comparing these. For example,

$$A_{SL} = \frac{L_{actual}}{L_{LP}} = \frac{\left\{\sum_{j=1}^{j=n} c_j \sum_{k=1}^{k=n} c_k \left[\frac{\left(\sigma_{VM,jk} - \sigma_{VM,\lim}\right)^c}{r_k^h}\right]\right\}_{actual}^{1/e}}{\left\{\sum_{j=1}^{j=n} c_j \sum_{k=1}^{k=n} c_k \left[\frac{\left(\sigma_{VM,jk}\right)_{LP}^c}{r_k^h}\right]\right\}_{LP}^{1/e}} = \frac{I_{actual}}{I_{LP}} \tag{8.43}$$

where I is called the life integral.

The accurate evaluation of I for each condition depends on the boundaries specified for the stress volume. It was shown that earlier, because only Hertz stresses were considered in their analysis, Lundberg and Palmgren were able to assume that the stressed volume was proportional to $\pi a d_r z_0$, where z_0 is the depth to the maximum orthogonal shear stress τ_0. In the analysis of the stress–life factor, von Mises stress is used in lieu of τ_0, and the effective stress is integrated over the appropriate volume. That volume is defined by the elements for which the effective stress is greater than zero; that is, $\sigma_{VM,i} - \sigma_{VM,limit} > 0$. It can be demonstrated using the Lundberg–Palmgren analysis that

$$L \propto \frac{1}{\tau_0^{c/e}} = \frac{1}{\tau_0^{9.3}} \tag{8.44}$$

Considering the equivalent integrated life, Harris and Yu [57] showed that

$$L_{ij} \propto \frac{1}{\tau_{ij}^{9.39}} \tag{8.45}$$

Moreover, they determined that all effective stresses

$$\sigma_{VM,i} - \sigma_{VM,limit} < 0.6 \left(\sigma_{VM,i} - \sigma_{VM,lim} \right)_{max}$$

influence life less than 1%. For simple Hertz loading, the life-influencing zone is illustrated in Figure 8.32. As compared with the Lundberg–Palmgren stressed volume proportionality for which $z_0/b \approx 0.5$, for Hertz loading, the critical stressed volume stretches down to $z/b \approx 1.6$.

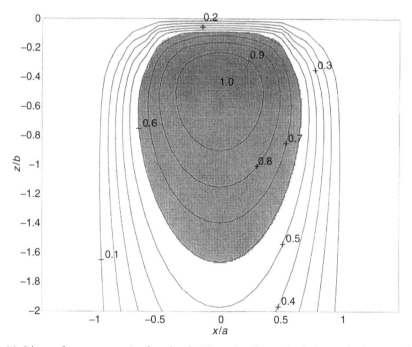

FIGURE 8.32 Lines of constant τ_{yz}/τ_0 for simple Hertz loading—shaded area indicates effective life-influencing stresses.

The critical stressed volume is different for each rolling element–raceway contact combination of applied and residual stress, and it should be used in the evaluation of the life integrals in Equation 8.43.

8.9.12 Fatigue Limit Stress

To evaluate the life integrals, the value of the fatigue limit stress must be known for the bearing component material. This can be determined by endurance testing of bearings or selected components. The test programs reported in Refs. [32,33] were extended to cover 129 bearing applications including additional materials. The analytical models to predict bearing application performance and ball/v-ring test performance were refined, and performance analyses were again conducted, using the von Mises stress as the fatigue failure-initiating criterion. Based on this subsequent study by Harris [56], Table 8.12 gives resulting values of fatigue limit stress for various materials.

Böhmer et al. [58] established that the fatigue limits of steels decrease as a function of temperature. From their graphical data, the following relationships may be determined by curve-fitting for various bearing steels operating at temperatures exceeding 80°C (176°F):

$$\left(\frac{\sigma_{VM,\lim(T)}}{\sigma_{VM,\lim(80)}}\right)_{52100} = 1.165 - 2.035 \times 10^{-3} T \tag{8.46}$$

$$\left(\frac{\sigma_{VM,\lim(T)}}{\sigma_{VM,\lim(80)}}\right)_{M50} = 1.076 - 9.494 \times 10^{-4} T \tag{8.47}$$

$$\left(\frac{\sigma_{VM,\lim(T)}}{\sigma_{VM,\lim(80)}}\right)_{M50NiL} = 1.079 - 1.040 \times 10^{-3} T \tag{8.48}$$

Equation 8.46 through Equation 8.48 were used in the application performance analyses that generated Table 8.12.

8.9.13 ISO Standard

In Equation 8.21 as presented in Ref. [5], A_{ISO} is used to indicate the "systems approach" life modification factor. Some manufacturers, for example, as in Ref. [43], have substituted their own subscript for "ISO." In this text as in Ref. [34], the integrated stress–life factor has been designated A_{SL}. The ISO standard [5] specifies that A_{ISO} can be expressed as a function of σ_u/σ, the endurance stress limit divided by the real stress, which can include as many influencing

---

TABLE 8.12
Fatigue Limit Stress (von Mises Criterion) for Bearing Materials

Material	$\sigma_{VM,limit}$ MPa (psi)
AISI 52100 CVD steel HR$_c$ 58 minimum	684 (99,200)
SAE 4320/8620 case-hardening steel HR$_c$ 58 minimum	590 (85,500)
VIMVAR M50 steel HR$_c$ 58 minimum	717 (104,000)
VIMVAR M50NiL case-hardening steel HR$_c$ 58 minimum	579 (84,000)
440C stainless steel	400 (58,000)
Induction-hardened steel (wheel bearings)	450 (65,000)
Silicon nitride ceramic	1,220 (177,000)

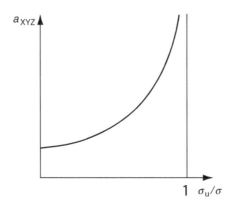

FIGURE 8.33 A_{ISO} vs. σ_u/σ for a given lubrication condition.

stress components as necessary. A_{ISO} vs. σ_u/σ is illustrated by the schematic diagram of Figure 8.33. While the diagram is constructed using normal stresses σ_u and σ, it can also be based on endurance strength in shear, which is the historical criterion for calculating rolling bearing fatigue life; for example, Lundberg and Palmgren [1,2] considered the range of the maximum orthogonal shear stress as the failure-initiating stress. It is noted from Figure 8.33 that A_{ISO}, and hence bearing fatigue life approaches infinity as the real stress σ approaches the endurance limit stress σ_u.

ISO [5] considers that the fatigue-initiating stress is substantially dependent on the internal load distribution in the bearing and the subsurface stresses associated with the loading in the most heavily loaded rolling element–raceway contact. To simplify the calculation of A_{ISO}, ISO introduces the following approximate equivalency:

$$A_{ISO} = f\left(\frac{\sigma_u}{\sigma}\right) \approx f\left(\frac{F_{lim}}{F_e}\right) \tag{8.49}$$

where F_{lim} is the statically applied load of the bearing at which the fatigue limit stress is just reached in the most heavily loaded rolling element–raceway contact. In the determination of F_{lim}, the following influences are considered:

- Bearing type, size, and internal geometry
- Profile of rolling elements and raceways
- Manufacturing quality of the bearing
- Fatigue limit stress of the bearing raceway material

As for the original Lundberg–Palmgren theory and life prediction methods [1,2], rolling element fatigue failure is not considered.

Specific means to calculate F_{lim} for high-quality ball and roller bearings manufactured from through-hardened 52100 steel are provided in an appendix to Ref. [5]. These are based on a maximum contact stress; that is, Hertz stress, of 1500 MPa. It is evident that the ISO standard [5] does not apply directly to bearings manufactured from other high-quality bearing steels.

Ioannides et al. [40] developed charts of A_{ISO} vs. $C_L A \cdot F_{lim}/F_e$ and κ for radial ball bearings, radial roller bearings, thrust ball bearings, and thrust roller bearings. These are provided herein as Figure CD8.16 through Figure CD8.19. Alternatively, A_{ISO} may be calculated using equations provided by the ISO standard [5]; for example,

TABLE 8.13
Constants and Exponents for A_{ISO} Equation 8.50

Bearing Type	Lubricant Film Adequacy	x_1	x_2	e_1	e_2	e_3	e_4
Radial ball	$0.1 \leq \Lambda < 0.4$	2.5671	2.2649	0.054381	0.83	1/3	−9.3
	$0.4 \leq \Lambda < 1$	2.5671	1.9987	0.19087	0.83	1/3	−9.3
	$1 \leq \Lambda \leq 4$	2.5671	1.9987	0.071739	0.83	1/3	−9.3
Radial roller	$0.1 \leq \Lambda < 0.4$	1.5859	1.3993	0.054381	1	0.4	−9.185
	$0.4 \leq \Lambda < 1$	1.5859	1.2348	0.19087	1	0.4	−9.185
	$1 \leq \Lambda \leq 4$	1.5859	1.2348	0.071739	1	0.4	−9.185
Thrust ball	$0.1 \leq \Lambda < 0.4$	2.5671	2.2649	0.054381	0.83	1/3	−9.3
	$0.4 \leq \Lambda < 1$	2.5671	1.9987	0.19087	0.83	1/3	−9.3
	$1 \leq \Lambda \leq 4$	2.5671	1.9987	0.071739	0.83	1/3	−9.3
Thrust roller	$0.1 \leq \Lambda < 0.4$	1.5859	1.3993	0.054381	1	0.4	−9.185
	$0.4 \leq \Lambda < 1$	1.5859	1.2348	0.19087	1	0.4	−9.185
	$1 \leq \Lambda \leq 4$	1.5859	1.2348	0.071739	1	0.4	−9.185

$$A_{ISO} = 0.1 \left[1 - \left(x_1 - \frac{x_2}{\kappa^{e_1}} \right)^{e_2} \left(\frac{C_L F_{\lim}}{F_e} \right)^{e_3} \right]^{e_4} \tag{8.50}$$

The constants x_1 and x_2 and the exponents e_1–e_4 are given in Table 8.13.

See Example 8.4 through Example 8.6.

8.10 CLOSURE

The Lundberg–Palmgren theory to predict fatigue life was a significant advancement in the state-of-the-art of ball and roller bearing technology, affecting the internal design and external dimensions for 40 years. The EHL theory, introduced by Grubin, and further advanced by scores of researchers, initially affected bearing microgeometry, but later, because of the possibility of increased endurance together with improved materials resulted in "downsizing" of ball and roller bearings. The Ioannides–Harris theory, in its ability to apply the total stress pattern to predict life in any bearing application, and in its use of a fatigue stress limit for rolling bearing materials carries the development to the next plateau by substantially increasing understanding of the significance of material quality and concentrated contact surface integrity. It is now apparent that a bearing, manufactured from material that is clean and homogeneous, which operates with its rolling/sliding contacts free from contaminants, and which is not overloaded may survive without fatigue. In fact, Palmgren [59] initially considered the existence of a fatigue limit stress; however, the rolling bearing sets that were tested in the development of the Lundberg–Palmgren theory failed rather completely under the test loading, and he abandoned the concept. During the early 1980s, when the Ioannides–Harris theory was under development, fatigue testing consumed substantial calendar time, often requiring $\frac{1}{2}$ a year and more with no bearing failures after more than 500 million revolutions.

This chapter converts the Ioannides–Harris theory into practice. The life theory is stress-based, as opposed to the factor-based, modified Lundberg–Palmgren theory (standard methods [3–5]) exemplified by Equation 8.16. Rather, the Ioannides–Harris theory utilizes the base Lundberg–Palmgren life equations together with a single factor A_{SL} that integrates the effect on fatigue life of all stresses acting on the bearing contact material. An accurate life prediction for any bearing application depends only on the successful evaluation of the appropriate stresses. With the application of modern computers and computational methods, these stresses are subjected to increasingly greater scrutiny. With the current availability of powerful, inexpensive, desktop and laptop computers, engineers worldwide have the capability to use rolling bearing performance analysis computer programs that can effectively employ the methods described in this text for such analysis.

REFERENCES

1. Lundberg, G. and Palmgren, A., Dynamic capacity of rolling bearings, *Acta Polytech. Mech. Eng. Ser. 1, Roy. Swed. Acad. Eng.*, 3(7), 1947.
2. Lundberg, G. and Palmgren, A., Dynamic capacity of roller bearings, *Acta Polytech. Mech. Eng. Ser. 2, Roy. Swed. Acad. Eng.*, 9(49), 1952.
3. American National Standards Institute, American National Standard (ANSI/ABMA) Std. 9–1990, Load ratings and fatigue life for ball bearings, July 17, 1990.
4. American National Standards Institute, American National Standard (ANSI/ABMA) Std.11–1990, Load ratings and fatigue life for roller bearings, July 17, 1990.
5. International Organization for Standards, International Standard ISO 281, Rolling bearings—dynamic load ratings and rating life, 2007.
6. Harris, T., Prediction of ball fatigue life in a ball/v-ring test rig, *ASME Trans., J. Tribol.*, 119, 365–374, July 1997.
7. Harris, T., How to compute the effects of preloaded bearings, *Prod. Eng.*, 84–93, July 19, 1965.
8. Jones, A. and Harris, T., Analysis of rolling element idler gear bearing having a deformable outer race structure, *ASME Trans., J. Basic Eng.*, 273–277, June 1963.
9. Harris, T. and Broschard, J., Analysis of an improved planetary gear transmission bearing, *ASME Trans., J. Basic Eng.*, 457–462, September 1964.
10. Harris, T., Optimizing the fatigue life of flexibly mounted, rolling bearings, *Lubr. Eng.*, 420–428, October 1965.
11. Harris, T., The effect of misalignment on the fatigue life of cylindrical roller bearings having crowned rolling members, *ASME Trans., J. Lubr. Technol.*, 294–300, April 1969.
12. Tallian, T., Sibley, L., and Valori, R., Elastohydrodynamic film effects on the load–life behavior of rolling contacts, ASME Paper 65-LUBS-11, *ASME Spring Lubr. Symp.*, NY, June 8, 1965.
13. Skurka, J., Elastohydrodynamic lubrication of roller bearings, ASME Paper 69-LUB-18, 1969.
14. Tallian, T., Theory of partial elastohydrodynamic contacts, *Wear*, 21, 49–101, 1972.
15. Harris, T., The endurance of modern rolling bearings, AGMA Paper 269.01, *Am. Gear Manufac. Assoc. Rol. Bear. Symp.*, Chicago, October 26, 1964.
16. Bamberger, E., et al., *Life Adjustment Factors for Ball and Roller Bearings*, AMSE Engineering Design Guide, 1971.
17. Schouten, M., *Lebensduur van Overbrengingen*, TH Eindhoven, November 10, 1976.
18. STLE, *Life Factors for Rolling Bearings*, E. Zaretsky, Ed., 1992.
19. Ville, F. and Nélias, D., Early fatigue failure due to dents in EHL contacts, Presented at the STLE Annual Meeting, Detroit, May 17–21, 1998.
20. Webster, M., Ioannides, E., and Sayles, R., The effect of topographical defects on the contact stress and fatigue life in rolling element bearings, *Proc. 12th Leeds–Lyon Symp. Tribol.*, 207–226, 1986.
21. Hamer, J., Sayles, R., and Ioannides E., Particle deformation and counterface damage when relatively soft particles are squashed between hard anvils, *Tribol. Trans.*, 32(3), 281–288, 1989.
22. Sayles, R., Hamer, J., and Ioannides, E., The effects of particulate contamination in rolling bearings—a state of the art review, *Proc. Inst. Mech. Eng.*, 204, 29–36, 1990.
23. Nélias, D. and Ville, F., Deterimental effects of dents on rolling contact fatigue, *ASME Trans., J. Tribol.*, 122, 1, 55–64, 2000.

24. Xu, G., Sadeghi, F., and Hoeprich, M., Dent initiated spall formation in EHL rolling/sliding contact, *ASME Trans., J. Tribol.*, 120, 453–462, July 1998.

25. Sayles, R. and MacPherson, P., Influence of wear debris on rolling contact fatigue, *ASTM Special Technical Publication 771*, J. Hoo, Ed., 255–274, 1982.

26. Tanaka, A., Furumura, K., and Ohkuna, T., Highly extended life of transmission bearings of "sealed-clean" concept, SAE Technical Paper, 830570, 1983.

27. Needelman, W. and Zaretsky, E., New equations show oil filtration effect on bearing life, *Pow. Transmis. Des.*, 33(8), 65–68, 1991.

28. Barnsby, R., et al., Life ratings for modern rolling bearings, ASME Paper 98-TRIB-57, presented at the ASME/STLE Tribology Conference, Toronto, October 26, 1998.

29. Tallian, T., On competing failure modes in rolling contact, *ASLE Trans.*, 10, 418–439, 1967.

30. Voskamp, A., Material response to rolling contact loading, ASME Paper 84-TRIB-2, 1984.

31. Ioannides, E. and Harris, T., A new fatigue life model for rolling bearings, *ASME Trans., J. Tribol.*, 107, 367–378, 1985.

32. Harris, T. and McCool, J., On the accuracy of rolling bearing fatigue life prediction, *ASME Trans., J. Tribol.*, 118, 297–310, April 1996.

33. Harris, T., Prediction of ball fatigue life in a ball/v-ring test rig, *ASME Trans., J. Tribol.*, 119, 365–374, July 1997.

34. Barnsby, R., et al., *Life Ratings for Modern Rolling Bearings—A Design Guide for the Application of International Standard ISO 281/2*, ASME Publication TRIB-Vol 14, New York, 2003.

35. Juvinall, R. and Marshek, K., *Fundamentals of Machine Component Design*, 2nd ed., Wiley, New York, 1991.

36. Thomas, H. and Hoersch, V., Stresses due the pressure of one elastic solid upon another, *Univ. Illinois, Bull.*, 212, July 15, 1930.

37. Nélias, D., et al., Experimental and theoretical investigation of rolling contact fatigue of 52100 and M50 steels under EHL or micro-EHL conditions, *ASME Trans., J. Tribol.*, 120, 184–190, April 1998.

38. Ahmadi, N., et al., The interior stress field caused by tangential loading of a rectangular patch on an elastic half space, *ASME Trans., J. Tribol.*, 109, 627–629, 1987.

39. Ai, X. and Cheng, H., The influence of moving dent on point EHL contacts, *Tribol. Trans.*, 37(2), 323–335, 1994.

40. Ioannides, E., Bergling, G., and Gabelli, A., An analytical formulation for the life of rolling bearings, *Acta Polytech. Scand.*, Mech. Eng. Series No. 137, Finnish Institute of Technology, 1999.

41. International Organization for Standards, International Standard ISO 4406, Hydraulic fluid power—fluids—method for coding level of contamination by solid particles, 1999.

42. International Organization for Standards, International Standard ISO 4372, Hydraulic fluid power—filters—multi-pass method for evaluating filtration performance, 1981.

43. SKF, *General Catalog 4000 US*, 2nd ed., 1997.

44. Nélias, D., *Contribution a L'etude des Roulements*, Dossier d'Habilitation a Diriger des Recherches, Laboratoire de Mécanique des Contacts, UMR-CNRS-INSA de Lyon No. 5514, December 16, 1999.

45. American National Standards Institute, American National Standard (ABMA/ANSI) Std 7–1972, Shaft and Housing Fits for Metric Radial Ball and Roller Bearings (Except Tapered Roller Bearings) 1972.

46. American National Standards Institute, American National Standard (ABMA/ANSI) Std 19.1–1987, Tapered Roller Bearings-Radial, Metric Design, October 19, 1987.

47. American National Standards Institute, American National Standard (ABMA/ANSI) Std 19.2–1994, Tapered Roller Bearings-Radial, Inch Design, May 12, 1994.

48. Timoshenko, S., *Strength of Materials, Part I, Elementary Theory and Problems*, Van Nostrand, 1955.

49. Society of Automotive Engineers, Residual stress measurements by X-ray diffraction, *SAE J784a*, 2nd ed., New York, 1971.

50. Koistinen, D., The distribution of residual stresses in carburized cases and their origins, *Trans. ASM*, 50, 227–238, 1958.

51. Gentile, A., Jordan, E., and Martin, A., Phase transformations in high-carbon high-hardness steels under contact loads, *Trans. AIME*, 233, 1085–1093, June 1965.
52. Bush, J., Grube, W., and Robinson, G., Microstructural and residual stress changes in hardened steel due to rolling contact, *Trans. ASM*, 54, 390–412, 1961.
53. Muro, H. and Tsushima, N., Microstructural, microhardness and residual stress changes due to rolling contact, *Wear*, 15, 309–330, 1970.
54. Voskamp, A., et al., Gradual changes in residual stress and microstructure during contact fatigue in ball bearings, *Metal. Tech.*, 14–21, January 1980.
55. Zaretsky, E., Parker, R., and Anderson, W., A study of residual stress induced during rolling, *J. Lub. Tech.*, 91, 314–319, 1969.
56. Harris, T., Establishment of a new rolling bearing fatigue life calculation model, Final Report U.S. Navy Contract N00421-97-C-1069, February 23, 2002.
57. Harris, T. and Yu, W.-K., Lundberg–Palmgren fatigue theory: considerations of failure stress and stressed volume, *ASME Trans., J. Tribol.*, 121, 85–89, 1999.
58. Böhmer, H.-J., et al., The influence of heat generation in the contact zone on bearing fatigue behavior, *ASME Trans., J. Tribol.*, 121, 462–467, July 1999.
59. Palmgren, A., The service life of ball bearings, *Zeitschrift des Vereines Deutscher Ingenieure*, 68(14), 339–341, 1924.

9 Statically Indeterminate Shaft–Bearing Systems

LIST OF SYMBOLS

Symbol	Description	Units
a	Distance to load point from right-hand bearing	mm (in.)
A	Distance between raceway groove curvature centers	mm (in.)
D	Rolling element diameter	mm (in.)
d_{m}	Pitch diameter	mm (in.)
$\mathfrak{D}_{\mathrm{o}}$	Outside diameter of shaft	mm (in.)
$\mathfrak{D}_{\mathrm{i}}$	Inside diameter of shaft	mm (in.)
E	Modulus of elasticity	MPa (psi)
F	Bearing radial load	N (lb)
f	r/D	
I	Section moment of inertia	mm^4 ($\mathrm{in.}^4$)
K	Load–deflection constant	$\mathrm{N/mm}^x$ ($\mathrm{lb/in.}^x$)
l	Distance between bearing centers	mm (in.)
$\mathfrak{M}$	Bearing moment load	$\mathrm{N \cdot mm}$ ($\mathrm{in. \cdot lb}$)
P	Applied load at a	N (lb)
Q	Rolling element load	N (lb)
$\mathfrak{R}$	Radius from bearing centerline to raceway groove center	mm (in.)
T	Applied moment load at a	$\mathrm{N \cdot mm}$ ($\mathrm{in. \cdot lb}$)
w	Load per unit length	N/mm (lb/in.)
x	Distance along the shaft	mm (in.)
y	Deflection in the y direction	mm (in.)
z	Deflection in the z direction	mm (in.)
α°	Free contact angle	rad, $^{\circ}$
γ	$\frac{D\cos\alpha}{d_m}$	
δ	Bearing radial deflection	mm (in.)
θ	Bearing angular misalignment	rad, $^{\circ}$
$\Sigma\rho$	Curvature sum	mm^{-1} ($\mathrm{in.}^{-1}$)
ψ	Rolling element azimuth angle	rad, $^{\circ}$

Subscripts

1, 2, 3	Bearing location	
a	Axial direction	
h	Bearing location	

j	Rolling element location
y	y Direction
z	z Direction
xy	xy Plane
xz	xz Plane

Superscript

k Applied load or moment

9.1 GENERAL

In some modern engineering applications of rolling bearings, such as high-speed gas turbines, machine tool spindles, and gyroscopes, the bearings must often be treated as integral to the system to be able to accurately determine shaft deflections and dynamic shaft loading as well as to ascertain the performance of the bearings. Chapter 1 and Chapter 3 detail methods of calculation of rolling element load distribution for bearings subjected to combinations of radial, axial, and moment loadings. These load distributions are affected by the shaft radial and angular deflections at the bearing. In this chapter, equations for the analysis of bearing loading as influenced by shaft deflections will be developed.

9.2 TWO-BEARING SYSTEMS

9.2.1 Rigid Shaft Systems

A commonly used shaft–bearing system involves two angular-contact ball bearings or tapered roller bearings mounted in a back-to-back arrangement as illustrated in Figure 9.1 and Figure 9.2. In these applications, the radial loads on the bearings are generally calculated independently using the statically determinate methods. It may be noticed from Figure 9.1 and Figure 9.2, however, that the point of application of each radial load occurs where the line defining the contact angle intersects the bearing axis. Thus, it can be observed that a back-to-back bearing mounting has a greater length between loading centers than does a face-to-face mounting. This means that the bearing radial loads will tend to be less for the back-to-back mounting.

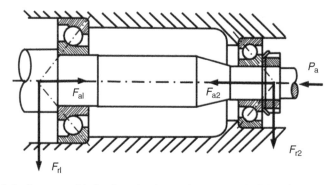

FIGURE 9.1 Rigid shaft mounted in back-to-back angular-contact ball bearings subjected to combined radial and thrust loadings.

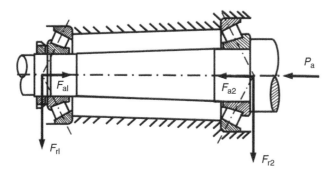

FIGURE 9.2 Rigid shaft mounted in back-to-back tapered roller bearings subjected to combined radial and thrust loadings.

The axial or thrust load carried by each bearing depends on the internal load distribution in the individual bearing. For simple thrust loading of the system, the method illustrated in Example 9.3 may be applied to determine the axial loading in each bearing. When each bearing must carry both radial and axial loads, although the system is statically indeterminate, for systems in which the shaft may be considered rigid, a simplified method of analysis may be employed. In Chapter 11 of the first volume of this handbook, it is demonstrated that a bearing subjected to combined radial and axial loading may be considered to carry an equivalent load defined by

$$F_e = XF_r + YF_a \tag{9.1}$$

Loading factors X and Y are functions of the free contact angle, which for this calculation is assumed invariant with rolling element azimuth location and unaffected by applied load. This condition is true for tapered roller bearings; however, as shown in Chapter 1, it is only approximated for ball bearings. Values for X and Y are usually provided for each ball bearing and tapered roller bearing in manufacturers' catalogs. Accordingly, assuming radial loads F_{r1} and F_{r2} are determined using statically determinate calculation methods, the bearing axial loads F_{a1} and F_{a2} may be approximated considering the following conditions:

If load condition (1) is defined by

$$\frac{F_{r2}}{Y_2} < \frac{F_{r1}}{Y_1}$$

and load condition (2) is defined by

$$\frac{F_{r2}}{Y_2} > \frac{F_{r1}}{Y_1} \quad P_a \geq \frac{1}{2}\left(\frac{F_{r2}}{Y_2} - \frac{F_{r1}}{Y_1}\right)$$

then,

$$F_{a1} = \frac{F_{r1}}{2Y_1} \tag{9.2}$$

$$F_{a2} = F_{a1} + P_a \tag{9.3}$$

If load condition (3) is defined by

$$\frac{F_{r2}}{Y_2} > \frac{F_{r1}}{Y_1} \quad P_a < \frac{1}{2}\left(\frac{F_{r2}}{Y_2} - \frac{F_{r1}}{Y_1}\right)$$

then,

$$F_{a2} = \frac{F_{r2}}{2Y_2} \tag{9.4}$$

$$F_{a1} = F_{a2} - P_a \tag{9.5}$$

See Example 9.1 and Example 9.2.

9.2.2 FLEXIBLE SHAFT SYSTEMS

In the more general two-bearing shaft system, flexure of the shaft induces moment loads $\mathfrak{M}_h$ at non-self-aligning bearing supports in addition to the radial loads F_h. This loading system, illustrated in Figure 9.3, is statically indeterminate in that there are four unknowns: F_1, F_2, $\mathfrak{M}_1$, and $\mathfrak{M}_2$; but, only two static equilibrium equations. For example,

$$\sum F = 0 \quad F_1 + F_2 - P = 0 \tag{9.6}$$

$$\sum M = 0 \quad F_1 l - \mathfrak{M}_1 + T - P(l - a) + \mathfrak{M}_2 = 0 \tag{9.7}$$

Considering the bending of the shaft, the bending moment at any section is given as follows:

$$EI \frac{d^2 y}{dx^2} = -\mathfrak{M} \tag{9.8}$$

where E is the modulus of elasticity, I is the shaft cross-section moment of inertia, and y is the shaft deflection at the section. For shafts that have circular cross-sections,

$$I = \frac{\pi}{64}\left(\mathfrak{D}_o^4 - \mathfrak{D}_i^4\right) \tag{9.9}$$

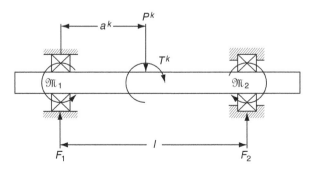

FIGURE 9.3 Statically indeterminate two-bearing shaft system.

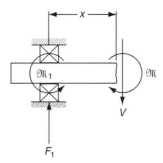

FIGURE 9.4 Statically indeterminate two-bearing shaft system forces and moments acting on a section to the left of the load application point.

For a cross-section at $0 \leq x \leq a$ illustrated in Figure 9.4,

$$EI \frac{d^2 y}{dx^2} = -F_1 x + \mathfrak{M}_1 \tag{9.10}$$

Integrating Equation 9.10 yields

$$EI \frac{dy}{dx} = -\frac{F_1 x^2}{2} + \mathfrak{M}_1 x + C_1 \tag{9.11}$$

Integrating Equation 9.11 yields

$$EIy = -\frac{F_1 x^3}{6} + \frac{\mathfrak{M}_1 x^2}{2} + C_1 x + C_2 \tag{9.12}$$

In Equation 9.11 and Equation 9.12, C_1 and C_2 are constants of integration. At $x = 0$, the shaft assumes the bearing deflection δ_{r1}. Also at $x = 0$, the shaft assumes a slope θ_1 in accordance with the resistance of the bearing to moment loading; hence,

$$C_1 = EI\theta_1$$

$$C_2 = EI\delta_{r1}$$

Therefore, Equation 9.11 and Equation 9.12 become

$$EI \frac{dy}{dx} = -\frac{F_1 x^2}{2} + \mathfrak{M}_1 x + EI\theta_1 \tag{9.13}$$

and

$$EIy = -\frac{F_1 x^3}{6} + \frac{\mathfrak{M}_t x^3}{2} + EI\theta_1 x + EI\delta_{r1} \tag{9.14}$$

For a cross-section at $a \leq x \leq l$ as shown in Figure 9.5,

$$EI \frac{d^2 y}{dx^2} = -F_1 x + \mathfrak{M}_1 + P(x - a) - T \tag{9.15}$$

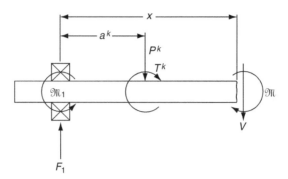

FIGURE 9.5 Statically indeterminate two-bearing shaft system forces and moments acting on a section to the right of the load application.

Integrating Equation 9.15 twice yields

$$EI\frac{dy}{dx} = -\frac{F_1 x^2}{2} + (\mathfrak{M}_1 - T)x + Px\left(\frac{x}{2} - a\right) + C_3 \tag{9.16}$$

$$EIy = -\frac{F_1 x^3}{6} + (\mathfrak{M}_1 - T)\frac{x^2}{2} + Px^2\left(\frac{x}{6} - \frac{a}{2}\right) + C_3 x + C_4 \tag{9.17}$$

At $x = l$, the slope of the shaft is θ_2 and the deflection is δ_{r2}, therefore,

$$EI\frac{dy}{dx} = \frac{F_1(l^2 - x^2)}{2} + (T - \mathfrak{M}_1)(l - x) + \frac{P}{2}[x(x - 2a) - l(l - 2a)] + EI\theta_2 \tag{9.18}$$

$$EIy = -\frac{F_1}{6}[l^2(2l - 3x) + x^3] + \frac{(\mathfrak{M}_1 - T)}{2}(l - x)^2 + \frac{P}{6}[x^2(x - 3a) \\ - l^2(3x + 3a - 2l) + 6xla] + EI[\delta_{r2} - \theta_2(l - x)] \tag{9.19}$$

At $x = a$, singular conditions of slope and deflection occur. Therefore at $x = a$, Equation 9.13 and Equation 9.18 are equivalent as are Equation 9.14 and Equation 9.19. Solving the resultant simultaneous equations yields

$$F_1 = \frac{P(l - a)^2(l + 2a)}{l^3} - \frac{6Ta(l - a)}{l^3} - \frac{6EI}{l^2}\left[\theta_1 + \theta_2 + \frac{2(\delta_{r1} - \delta_{r2})}{l}\right] \tag{9.20}$$

$$\mathfrak{M}_1 = \frac{Pa(l - a)^2}{l^2} + \frac{T(l - a)(l - 3a)}{l^2} - \frac{2EI}{l}\left[2\theta_1 + \theta_2 + \frac{3(\delta_{r1} - \delta_{r2})}{l}\right] \tag{9.21}$$

Substituting Equation 9.20 and Equation 9.21 in Equation 9.6 and Equation 9.7 yields

$$F_2 = \frac{Pa^2(3l - 2a)}{l^3} + \frac{6Ta(l - a)}{l^3} + \frac{6EI}{l^2}\left[\theta_1 + \theta_2 + \frac{2(\delta_{r1} - \delta_{r2})}{l}\right] \tag{9.22}$$

$$\mathfrak{M}_2 = \frac{Pa^2(l - a)}{l^2} + \frac{Ta(2l - 3a)}{l^2} + \frac{2EI}{l}\left[\theta_1 + 2\theta_2 + \frac{3(\delta_{r1} - \delta_{r2})}{l}\right] \tag{9.23}$$

In Equation 9.20 through Equation 9.23, slope θ_1 and δ_{r1} are considered positive and the signs of θ_2 and δ_{r2} may be determined from Equation 9.18 and Equation 9.19. The relative magnitudes of P and T and their directions will determine the sense of the shaft slopes at the bearings. To determine the reactions, it is necessary to develop equations relating bearing misalignment angles θ_h to the misaligning moments $\mathfrak{M}_h$ and bearing radial deflections δ_{rh} to loads F_h. This may be done by using the data of Chapter 1 and Chapter 3.

When the bearings are considered as axially free pin supports, Equation 9.20 and Equation 9.22 are identical to Equation 4.29 and Equation 4.30, given in the first volume of this handbook for a statically determinate system. That format is obtained by setting $\mathfrak{M}_h = \delta_{rh} = 0$ and solving Equation 9.21 and Equation 9.23 simultaneously for θ_1 and θ_2. Substitution of these values in Equation 9.20 and Equation 9.22 produces the resultant equations. If the shaft is very flexible and the bearings are very rigid with regard to misalignment, then $\theta_1 = \theta_2 = 0$. This substitution in Equation 9.20 through Equation 9.23 yields the classical solution for a beam with both ends built in. The various types of two-bearing support may be examined by using Equation 9.20 through Equation 9.23. If more than one load or torque is applied between the supports, then by the principle of superposition

$$F_1 = \frac{1}{l^3} \sum_{k=1}^{k=n} P^k \left(l - a^k \right)^2 \left(l + 2a^k \right) - \frac{6}{l^3} \sum_{k=1}^{k=n} T^k a^k \left(l - a^k \right) - \frac{6EI}{l^2} \left[\theta_1 + \theta_2 + \frac{2(\delta_{r1} - \delta_{r2})}{l} \right] \quad (9.24)$$

$$\mathfrak{M}_1 = \frac{1}{l^3} \sum_{k=1}^{k=n} P^k a^k \left(l - a^k \right)^2 + \frac{1}{l^2} \sum_{k=1}^{k=n} T^k \left(l - a^k \right) \left(l - 3a^k \right) - \frac{2EI}{l} \left[2\theta_1 + \theta_2 + \frac{3(\delta_{r1} - \delta_{r2})}{l} \right] \quad (9.25)$$

$$F_2 = \frac{1}{l^3} \sum_{k=1}^{k=n} P^k \left(a^k \right)^2 \left(3l - 2a^k \right) + \frac{6}{l^3} \sum_{k=1}^{k=n} T^k a^k \left(l - a^k \right) + \frac{6EI}{l^2} \left[\theta_1 + \theta_2 + \frac{2(\delta_{r1} - \delta_{r2})}{l} \right] \quad (9.26)$$

$$\mathfrak{M}_2 = \frac{1}{l^3} \sum_{k=1}^{k=n} P^k \left(a^k \right)^2 \left(l - a^k \right) + \frac{1}{l^2} \sum_{k=1}^{k=n} T^k a^k \left(2l - 3a^k \right) + \frac{2EI}{l} \left[\theta_1 + 2\theta_2 + \frac{3(\delta_{r1} - \delta_{r2})}{l} \right] \quad (9.27)$$

See Example 9.3.

9.3 THREE-BEARING SYSTEMS

9.3.1 Rigid Shaft Systems

When the shaft is rigid and the distance between bearings is small, the influence of the shaft deflection on the distribution of loading among the bearings may be neglected. An application of this kind is illustrated in Figure 9.6.

In this system, the angular-contact ball bearings are considered as one double-row bearing. The thrust load acting on the double-row bearing is the thrust load P_a applied by the bevel gear. To calculate the magnitude of the radial loads F_r and F_{r3}, the effective point of application of F_r must be determined. F_r acts at the center of the double-row bearing only if $P_a = 0$. If a thrust load exists, the line of action of F_r is displaced toward the pressure center of the rolling element row that supports the thrust load. This displacement may be neglected only if the distance l between the center of the double-row ball bearing set and the roller bearing is large compared with the distance b. Using the X and Y factors (see Equation 9.1), pertaining to the single-row bearings, Figure 9.7 gives the relative distance b_1/b as a function of the parameter $F_a Y/F_r$ $(1 - X)$. The X and Y factors for the load condition $F_a/F_r > e$ must be selected from the bearing catalog.

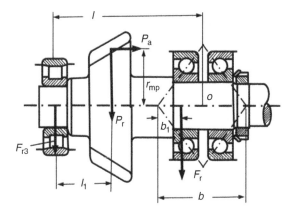

FIGURE 9.6 Example of three-bearing shaft system with a rigid shaft.

See Example 9.4.

9.3.2 Nonrigid Shaft Systems

The generalized loading of a three-bearing-shaft support system is illustrated in Figure 9.8a.

This system may be reduced to the two systems of Figure 9.8b and analyzed according to the methods given previously for a two-bearing nonrigid shaft system provided that

$$F_2' + F_2'' = F_2 \tag{9.28}$$

$$\mathfrak{M}_2' - \mathfrak{M}_2'' = \mathfrak{M}_2 \tag{9.29}$$

Hence, from Equation 9.24 through Equation 9.27,

$$F_1 = \frac{1}{l_1^3} \sum_{k=1}^{k=n} P_1^k (l_1 - a_1^k)^2 (l_1 + 2a_1^k) - \frac{6}{l_1^3} \sum_{k=1}^{k=n} T_1^k a_1^k (l_1 - a_1^k) - \frac{6EI_1}{l_1^2} \left[\theta_1 + \theta_2 + \frac{3(\delta_{r1} - \delta_{r2})}{l_1} \right] \tag{9.30}$$

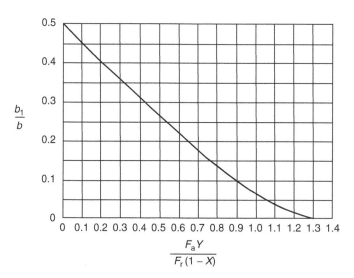

FIGURE 9.7 b_1/b vs. $F_a Y/F_r (1 - X)$ for the double-row bearing in a three-bearing rigid shaft system.

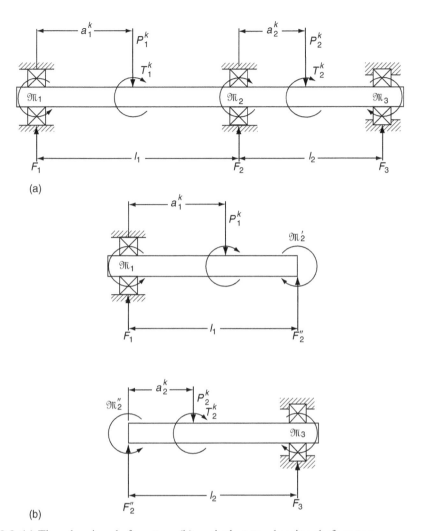

FIGURE 9.8 (a) Three-bearing shaft system; (b) equivalent two-bearing shaft system.

$$\mathfrak{M}_1 = \frac{1}{l_1^2} \sum_{k=1}^{k=n} P_1^k a_1^k (l_1 - a_1^k)^2 + \frac{1}{l_1^2} \sum_{k=1}^{k=n} T_1^k (l_1 - a_1^k)(l_1 - 3a_1^k) - \frac{2EI_1}{l_1}\left[2\theta_1 + \theta_2 + \frac{3(\delta_{r1} - \delta_{r2})}{l_1} \right]$$

(9.31)

$$
\begin{aligned}
F_2 = {} & \frac{1}{l_1^3} \sum_{k=1}^{k=n} P_1^k (a_1^k)^2 (3l_1 - 2a_1^k) + \frac{1}{l_2^3} \sum_{k=1}^{k=n} P_2^k (l_2 - a_2^k)(l_2 + 2a_2^k) \\
& + \frac{6}{l_1^3} \sum_{k=1}^{k=n} T_1^k a_1^k (l_1 - a_1^k) - \frac{6}{l_2^3} \sum_{k=1}^{k=n} T_2^k a_2^k (l_2 - a_2^k) \\
& + 6E\left[\frac{I_1}{l_1^2}(\theta_1 + \theta_2) - \frac{I_2}{l_2^2}(\theta_2 + \theta_3) \right] \\
& + 12E\left[\frac{I_1}{l_1^3}(\delta_{r1} - \delta_{r2}) - \frac{I_2}{l_2^3}(\delta_{r2} - \delta_{r3}) \right]
\end{aligned}
$$

(9.32)

$$\mathfrak{M}_2 = \frac{1}{l_1^2} \sum_{k=1}^{k=n} P_1^k (a_1^k)^2 (l_1 - a_1^k) - \frac{1}{l_2^2} \sum_{k=1}^{k=n} P_2^k a_2^k (l_2 - a_2^k)$$

$$+ \frac{1}{l_1^2} \sum_{k=1}^{k=n} T_1^k a_1^k (2l_1 - 3a_1^k) - \frac{1}{l_2^2} \sum_{k=1}^{k=n} T_2^k (l_2 - a_2^k)(l_2 - 3a_2^k)$$

$$+ 2E \left[\frac{I_1}{l_1} (\theta_1 + 2\theta_2) + \frac{I_2}{l_2} (2\theta_2 + \theta_3) \right]$$

$$+ 6E \left[\frac{I_1}{l_1^2} (\delta_{r1} - \delta_{r2}) + \frac{I_2}{l_2^2} (\delta_{r2} - \delta_{r3}) \right]$$

$$(9.33)$$

$$F_3 = \frac{1}{l_2^3} \sum_{k=1}^{k=n} P_2^k (a_2^k)^2 (3l_2 - 2a_2^k) + \frac{6}{l_2^3} \sum_{k=1}^{k=n} T_2^k a_2^k (l_2 - a_2^k) + \frac{6EI_2}{l_2^2} \left[\theta_2 + \theta_3 + \frac{2(\delta_{r2} - \delta_{r3})}{l_2} \right] \quad (9.34)$$

$$\mathfrak{M}_3 = \frac{1}{l_2^2} \sum_{k=1}^{k=n} P_2^k (a_2^k)^2 (l_2 - a_2^k) + \frac{1}{l_2^2} \sum_{k=1}^{k=n} T_2^k a_2^k (2l_2 - 3a_2^k) + \frac{2EI_2}{l_2} \left[\theta_2 + 2\theta_3 + \frac{3(\delta_{r2} - \delta_{r3})}{l_2} \right]$$

$$(9.35)$$

An example of the utility of the generalized equations Equation 9.30 through Equation 9.35 is the system illustrated in Figure 9.9. For that system, it is assumed that moment loads are zero and that the differences between bearing radial deflections are negligibly small. Hence, Equation 9.30 through Equation 9.35 become

$$F_1 = \frac{P(l_1 - a)^2 (l_1 + 2a)}{l_1^3} - \frac{6EI}{l_1^2} (\theta_1 + \theta_2) \quad (9.36)$$

$$2\theta_1 + \theta_2 = \frac{Pa(l_1 - a)^2}{2EIl_1} \quad (9.37)$$

$$F_2 = \frac{Pa^2 (3l_1 - 2a)}{l_1^3} + 6EI \left[\frac{(\theta_1 + \theta_2)}{l_1^2} - \frac{(\theta_2 + \theta_3)}{l_2^2} \right] \quad (9.38)$$

$$\frac{(\theta_1 + 2\theta_2)}{l_1} + \frac{(2\theta_2 + \theta_3)}{l_2} = -\frac{Pa^2 (l_1 - a)}{2EIl_1^3} \quad (9.39)$$

$$F_3 = \frac{6EI(\theta_2 + \theta_3)}{l_2^2} \quad (9.40)$$

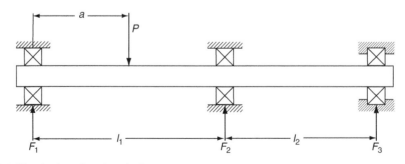

FIGURE 9.9 Simple three-bearing shaft system.

$$\theta_2 + 2\theta_3 = 0 \tag{9.41}$$

Equation 9.37, Equation 9.39, and Equation 9.4 can be solved for θ_1, θ_2, and θ_3. Subsequent substitution of these values in Equation 9.36, Equation 9.38, and Equation 9.40 yields the following result:

$$F_1 = \frac{P(l_1 - a)[2l_1(l_1 + l_2) - a(l_1 + a)]}{2l_1^2(l_1 + l_2)} \tag{9.42}$$

$$F_2 = \frac{Pa[(l_1 + l_2)^2 - a^2 - l_2^2]}{2l_1^2 l_2} \tag{9.43}$$

$$F_3 = \frac{-Pa(l_1^2 - a^2)}{2l_1 l_2(l_1 + l_2)} \tag{9.44}$$

9.3.2.1 Rigid Shafts

When the distances between bearings are small or the shaft is otherwise very stiff, the bearing radial deflections determine the load distribution among the bearings. From Figure 9.10, it can be seen that by considering similar triangles

$$\frac{\delta_{r1} - \delta_{r2}}{l_1} = \frac{\delta_{r2} - \delta_{r3}}{l_2} \tag{9.45}$$

This identical relationship can be obtained from Equation 9.30 through Equation 9.35 by setting shaft cross-section moment of inertia I to an infinitely large value. For a radially loaded bearing with rigid rings, the maximum rolling element load is directly proportional to the applied radial load F_r, and the maximum rolling element deflection determines the bearing radial deflection. Since rolling element load $Q = K\delta^n$, therefore,

$$F_r = K\delta_r^n \tag{9.46}$$

Rearranging Equation 9.46,

$$\delta_r = \left(\frac{F_r}{K}\right)^{1/n} \tag{9.47}$$

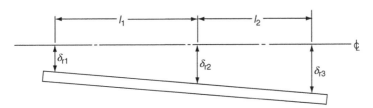

FIGURE 9.10 Deflection of a three-bearing shaft system with a rigid shaft.

Substitution of Equation 9.47 in Equation 9.45 yields

$$\left(\frac{F_{r1}}{K_1}\right)^{1/n} - \left(\frac{F_{r2}}{K_2}\right)^{1/n} = \frac{l_1}{l_2}\left[\left(\frac{F_{r2}}{K_2}\right)^{1/n} - \left(\frac{F_{r3}}{K_3}\right)^{1/n}\right] \tag{9.48}$$

Equation 9.48 is valid for bearings that support a radial load only. More complex relationships are required in the presence of simultaneous applied thrust and moment loading. Equation 9.48 can be solved simultaneously with the equilibrium equations to yield values of F_{r1}, F_{r2}, and F_{r3}.

See Example 9.5.

9.4   MULTIPLE-BEARING SYSTEMS

Equation 9.30 through Equation 9.35 may be used to determine the bearing reactions in a multiple-bearing system such as that shown in Figure 9.11 with a flexible shaft. It is evident that the reaction at any bearing support location h is a function of the loading existing at and in between the bearing supports located at $h-1$ and $h+1$. Therefore, from Equation 9.30 through Equation 9.35, the reactive loads at each support location h are given as follows:

$$
\begin{aligned}
F_h ={}& \frac{1}{l_{h-1}^3}\sum_{k=1}^{k=p}P_{h-1}^k(a_{h-1}^k)^2(3l_{h-1}-2a_{h-1}) \\
&+\frac{1}{l_h^3}\sum_{k=1}^{k=q}P_h^k(l_h-a_h^k)^2(l_h+2a_h^k) \\
&+\frac{6}{l_{h-1}^3}\sum_{k=1}^{k=r}T_{h-1}^k a_{h-1}^k(l_{h-1}-a_{h-1}^k) \\
&-\frac{6}{l_h^3}\sum_{k=1}^{k=s}T_h^k a_h^k(l_h-a_h^k) \\
&+6E\left\{\frac{I_{h-1}}{l_{h-1}^2}\left[\theta_{h-1}+\theta_h+\frac{2}{l_{h-1}}(\delta_{r,h-1}-\delta_{r,h})\right]\right. \\
&\left.-\frac{I_h}{l_h^2}\left[\theta_h+\theta_{h+1}+\frac{2}{l_h}(\delta_{r,h}-\delta_{r,h+1})\right]\right\}
\end{aligned}
\tag{9.49}
$$

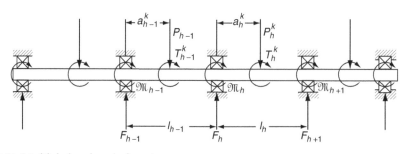

FIGURE 9.11 Multiple-bearing shaft system.

$$\mathfrak{M}_h = \frac{1}{l_{h-1}^2} \sum_{k=1}^{k=p} P_{h-1}^k (a_{h-1}^k)^2 (l_{h-1} - a_{h-1}^k) - \frac{1}{l_h^2} \sum_{k=1}^{k=q} P_h^k a_h^k (l_h - a_h^k)^2$$

$$+ \frac{1}{l_{h-1}^2} \sum_{k=1}^{k=r} T_{h-1}^k a_{h-1}^k (2l_{h-1} - 3a_{h-1}^k)$$

$$- \frac{1}{l_h^2} \sum_{k=1}^{k=s} T_h^k (l_h - a_h^k)(l_h - 3a_h^k)$$ (9.50)

$$+ 2E \left\{ \frac{I_{h-1}}{l_{h-1}} \left[\theta_{h-1} + 2\theta_h + \frac{3}{l_{h-1}} (\theta_{r,h-1} - \theta_{r,h}) \right] \right.$$

$$\left. + \frac{I_h}{l_h} \left[2\theta_h + \theta_{h+1} + \frac{3}{l_h} (\delta_{r,h} - \delta_{r,h+1}) \right] \right\}$$

For a shaft–bearing system of n supports, that is, $h = n$, Equation 9.49 and Equation 9.50 represent a system of $2n$ equations. In the most elementary case, all bearings are considered as sufficiently self-aligning such that all $\mathfrak{M}_h$ equal zero; furthermore, all $\delta_{r,h}$ are considered negligible compared with shaft deflection. Equation 9.49 and Equation 9.50 thereby degenerate to the familiar equation of "three moments."

It is evident that the solution of Equation 9.49 and Equation 9.50 to obtain bearing reactions $\mathfrak{M}_h$ and F_h depends on relationships between radial load and radial deflection and moment load and misalignment angle for each radial bearing in the system. These relationships have been defined in Chapter 1 and Chapter 3. Thus, for a very sophisticated solution to a shaft–bearing problem as illustrated in Figure 9.12 one could consider a shaft that has two degrees of freedom with regard to bending, that is, deflection in two of three principal directions, supported by bearings h and accommodating loads k. At each bearing location h, one must establish the following relationships:

$$\delta_{y,h} = f_1(F_{x,h}, F_{y,h}, F_{z,h}, \mathfrak{M}_{xy,h}, \mathfrak{M}_{xz,h})$$ (9.51)

$$\delta_{z,h} = f_2(F_{x,h}, F_{y,h}, F_{z,h}, \mathfrak{M}_{xy,h}, \mathfrak{M}_{xz,h})$$ (9.52)

$$\theta_{xy,h} = f_3(F_{x,h}, F_{y,h}, F_{z,h}, \mathfrak{M}_{xy,h}, \mathfrak{M}_{xz,h})$$ (9.53)

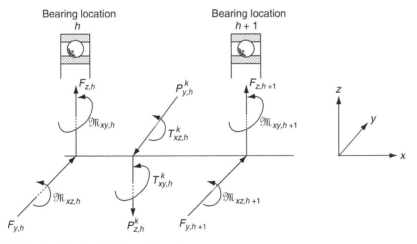

FIGURE 9.12 System loading in three dimensions.

$$\theta_{xz,h} = f_4(F_{x,h}, F_{y,h}, F_{z,h}, \mathfrak{M}_{xy,h}, \mathfrak{M}_{xz,h}) \tag{9.54}$$

To accommodate the movement of the shaft in two principal directions, the following expressions will replace Equation 3.72 and Equation 3.73 for each ball bearing (see Ref. [1]):

$$S_{xj} = BD \sin \alpha^\circ + \delta_x + \theta_{xz} \mathfrak{R}_i \sin \psi_j + \theta_{xy} \mathfrak{R}_i \cos \psi_j \tag{9.55}$$

$$S_{zj} = BD \cos \alpha^\circ + \delta_y \sin \psi_j + \delta_j \cos \psi_j \tag{9.56}$$

9.5 CLOSURE

For most rolling bearing applications, it is sufficient to consider the shaft and housing as rigid structures. As demonstrated in Example 9.3, however, when the shaft is considerably hollow and the span between bearing supports is sufficiently great, the shaft bending characteristics cannot be considered separately from the bearing deflection characteristics with the expectation of accurately ascertaining the bearing loads or the overall system deflection characteristics. In practice, the bearings may be stiffer than might be anticipated by the simple deflection formulas or even stiffer than a more elegant solution that employs accurate evaluation of load distribution might predict for the assumed loading. The penalty for increased stiffness will be paid in shortened bearing life since the improved stiffness is obtained at the expense of induced moment loading.

It is of interest to note that the accurate determination of bearing loading in integral shaft–bearing–housing systems involves the solution of many simultaneous equations. For example, in a high-speed shaft supported by three ball bearings, each of which has a complement of 10 balls, the shaft being loaded so as to cause each bearing to experience five degrees of freedom in deflection requires the solution of 142 simultaneous equations, most of which are nonlinear in the variables to be determined. Most likely, the system would include some roller bearings, these having complements of 20 or more rollers per row, thus adding to the number of equations to be solved simultaneously. Furthermore, the bearing outer rings and inner rings may be flexibly supported as in aircraft power transmissions, adding to the complexity of the analytical system and the difficulty of obtaining a solution using numerical analysis techniques such as the Newton–Raphson method for simultaneous, nonlinear equations.

REFERENCE

1. Jones, A., A general theory of elastically constrained ball and radial roller bearings under arbitrary load and speed conditions, *ASME Trans.*, *J. Basic Eng.*, 82, 309–320, 1960.

10 Failure and Damage Modes in Rolling Bearings

10.1 GENERAL

Although ball and roller bearings appear to be relatively simple mechanisms, their internal operations are relatively complex as witnessed by the number of pages devoted by *Rolling Bearing Analysis, 5th Ed.* to the evaluation of their design and operation. It has been established in these pages that rolling bearings can perform in many applications without interruption of successful operation, provided:

- The bearing selected for the given application is of correct design and sufficient size.
- The bearing is properly mounted on the shaft and in the housing.
- The bearing lubrication system is of proper design; the lubricant film thicknesses generated are sufficient to adequately separate the rolling contact surfaces; and the amount of lubricant supplied is sufficient.
- Lubrication of rolling element–cage and cage–bearing ring land interfaces is adequate.
- The bearing is operated at speeds consistent with the lubrication method such that overheating is prevented.
- The bearing is protected from the ingress of contaminants.

It has also been established that, in many applications, it is possible to accommodate these conditions.

In some applications, however, the conditions for application design functional performance and endurance are not met due to:

- Extreme operating conditions of heavy or complex loading, very high speed or accelerations, and very high or very low operating temperatures to cite a few

and, perhaps,

- Insufficient attention to proper machinery assembly and operating practice.

Operation under such conditions will very frequently lead to early bearing failure, and possibly early machinery failure.

As implied above, rolling bearing failure can be defined as not meeting the design requirements of the application. Thus, failure can manifest itself as:

1. Excessive deflection
2. Excessive vibration or noise

3. Unacceptably high friction torque and temperature, or
4. Bearing seizure

Actually, conditions 1–3, singly or in combination, may lead to the last. Very likely, conditions 1–3 are the result of damage to the rolling contact surfaces. The likelihood of such damage in a given application can, in the best circumstance, be predicted and consequently avoided using the analytical techniques contained in this text. In the worst instance, the reason(s) for early failure can be found through such analyses.

The purpose of this chapter is to elucidate the various types of damage and failure that may occur in rolling bearing applications and to connect these to the physical phenomena that cause them.

10.2 BEARING FAILURE DUE TO FAULTY LUBRICATION

10.2.1 INTERRUPTION OF LUBRICANT SUPPLY TO BEARINGS

Most ball and roller bearing failures are caused by interruption of the lubricant supply to the bearing or inadequate delivery of the lubricant to the rolling element–raceway contacts in the first place. In the aircraft gas turbine engine mainshaft application, in which engine failure is considered life-critical, ball and roller bearing cages are coated with silver. In the event of temporary loss of lubricant supplied to the bearings, some silver is transferred to the rolling element surfaces, providing increased lubricity and lower coefficient of friction than steel-on-steel. Also, in the latter instance, bearings that have silicon nitride rolling elements experience lower friction in both the rolling element–raceway and rolling element–cage contacts than do bearings that have steel rolling elements.

10.2.2 THERMAL IMBALANCE

During operation of ball and roller bearings, it is important that the temperature gradient between the bearing inner and outer raceways is maintained such that radial preloading does not occur. This condition leads to increased rolling element–raceway loading, increased friction, and increased temperatures. If the rate of heat dissipation from the bearing outer ring is greater than that from the inner ring, a temperature excursion occurs, resulting in bearing seizure. Heat generation in other components of an application is frequently greater than that generated by bearing operation; for example, the heat generated by the windings in an electric motor. In this case, it is important that the paths for heat transfer are designed such that the temperature gradient across the bearing does not result in a thermal excursion.

High friction is also caused by excessive amounts of sliding in a bearing. This condition can occur as a result of rolling–raceway contacts that operate in the boundary lubrication regime. In other words, the lubricant film thicknesses formed in the rolling element–raceway contacts do not sufficiently separate the rolling/sliding components, allowing the interaction of surface asperities on the contacting bodies. High friction also occurs in solid-film-lubricated bearings; for example, bearings lubricated with molybdenum disulfide.

The first stage of excessive friction heat generation is lubricant oxidation and degradation. In this case, the lubricant changes to darker colors, eventually becoming black and having even greater friction; see Figure 10.1. Lubricant overheating and oxidation can also lead to chemical deposits on, and discoloration of, rolling elements as shown in Figure 10.2 as well as rings as illustrated in Figure 10.3 and cage in Figure 10.4.

As the bearing component temperatures increase, the hardness of bearing ring and rolling element steels decreases, giving rise to loss of elasticity and resulting in plastic deformations

FIGURE 10.1 Grease-lubricated ball bearing showing lubricant oxidation. (Courtesy NTN.)

(see Figure 10.5 through Figure 10.7). Ultimately, heat imbalance failure leads to breakage of bearing components and bearing seizure as illustrated in Figure 10.8 through Figure 10.10. Bearing seizure is obviously a complete loss of bearing function and, most likely, machinery function. This type of failure can be catastrophic in life-critical applications; for example, automobile wheel bearings and helicopter power transmission bearings to name a few.

FIGURE 10.2 Hard organic coating on balls formed by grease polymerization due to high temperature caused by sliding under high contact stresses.

FIGURE 10.3 Discoloration and oxidation of bearing ring due to overheating of bearing during operation.

10.3 FRACTURE OF BEARING RINGS DUE TO FRETTING

For applications involving shaft rotation, bearing inner rings are usually press-fitted or shrink-fitted on the shaft to prevent ring rotation about the shaft due to operation under applied loading. For outer ring rotation applications such as automobile wheel bearings, the bearing outer ring is usually mounted in its housing with an interference fit to prevent ring rotation relative to the housing during bearing operation. The inner ring rotation about the shaft or the outer ring rotation relative to the housing is generally a small intermittent motion occasioned by the circumferential spacing of the rolling elements. If the interference fitting is insufficient to prevent this motion, a condition called fretting occurs. Fretting is a chemical

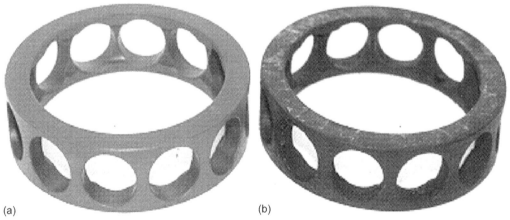

(a)   (b)

FIGURE 10.4 Machine tool ball bearing phenolic cage: (a) original color and (b) discolored and oxidized due to overheating of bearing during operation.

FIGURE 10.5 Tapered roller bearing—deformed cone, rollers, and cage due to heat imbalance failure.

attack on the surfaces in relative motion, and it entails localized removal of material called fretting corrosion or fretting wear. Figure 10.10 and Figure 10.11 show bearing rings with fretting. This corrosion or wear can result in ring cracking as shown in Figure 10.12. Hence, fretting corrosion on bearing ring surfaces can lead to loss of bearing function and potential catastrophic failure.

FIGURE 10.6 Spherical roller bearing—deformed inner raceways and rollers due to heat imbalance failure.

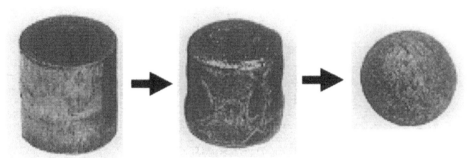

FIGURE 10.7 Cylindrical roller bearing—transformation of rollers into balls due to heat imbalance failure.

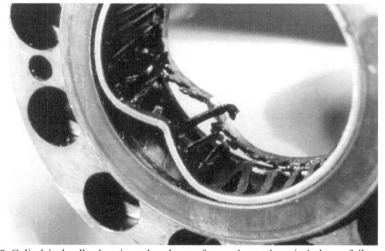

FIGURE 10.8 Cylindrical roller bearing—breakage of cage due to heat imbalance failure.

FIGURE 10.9 Deep-groove ball bearing—breakage of cage and balls due to heat imbalance failure.

FIGURE 10.10 Fretting corrosion in the bore of a bearing inner ring.

10.4 BEARING FAILURE DUE TO EXCESSIVE THRUST LOADING

Excessive thrust loading in a ball bearing can cause the balls to ride over the ring land as shown in Figure 10.13. This causes the raceway area to be truncated resulting in much higher contact stress, much higher surface friction shear stress, and rapid bearing failure due to

FIGURE 10.11 Fretting corrosion on the outside diameter of a bearing outer ring.

FIGURE 10.12 Cracking of a ball bearing outer ring due to fretting corrosion.

overheating. Figure 10.14 provides a postmortem view of the inner and outer raceway patterns.

Excessive thrust loading in tapered roller and spherical roller bearings results in greatly magnified roller–raceway loading and early subsurface-initiated fatigue failure (see later sections).

10.5 BEARING FAILURE DUE TO CAGE FRACTURE

In Chapter 7 of the first volume of this handbook, it was indicated that interference fitting of the inner ring on the shaft or the outer ring in the housing in a radial ball or roller

FIGURE 10.13 Ball bearing inner ring with rolled over left side land due to very heavy applied thrust loading.

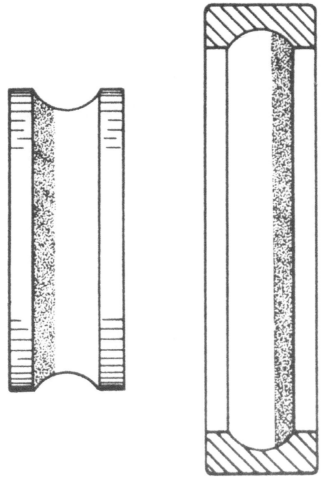

FIGURE 10.14 Postmortem diagram of inner and outer raceway in a ball bearing operated with excessive thrust loading.

bearing results in the loss of radial clearance. Also, if the inner raceway runs hotter than the outer raceway, then radial clearance is also reduced. If radial clearance is lost completely and radial interference occurs during bearing operation, loading between the bearing cage and the rolling elements may become excessive and cause breakage of the cage. This is illustrated in Figure 10.15 and Figure 10.16. In the event of cage fracture, fragments may break off and wedge themselves between the rolling elements and raceways, causing increased friction, overheating, and bearing seizure. This can be a catastrophic-type failure. Postmortem examination of the bearing raceways in such a case would reveal that the inner raceway was somewhat wider than the design, and the outer raceway extends a complete 360° as shown in Figure 10.17. This indicates excessive radial preload as the cause of bearing failure.

Cage fracture can also occur due to excessive misalignment in the bearing during operation. This places high fore-and-aft axial loading on the cage causing the breakage. The postmortem loading patterns of the raceways are shown in Figure 10.18.

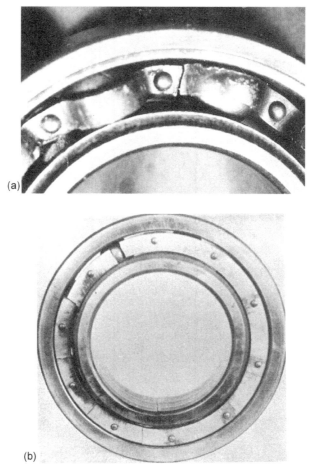

FIGURE 10.15 Fractured steel cages in deep-groove ball bearings: (a) ribbon-type cage and (b) machined and riveted cage.

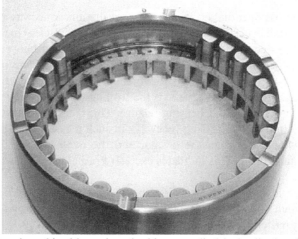

FIGURE 10.16 Fractured machined brass in a double-row cylindrical roller bearing.

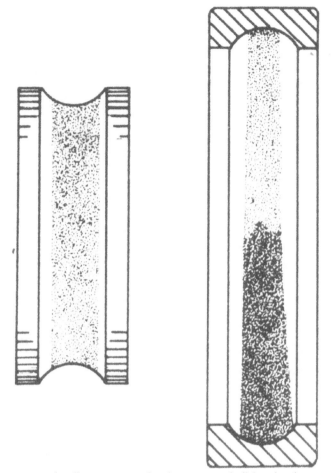

FIGURE 10.17 Postmortem loading patterns of a deep-groove ball bearing inner and outer raceways indicating heavy radial preloading that occurred in the bearing.

10.6 INCIPIENT FAILURE DUE TO PITTING OR INDENTATION OF THE ROLLING CONTACT SURFACES

10.6.1 CORROSION PITTING

Operation of a properly operating rolling bearing entails only a small amount of friction torque. As implied in Section 10.2.2, application design must be such as to accommodate the dissipation of both the heat generated by the application and the friction heat generated by the bearings without significant temperature rise. It was shown in Chapter 2 that, in most rolling element–raceway contacts, a combination of rolling and sliding motions occurs. It was further shown that sliding motion is the major cause of rolling bearing friction. Interruption of the rolling contact surfaces by corrosion pits or indentations exacerbates this condition. Figure 10.19 and Figure 10.20 illustrate corrosion pitting and oxidation (rusting) of rolling contact surfaces.

Figure 10.21 demonstrates the corrosion of a tapered roller bearing cone raceway due to moisture in the lubricant. Each of these conditions represents an interruption in the smooth surface of the rolling contact surfaces.

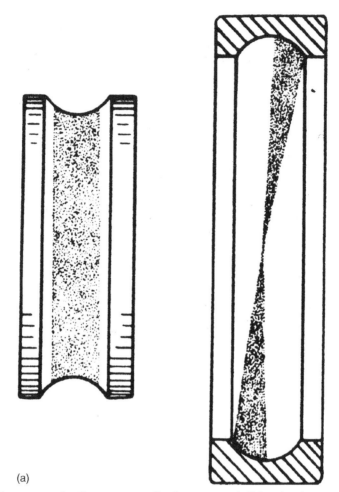

(a)

FIGURE 10.18 Postmortem loading patterns of a deep-groove ball bearing inner and outer raceways indicating significant misalignment that occurred in the bearing: (a) outer ring axis misaligned relative to the shaft axis. (continued)

10.6.2 True Brinnelling

Brinnelling in a rolling bearing is defined as the plastic deformation caused by either sudden impact loading during bearing operation or heavy loading while the bearing is not rotating. Figure 10.22 demonstrates such indentations, typically located at rolling element circumferential spacing.

10.6.3 False Brinnelling in Bearing Raceways

False brinnelling as illustrated in Figure 10.23 is actually fretting wear that occurs in the bearing raceways. It is caused by vibration that occurs during transportation of the bearing before installation or of the assembled application. It is also caused in the application by a vibrating load that results in small amplitude oscillations. The lubricant is driven from the contacts and fretting wear results. As seen in Figure 10.23, the indentations are wider than those caused by true brinnelling. Figure 10.24 and Figure 10.25 also depict false brinnelling on bearing raceways.

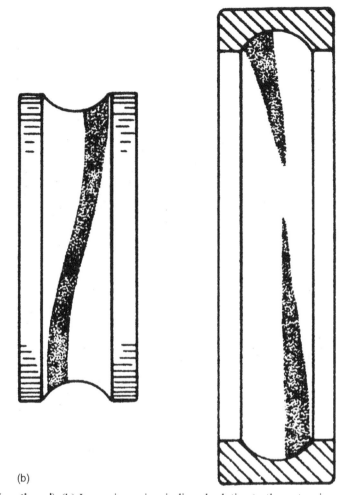

(b)

FIGURE 10.18 (continued) (b) Inner ring axis misaligned relative to the outer ring axis.

10.6.4 PITTING DUE TO ELECTRIC CURRENT PASSING THROUGH THE BEARING

In applications involving electric motors, if the bearing is not electrically insulated from the application, electric current may pass through the bearing. This current passage will form clusters of tiny pits in the rolling surface. Continued operation leads to corrugation of the surfaces called fluting, as shown in Figure 10.26 and Figure 10.27; the spacing of the corrugations is a function of the bearing internal speeds and the frequency of the electrical current. Rolling elements may also experience electrical pitting as shown in Figure 10.28. Figure 10.29 shows the special morphology surrounding a pit caused by electrical arcing through a bearing.

10.6.5 INDENTATIONS CAUSED BY HARD PARTICLE CONTAMINANTS

Disruptions or dents in the rolling contact surfaces can also be caused by hard particle contaminants that gain ingress past seals or shields into the lubricant and into the free space within the bearing boundaries. Such particles become trapped between the rolling

FIGURE 10.19 Spherical roller bearing—corrosion pitting of a roller.

elements and raceways and get rolled over. This results in relatively deep impressions in the rolling surfaces as illustrated in Figure 10.30 through Figure 10.32.

10.6.6   EFFECT OF PITTING AND DENTING ON BEARING FUNCTIONAL PERFORMANCE AND ENDURANCE

As indicated in Chapter 8, when a ball or roller bearing operates with lubricant films that have thicknesses at least four times the composite rms roughness of the opposing rolling contact surfaces, extremely long life generally results. Raceways in medium-size, modern, deep-groove ball bearings are typically manufactured with raceways having surface roughness $R_a = 0.05 \, \mu m$ (2 μin.) or less; the equivalent rms roughness value is $0.0625 \, \mu m$ (2.5 μin.).

FIGURE 10.20 Spherical roller bearing—oxidation (rusting) of the inner raceways.

FIGURE 10.21 Tapered roller bearing—corrosion on cone raceway caused by moisture in the lubricant.

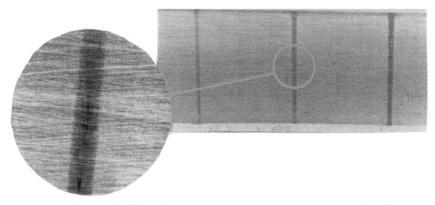

FIGURE 10.22 Tapered roller bearing—brinnelling on cup raceway. (Courtesy of the Timken Company.)

FIGURE 10.23 Tapered roller bearing—false brinnelling on cup raceway. (Courtesy of the Timken Company.)

FIGURE 10.24 Deep-groove ball bearing—false brinnelling on inner raceway.

Therefore, the ideal lubricant film thickness would be approximately $0.25\,\mu m$ ($10\,\mu in.$). (R_a for balls is only a small fraction of that for the raceways and does not significantly affect the calculation.) Larger bearings and roller bearings generally have "rougher" finishes; for example, $R_a = 0.25\,\mu m$ ($10\,\mu in.$) can be representative of roller bearing raceways and rollers.

FIGURE 10.25 Tapered roller bearing—false brinnelling on cone raceway. (Courtesy of the Timken Company.)

FIGURE 10.26 Tapered roller bearing—cone raceway fluting caused by electrical arcing. (Courtesy of the Timken Company.)

FIGURE 10.27 Cylindrical roller bearing—inner raceway fluting caused by electrical arcing.

FIGURE 10.28 Tapered roller bearing—pitting of tapered rollers caused by electrical arcing. (Courtesy of the Timken Company.)

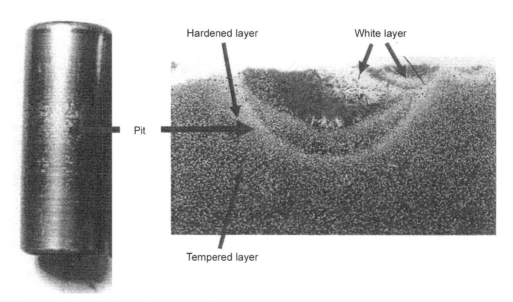

FIGURE 10.29 Morphology of a pit caused by electrical arcing.

This would give a composite rms roughness of $0.442\,\mu m$ ($17.68\,\mu in.$). In this case, an ideal lubricant film thickness would be approximately $1.77\,\mu m$ ($70.7\,\mu in.$).

To determine the effectiveness of the lubricant film in separating the rolling contact surfaces in the presence of a pit or dent, the depth of the dent or depression needs to be considered. Figure 10.33 is an elevation view of a section taken through a dent on a raceway surface. The depth of such a dent is typically in the order of several micrometers. This means that the lubricant film will tend to collapse into the depression. Figure 10.34 is a photograph taken through a transparent disk on a ball–disk friction testing rig (see Chapter 11). It depicts

FIGURE 10.30 Severe denting of the inner raceway of a deep-groove ball bearing.

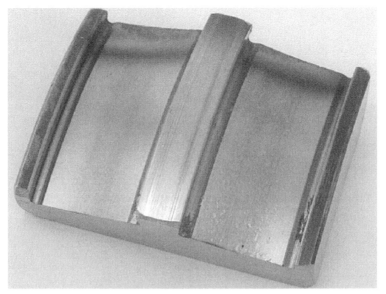

FIGURE 10.31 Denting of one inner raceway in a double-row spherical roller bearing.

FIGURE 10.32 Denting of rolling elements: (a) ball, (b) spherical roller, and (c) tapered roller.

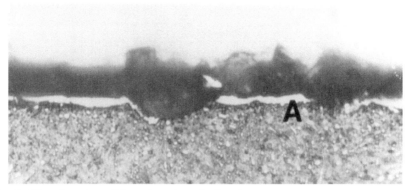

FIGURE 10.33 Elevation view of dented section of a bearing raceway.

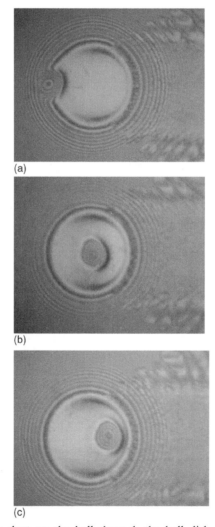

(a)

(b)

(c)

FIGURE 10.34 Passage of a dent on the ball through the ball–disk contact of a ball–disk testing machine: (a) the dent is entering the contact, (b) the dent is in the center of the contact, and (c) the dent is preparing to exit the contact. (Courtesy of Wedeven Associates, Inc.)

the passage of a dent through the oil-lubricated ball–disk contact. It shows how the lubricant film thickness and pressure distribution over the contact are altered by the dent. Extremely high-pressure ridges form in the vicinity of the depression. In a bearing, these high pressures greatly affect the surface and subsurface stresses in both the rolling element and raceway material, providing initiation points for fatigue failures. Figure 10.35a shows a dent in a raceway. The depression in the material is surrounded by a ridge that acts as a stress riser. Figure 10.35b shows a fatigue spall starting at the ridge on the trailing edge of a dent. Hence, corrosion and oxidation pits, true and false brinnelling, and hard particle contamination dents act as locations for incipient fatigue. This can cause bearing endurance to be shorter than that designed and may also lead to rapid failure of the bearing.

10.7   WEAR

10.7.1   DEFINITION OF WEAR

According to Tallian [1]:

> Wear (of a contact component) is defined as the removal of component surface material in the form of loose particles by the application of high tractive forces in asperity dimensions during service.

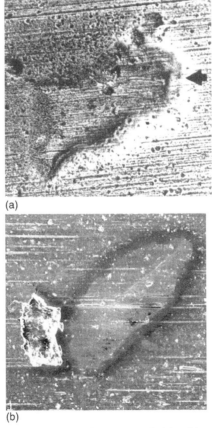

(a)

(b)

FIGURE 10.35 Dents in a raceway: (a) depression surrounded by ridge and (b) fatigue spall formed on the trailing edge of the dent.

The result of wear is continuing loss of the geometric accuracy of the rolling contact surfaces, and gradual deterioration of bearing function; for example, increased deflection, increased friction and temperature, increased vibration, and so forth. In ball and roller bearings, wear is considered preventable by proper attention to bearing and application design, manufacturing accuracy, lubrication adequacy, and prevention of ingress of contaminants. Therefore, no effort is made to estimate the life of rolling bearings as occasioned by wear.

According to some bearing practitioners, the term wear is used loosely to include all modes of surface material removal, including pitting and spalling. Herein, the latter modes of material loss are not included in the definition of wear of rolling bearing material.

10.7.2 Types of Wear

Mild wear is frequently called simply wear. Distinction is often made between two types of mild wear as follows:

1. Adhesive or two-body wear occurring at the interface of the contacting surfaces.
2. Abrasive or three-body wear occurring due to extraneous hard particles acting at the interface of the contacting surfaces.

Tallian [1] indicates that the worn surface to the naked eye appears "featureless, matte, and nondirectional" and characteristic finishing marks of the original manufactured surface are worn away. He further states that the characteristic appearances of other defined modes of material removal such as fretting, micropitting, and galling are distinctly not present. In any case, mild wear by itself, is not a mode of bearing failure, nor does it lead significantly to rapid bearing failure.

Severe wear or galling is defined as the transfer of component surface material in visible patches from a location on one surface to a location on the contacting surface, and possibly back onto the original surface. This transfer of material takes place because of high-friction shear forces due to sliding over the asperities of the surfaces. In rolling bearings, this severe wear phenomenon is also called smearing. It is a welding phenomenon entailing adhesive bonding between material portions of the contacting surfaces. Figure 10.36 through Figure 10.38 show bearing components with smearing on rolling contact surfaces. Smearing indicates

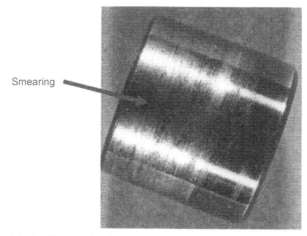

Smearing

FIGURE 10.36 Cylindrical roller bearing inner raceway with smearing damage.

FIGURE 10.37 Asymmetrical roller with smearing damage from spherical roller thrust bearing.

increased bearing friction and can lead to less-than-expected bearing endurance. Figure 10.39 shows an enlargement of a smeared area.

10.8 MICROPITTING

Tallian [1] narrowly defines surface distress as the plastic flow of surface material due to the application of "high normal forces in asperity dimensions." This surface distress results in micropitting, illustrated in Figure 10.40. The implication in this definition is that surface distress and micropitting occur during simple rolling motion; that is, in the absence of sliding. In any case, micropitting appears to be a severe form of surface distress.

10.9 SURFACE-INITIATED FATIGUE

When the repeatedly cycled stress on a surface in rolling contact with another exceeds the endurance strength of the material, fatigue cracking of the surface will occur. The crack will propagate until a large pit or spall occurs in the surface as shown in Figure 10.41. Some salient characteristics of such a spall are:

FIGURE 10.38 Smearing damage on the cone raceway of a tapered roller bearing.

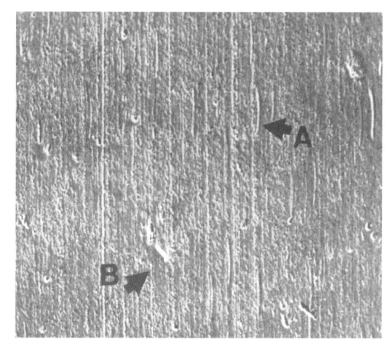

FIGURE 10.39 Enlarged photograph of smearing on a bearing raceway. Movement of metal is apparent.

- It is relatively shallow in depth.
- It commences at the trailing edge of the contact.
- The starting point of the arrowhead is frequently distinguishable if the fatigue spall is detected before significant propagation.

FIGURE 10.40 Extreme surface distress (micropitting) on a ball bearing inner raceway.

Rolling direction ➡

FIGURE 10.41 Surface-initiated fatigue spall in a bearing raceway.

Figure 10.42 and Figure 10.43 show bearings in advanced stages of surface-initiated fatigue.

In a properly designed, manufactured, application-selected, mounted, and lubricated rolling bearing, the potential for the occurrence of surface-initiated fatigue is virtually nil. Therefore, notwithstanding Tallian's definition of surface distress, a condition of sliding in marginally lubricated rolling element–raceway contacts is usually present when surface-initiated fatigue occurs. Furthermore, the surface friction shear stresses during sliding are

FIGURE 10.42 Thrust ball bearing raceway and balls with advanced stage of surface-initiated fatigue.

FIGURE 10.43 Cylindrical roller bearing inner raceway with advanced stage of surface-initiated fatigue.

most likely increased by the presence of depressions in the contacting surfaces caused by the aforementioned conditions of:

- Corrosion or electric arc pitting
- True brinnelling
- False brinnelling
- Denting due to hard particle contaminants
- Micropitting

Another mode of rolling contact surface failure is caused by hydrogen ions, which attack the surface material, resulting in pitting or spalling of the surface. Figure 10.44 depicts the spalled surface of a bearing ball caused by hydrogen embrittlement. This failure mode, which is relatively rare, is generally associated with rolling contact surface operation at steady-state temperature above that at which degradation of the mineral oil lubricant commences. It is generally associated with significant differential or gross sliding between rolling contact surfaces subjected to high Hertz stress, the surfaces incompletely or marginally separated by a mineral oil lubricant film. The high temperature that results in the contact causes the chemical breakdown of the lubricant, releasing hydrogen ions. Hydrogen embrittlement is also associated with an environment surrounding the bearing, which does not allow the hydrogen ions to easily dissipate from the vicinity of the bearing; for example, in a well-sealed application.

The essentially circular shapes of the spalled areas of Figure 10.44 are probably associated with the axisymmetric residual stress distribution existing in the bearing ball after heat treatment and surface finishing. As illustrated in Figure 10.45, the hydrogen ions penetrate the steel from the surface of the component, resulting in cracks. These in turn propagate weakening the material until spalling occurs. Many researchers have investigated the occur-

FIGURE 10.44 Spalling of bearing ball surface due to hydrogen embrittlement.

rence of hydrogen embrittlement failure. In all these reported experiments, hydrogen was introduced into the application in the presence of excessive contact stresses, and in most cases, in the presence of elevated temperatures. In none of these cases did the production of hydrogen ions result from lubricant chemical breakdown.

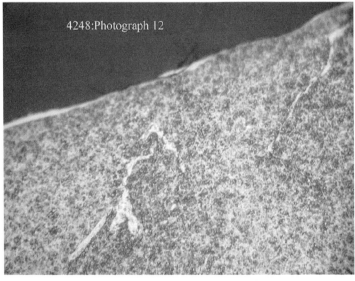

FIGURE 10.45 Cracking of steel ball surface due to penetration of hydrogen ions.

10.10   SUBSURFACE-INITIATED FATIGUE

As stated at the beginning of this chapter, each of the indicated modes of rolling bearing damage and failure is considered avoidable. Under very heavy loading, however, even though an accurately manufactured and properly mounted bearing is well-lubricated, it is possible for bearing failure to occur due to subsurface-initiated fatigue. The life of a rolling bearing from start of operation to occurrence of the first subsurface-initiated spall is the basis specified in the ISO [2] standard and supporting national standards for the calculation of fatigue endurance. See Chapter 11 of the first volume of this handbook. Figure 10.46 illustrates a ball bearing raceway with a subsurface-initiated spall. It is apparent that the depth is not shallow. Figure 10.47 shows spalling that has propagated in a spherical roller bearing raceway, while Figure 10.48 indicates spalling in a tapered roller bearing raceway due to edge loading.

Fatigue spalling is not considered a catastrophic-type failure. Depending on the type and quantity of lubrication, the bearing will continue to rotate, however, with ever-increasing friction (see Refs. [3,4]). After some time, depending on the magnitude of loading, operational speed, and lubrication effectiveness, the bearing will experience either excessive vibration or surface friction heat generation causing the bearing to seize.

10.11   CLOSURE

This chapter detailed the various common modes of failure to which ball and roller bearings may succumb. It is seen that most of these involve situations caused by bearing operations outside of recommended practice. As stated previously herein and in the first volume of this handbook, rolling bearings are rated according to their ability to resist or avoid subsurface-initiated rolling contact fatigue. The data of Figure 10.49, based on returns of failed bearings to manufacturers, show that the latter comprises only a small fraction of common failure

FIGURE 10.46 Subsurface-initiated fatigue spall in a ball bearing raceway.

FIGURE 10.47 Subsurface-initiated fatigue spall and propagation in a spherical roller bearing raceway.

FIGURE 10.48 Subsurface-initiated fatigue spalls due to edge loading in a tapered roller bearing cone raceway.

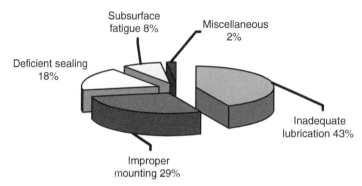

FIGURE 10.49 Frequency of occurrence of bearing failure modes.

types. Therefore, with regard to a given bearing application, attention to proper bearing selection, proper lubricant selection and adequate means of delivery, proper mounting techniques, avoidance of contamination, and general adherence to good operating practice will enable the achievement of the bearing design life.

REFERENCES

1. Tallian, T., *Failure Atlas for Hertz Contact Machine Elements*, 2nd Ed., ASME Press, 1999.
2. International Organization for Standards, International Standard ISO 281, *Rolling Bearings— Dynamic Load Ratings and Rating Life*, 2006.
3. Kotzalas, M. and Harris, T., Fatigue failure progression in ball bearings, *ASME Trans., J. Tribol.*, 123(2), 283–242, April 2001.
4. Kotzalas, M. and Harris, T., Fatigue failure and ball bearing friction, *Tribol. Trans.*, 43(1), 137–143, 2000.

11 Bearing and Rolling Element Endurance Testing and Analysis

LIST OF SYMBOLS

Symbol	Description	Units
A_1	Reliability-life factor	
A_{SL}	Stress-life factor	
C	Basic dynamic capacity	N (lb)
$f(x)$	Probability density function	
$F(x)$	Cumulative distribution function	
F_e	Equivalent applied load	N (lb)
h	Lubricant film thickness	μm (μin.)
h	Hazard rate	
H	Cumulative hazard rate	
i	Failure order number	
k	Number of samples	
k_p	$-\ln(1-p)$	
l	Number of subgroups in a sudden death test	
m	Sample size of a sudden death subgroup	
n	Sample size	
p	Probability value	
$q(l,m,p)$	Pivotal function used for sudden death test analysis	
r	Number of failures in a censored sample	
R	Ratio of upper to lower confidence limits for β	
$R_{0.50}$	Median ratio of upper to lower confidence limits for $x_{0.10}$	
$\mathcal{S}$	Probability of Survival	
$t_1(r,n,k)$	Pivotal function for testing for differences among k estimates of $x_{0.10}$	
$u(r,n,p)$	Pivotal function for setting confidence limits on x_p	
$u_1(r,n,p,k)$	Pivotal function for setting confidence limits on x_p using k data samples	
$v(r,n)$	Pivotal function for setting confidence limits on β	
$v_1(r,n,k)$	Pivotal function for setting confidence limits on β using k data samples	
$w(r,n,k)$	Pivotal function for testing whether k Weibull populations have a common β	

x	Random variable	
x_p	pth percentile of the distribution of the random variable x	
β	Weibull shape parameter	
η	Weibull scale parameter	
Λ	Lubricant film parameter	
κ	Ratio of actual lubricant viscosity to viscosity required	
σ	rms surface roughness	μm (μin.)

11.1 GENERAL

Similar to the lives of light bulbs and humans, ball and roller bearing lives, specifically rolling contact (RC) fatigue lives, are probabilistic in nature, as shown in Figure 11.1. They do not achieve a specific, uniquely predictable life, notwithstanding functioning in the same environment. As indicated in Chapter 10, with proper attention to bearing design, manufacture, and application, all modes of rolling bearing failure can be avoided with the exception of RC fatigue when contact stresses due to application loading exceed the bearing's endurance strength. In that situation, the life of any one bearing operating in the application can differ significantly from that of an apparently identical unit. To estimate rolling bearing fatigue life in a given application, statistical procedures have been established for the analysis of measured endurance data.

Using the method introduced by Weibull [1] to analyze the experimental data accumulated on fatigue lives of more than 1500 ball and roller bearings, Lundberg and Palmgren [2,3] developed formulas and methods to enable the calculation of load and life ratings for rolling bearings. The analysis by Lundberg and Palmgren was based on the influence of rolling element–raceway contact normal stresses (Hertz stresses) on RC fatigue life of the bearing raceways. Chapter 11 in the first volume of this handbook discusses these analytical methods

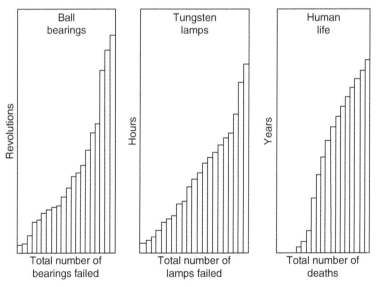

FIGURE 11.1 Comparison of rolling bearing fatigue life distribution with the distribution of the service lives of light bulbs and life spans of human beings.

in detail. The spread of rolling bearing fatigue lives recorded by Lundberg and Palmgren is illustrated for a typical test group in Figure 11.2.

From Figure 11.2, it is noted that two points on the curve are typically determined:

- L_{10}, the life that 90% of the bearings will survive and exceed.
- L_{50}, the median life that 50% of the bearings will survive and exceed.

It should be observed that L_{10} is the bearing rating life, the life on which bearing selection is typically based.

More recently, Ioannides and Harris [4], as detailed in Chapter 8, extended the Lundberg–Palmgren analysis to include the influence on bearing fatigue life of all stresses in the material in the vicinity of each contact as well as the concept of a material fatigue endurance limit stress. The combination of the two techniques resulted in the current ISO [5] standard load and fatigue life rating equation:

$$L_n = A_1 A_{SL} \left(\frac{C}{F_e}\right)^p \qquad (8.23)$$

In Equation 8.23, exponent $p = 3$ for ball bearings and 10/3 for roller bearings.

The Lundberg and Palmgren experimental effort [2,3] was based on bearings manufactured from 52100 steel during the 1930s and 1940s. Modern bearing manufacturing methods have since modified and improved rolling bearing geometries. Moreover, modern 52100 steel has been considerably improved in cleanliness and homogeneity since the time of Lundberg and Palmgren, and modern bearings are manufactured from a variety of steels and even ceramics. To evaluate the effects of new materials, material processing methods, and perhaps

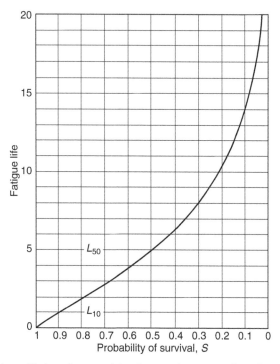

FIGURE 11.2 Distribution of fatigue lives resulting from endurance testing of a group of rolling bearings.

modified geometries on rolling bearing life, it is yet necessary to conduct fatigue endurance tests. These can be accomplished through testing of complete bearings or, possibly, using components such as balls or rollers. The spread in experimental fatigue data and the limitations of the statistical analysis techniques require that many test units be operated for a long time to obtain valid estimates of bearing life. It is obviously less expensive to endurance test balls or rollers in lieu of complete bearings; however, the accuracy of extrapolation of the component test results to complete bearings is always subject to question.

This chapter discusses the concepts, methods, and philosophies of conducting endurance tests on complete bearing assemblies as well as on elemental RC configurations.

11.2 LIFE TESTING PROBLEMS AND LIMITATIONS

11.2.1 Acceleration of Endurance Testing

The ability to use life test data collected on a particular bearing type and size under a specific set of operating conditions to predict general bearing performance requires a systematic relationship between applied load and life. This relationship, given by Equation 8.23, provides the means to use experimental life data collected under one set of test conditions to establish projected bearing performance under a wide range of operating conditions.

The time to initiation of a RC fatigue spall in a typical application is several years; for example, 10 years or more, assuming that applied loading is sufficiently heavy to cause fatigue of RC surfaces. It is therefore obvious that any practical laboratory test sequence must be conducted under accelerated conditions if the necessary data are to be accumulated within a reasonable time. Several potential methods exist for acceleration of testing. RC damage modes exist, however, that compete in individual bearings to produce the final failure. Care must be exercised to ensure that the method of test acceleration does not alter the desired failure mode of RC fatigue. Generally, two methods have been used to accelerate endurance testing: (1) increasing the level of applied load and (2) increasing the operating speed.

11.2.2 Acceleration of Endurance Testing through Very Heavy Applied Loading

The experimental results obtained under increased load levels can be rather easily extrapolated to other conditions by using the basic load–life relationship. Thus, this is the most widely used method of test acceleration. It is important, however, that consistency be maintained with the basic assumptions used in the development of the life formulas. Key among these is that the stresses generated at and below the RC surfaces should remain within the elastic regime. As indicated by Valori et al. [7], exceeding elastic limits of the bearing raceway and rolling element materials will produce deviations from the basic load–life relationship. Testing conducted in the material plastic regime produces results that are inconsistent with bearing operating practice and cannot be reliably extrapolated. The practical maximum Hertz stress limit for bearing endurance testing is 3300 MPa (475 psi).

11.2.3 Avoiding Test Operation in the Plastic Deformation Regime

Endurance testing of some bearings requires special consideration. For example, the outer raceway of a self-aligning ball bearing is a spherical surface producing circular point contacts between balls and raceways. Under very heavy applied loading, these contacts develop stresses in the plastic regime, more rapidly than considered by the dynamic capacity calculation. Johnston et al. [8] indicated that applied load should be no greater than $C/8$ to prevent substantial plastic deformation during testing of these bearings. Similarly, it is anticipated that some bearing types that have nonstandard internal geometries could also experience significant plastic deformations at lower than expected loads.

Cylindrical roller, tapered roller, and needle roller bearings are designed to operate with line contact between rollers and raceways under most applied loading. Spherical roller bearings, however, may operate with point contact until the applied load is sufficiently great to cause operation of the most heavily loaded rolling element in modified line contact (see Chapter 6 and Chapter 11 of the first volume of this handbook). When the load becomes too heavy, all roller bearings will tend to experience edge loading and plastic deformations, at least in the most heavily loaded roller–raceway contacts. Accordingly, in endurance testing of roller bearings, roller and raceway geometries must be profiled in the axial direction to ensure that stress concentrations do not occur at the contact extremities with attendant plastic deformations. The profiles used in standard design roller bearings are often insufficient for the heavy loads used in an accelerated life test series. Edge loading will tend to produce fatigue lives that are substantially less than lives experienced in field applications. Hence, endurance test results generated under conditions involving edge loading could not be accurately extrapolated to normal applications.

11.2.4 Load–Life Relationship of Roller Bearings

Even when edge stresses do not occur, roller bearing life under heavy loading does not tend to follow the standard load–life relationship. Lundberg and Palmgren [3] showed that test series conducted on cylindrical roller bearings indicated a load–life exponent of 4 in lieu of the standard 10/3. Actually, the exponent 10/3 was chosen as a compromise to accommodate the combination of line and point contacts that occurs in the operation of spherical roller bearings. Interpretation of the endurance test data for roller bearings needs to take this consideration into account.

11.2.5 Acceleration of Endurance Testing through High-Speed Operation

If all operating parameters remained unchanged, endurance testing a bearing at higher rotational speed would shorten the duration of testing by generating a more rapid accumulation of test cycles. Unfortunately, the objective of a shorter test duration is generally not achieved because fatigue endurance is usually conducted with oil film lubrication, the lubricant generally delivered to the bearing in copious quantities. Referring to Equation 4.57 for line contacts and Equation 4.60 for point contacts, it can be seen that the minimum lubricant film thickness is a function of approximately the 0.7 power of rotational speed. Hence, as speed increases so does lubricant film thickness. In Chapter 8, it was demonstrated that as the ability of the lubricant films to separate the RC surfaces increases, fatigue life increases at a greater rate. Therefore, increasing the rotational speed will most likely increase rather than decrease the duration of testing.

Associated with the effect of lubricant film thickness on fatigue life is the delivery of lubricant to the rolling element–raceway contacts in sufficient quantity to enable complete lubricant films to be generated. As the bearing speed increases, this becomes more difficult because the rapidity of oil reflux to the contacts may not keep pace with rolling element passage. As indicated in Chapter 4, this is called lubricant starvation. This is a particular consideration when endurance testing is to be performed with grease-lubricated bearings.

Also, as the operating speed is increased, other life-influencing effects occur. Under high-speed operating conditions, rolling element centrifugal force increases. This means that the maximum contact stresses may occur at the outer raceway, rather than the inner raceway, causing that component to experience spalling first, and changing the expected failure mode. Concurrently, as illustrated in Chapter 1, for the high-speed operation of radially loaded bearings, the number of rolling elements under load decreases, increasing the contact stresses on the inner raceway, but changing the mode of operation.

As discussed in Chapter 1, operation of thrust-loaded ball bearings at high speed causes ball–outer raceway contact angles to decrease and ball–inner raceway contact angles to increase. This changes the friction characteristics of the bearing, which also influence fatigue endurance.

A parameter often used to express the severity of bearing speed conditions is dN, the product of the bearing bore measured in millimeters and the shaft speed in revolutions per minute. It is usual to consider high-speed bearing applications as those that have $dN \geq 1$ million. For bearing operations under high-speed conditions, sophisticated analytical techniques, such as those presented in this text, are required to reliably calculate bearing rating lives for comparison with endurance test data.

Using high speed to accelerate bearing endurance test programs has other limitations. Standard rolling bearings have functional speed limits because of the stamped metal or molded plastic cage designs, which are not adequate for high-speed operations. Excessive heat generation rates may occur at the rolling element–raceway contacts, which have been designed primarily for maximum load-carrying capabilities at lower speeds, and component precision may be altered due to the dynamic loading occurring during high-speed operations.

System operating effects can also produce significant life effects on high-speed bearings. For example, insufficient cooling or the inadequate distribution of the cooling medium can create thermal gradients in the bearings that alter internal clearances and component geometries. Higher operating temperatures are generated at higher speeds. The test lubricants used must then be capable of sustained extended exposure to these temperatures without suffering degradation. The conduct of high-speed life tests requires extra care to ensure that the failures obtained are fatigue-related and not precipitated by some speed-related performance malfunction.

11.2.6 TESTING IN THE MARGINAL LUBRICATION REGIME

In Chapter 8, the means to quantify the lubrication-associated effect of speed on bearing fatigue endurance is demonstrated. Particularly in the regime of marginal lubrication, the effect is complex owing to the interactions of rolling component surface finishes and chemistry, lubricant chemical and mechanical properties, lubrication adequacy, contaminant types, and contamination levels. Testing at speeds slow enough to cause operation in the marginal lubrication regime may indeed reduce the fatigue life in revolutions survived; however, as with high-speed testing, the duration of testing may increase; in this case, due to the slower speed of accumulating stress cycles. Furthermore, the above-indicated side effects must be considered in the evaluation of test results.

11.3 PRACTICAL TESTING CONSIDERATIONS

11.3.1 PARTICULATE CONTAMINANTS IN THE LUBRICANT

An individual bearing may fail for several reasons; however, the results of an endurance test series are only meaningful when the test bearings fail by fatigue-related mechanisms. The experimenter must control the test process to ensure that this occurs. Some of the other failure modes that can be experienced are discussed in detail by Tallian [6]. The following paragraphs deal with a few specific failure types that can affect the conduct of a life test sequence.

In Chapter 8, the influence of lubrication on contact fatigue life was discussed from the standpoint of elastohydrodynamic lubrication (EHL) film generation. There are also other lubrication-related effects that can affect the outcome of the test series. The first is particulate contaminants in the lubricant. Depending on bearing size, operating speed, and lubricant rheology, the overall thickness of the lubricant film developed at the rolling element–raceway contacts may fall between 0.05 and 0.5 μm (2 and 20 μin.). Solid particles larger than the film can be mechanically trapped in the contact regions and damage the raceway and rolling element surfaces, leading to substantially shortened endurances. This has been amply demonstrated by Sayles and MacPherson [9] and others.

Therefore, filtration of the lubricant to the desired level is necessary to ensure meaningful test results. The desired level is determined by the application which the testing purports to approximate. If this degree of filtration is not provided, effects of contamination must be considered when evaluating test results. Chapter 8 discusses the effect of various degrees of particulate contamination, and hence filtration, on bearing fatigue life.

11.3.2 MOISTURE IN THE LUBRICANT

The moisture content in the lubricant is another important consideration. It has long been apparent that quantities of free water in the oil cause corrosion of the RC surfaces and thus have a detrimental effect on bearing life. It has been further shown by Fitch [10] and others, however, that water levels as low as 50–100 parts per million (ppm) may also have a detrimental effect, even with no evidence of corrosion. This is due to hydrogen embrittlement of the rolling element and raceway material (see also Chapter 8). Moisture control in test lubrication systems is thus a major concern, and the effect of moisture needs to be considered during the evaluation of life test results. A maximum of 40 ppm is considered necessary to minimize life reduction effects.

11.3.3 CHEMICAL COMPOSITION OF THE LUBRICANT

Most commercial lubricants contain a number of proprietary additives developed for specific purposes; for example, to provide antiwear properties, to achieve extreme pressure and thermal stability, and to provide boundary lubrication in case of marginal lubricant films. These additives can also affect bearing endurance, either immediately or after experiencing time-related degradation. Care must be taken to ensure that the additives included in the test lubricant do not suffer excessive deterioration as a result of accelerated life test conditions. Also, for consistency of results and comparing life test groups, it is a good practice to use one standard test lubricant from a particular producer for the conduct of all general life tests.

11.3.4 CONSISTENCY OF TEST CONDITIONS

11.3.4.1 Condition Changes over the Test Period

The statistical nature of RC fatigue requires many test samples to obtain a reasonable estimate of life; therefore, a bearing life test sequence generally occurs over a long time. A major job of the experimenter is to ensure the consistency of the applied test conditions throughout the test period. The process is not simple because subtle changes can occur during this period. Such changes might be overlooked until their effects become major, and it is too late to salvage the collected data. The test may then have to be redone under better controls.

11.3.4.2 Lubricant Property Changes

An example of the above is that the stability of the additive packages in a test lubricant can be a source of changing test conditions. Some lubricants are known to suffer additive depletion after an extended period of operation. The degradation of the additive package can alter the RC surface friction conditions, altering bearing life. Generally, the normal chemical tests used to evaluate lubricants do not determine the conditions of the additive content. Therefore, if a lubricant is used for endurance testing over a long period of time, a sample of the fluid should be returned to the producer at regular intervals, for example annually, for a detailed evaluation of its condition.

11.3.4.3   Control of Temperature

Adequate temperature controls must also be employed during the test period. The thickness of the EHL film is sensitive to the contact temperature. Referring to Equation 4.57 for line contacts and Equation 4.60 for point contacts, it can be seen that the minimum lubricant film thickness is a function of approximately the 0.7 power of lubricant viscosity, which is highly sensitive to temperature. Most test machines are located in standard industrial environments where rather wide fluctuations in ambient temperature are experienced over a period of one year. In addition, the heat generation rates of individual bearings can vary as a result of the combined effects of normal manufacturing tolerances. Both these conditions produce variations in operating temperature levels in a lot of bearings and affect the validity of the life data. Means must be provided to monitor and control the operating temperature level of each bearing to achieve a degree of consistency. A tolerance level of $\pm 3°C$ (5.4°F) is normally considered adequate for the endurance test period.

11.3.4.4   Deterioration of Bearing Mounting Hardware

The condition of the hardware involved in mounting and dismounting of bearings requires constant monitoring. The heavy loads used for life testing require heavy interference fits between bearing inner rings and shafts. Repeated mounting and dismounting of bearings can damage the shaft surface, which can in turn alter the geometry of the mounted ring. The shaft surface and the housing bore are also subject to deterioration from fretting corrosion (see Chapter 10). This can produce significant variations in the geometry of the mounting surfaces, which can alter the internal bearing geometry and, thereby, reduce bearing life.

11.3.4.5   Failure Detection

Fatigue theory considers failure as the initiation of the fatigue crack in the bulk material. To be detectable in a practical manner, the crack must propagate to the surface and produce a spall of sufficient magnitude to produce a marked effect on a bearing operating parameter; for example, vibration, noise, and temperature. The ability to detect early signs of failure varies with the complexity of the test system, the type of bearing under evaluation, and other test conditions. There is no single method that can consistently provide the failure discrimination necessary for all types of bearing tests. It is, therefore, necessary to select a method and system that will repeatedly terminate machine operation upon the consistent occurrence of a minimal degree of damage.

Considering the above, failure propagation rate is important. If the degree of damage at test termination is consistent among test elements, the only variation between the experimental and theoretical lives is the lag in failure detection. In standard through-hardened bearing steels, the failure propagation rate is quite rapid under endurance test conditions, and this is not a major factor, considering the typical dispersion of endurance test data and the degree of confidence obtained from statistical analysis. Care must be exercised when evaluating these latter results and particularly when comparing the experimental lives with those obtained from standard steel lots.

Post-test analysis is a detailed examination of all tested bearings using:

1. High magnification optical inspection
2. Higher magnification electron microscopy
3. Metallurgical examination
4. Dimensional examination
5. Chemical evaluation as required

The characteristics of the failures are examined to establish their origins, and the residual surface conditions are evaluated for indications of extraneous effects that may have influenced bearing life. This technique enables the experimenter to ensure that the data are indeed valid. Tallian's "Damage Atlas" [11], containing numerous photographs of the various failure modes, can provide valuable assistance in this effort.

11.3.4.6 Concurrent Test Analysis

Whenever a bearing is removed from the test machine, the experimenter should conduct a preliminary evaluation. Herein, the bearing is examined optically at magnifications up to 30× for indications of improper or out-of-control test parameters. Examples of indications that may be observed are given in Chapter 10. Figure 10.46 illustrates the appearance of a typical, subsurface fatigue-initiated spall on a ball bearing raceway. Figure 10.48 shows spalling of a tapered roller bearing that most likely resulted from bearing misalignment. Figure 10.12 illustrates a spalling failure on a ball bearing outer ring that resulted from fretting corrosion on the outer diameter of the ring. Figure 10.41 illustrates a more subtle form of test alteration, where the spalling failure originated from the presence of a debris dent on the raceway surface. The last three failures are not valid subsurface-initiated fatigue spalls and indicate the need to correct the test methods. Furthermore, the data points need to be eliminated from the failure data to obtain a valid estimate of the experimental bearing life.

11.4 TEST SAMPLES

11.4.1 Statistical Requirements

The statistical techniques used to evaluate the failure data require that the bearings be statistically similar assemblies. Therefore, the individual components must be manufactured in the same processing lot from one heat of material. Generally, it is prudent to manufacture the total bearing assembly in this manner; however, when highly experimental materials or processes are considered, this is often not cost effective or even possible. In those cases, the critical element in a bearing assembly from a fatigue point of view can be used as the test element with the other components manufactured from standard material. The effects of failures occurring on the other parts can be eliminated during analysis of the test data. There is some risk in this approach because it is possible that too many failures might occur on these nontest parts, rendering it impossible to calculate an accurate life estimate for the material under evaluation. This risk is generally small because an initial result indicating the superior performance of an experimental process is usually sufficient to justify continued development effort even without a firm numerical life estimate. Additional life tests would, however, be required to establish the magnitude of the expected lot-to-lot variation before adopting a new material or implementing a new manufacturing process.

11.4.2 NUMBER OF TEST BEARINGS

Statistical analysis provides a numerical estimate of the value of the experimental life enclosed by upper-boundary and lower-boundary estimates at specified confidence limits. The precision of the experimental life estimate can be defined by the ratio of these upper and lower confidence limits; the experimental aim is to minimize this spread. The magnitude of the confidence interval decreases as the size of the test lot increases; however, the cost of conducting the test also increases with lot test size. Therefore, the degree of precision required in the test result should be established during the test planning stage to define the size of the test lot to achieve the required results.

11.4.3 TEST STRATEGY

The usual method of performing endurance tests is to use one large group of bearings, running each bearing to failure. This process is time-consuming, but it provides the best experimental estimates of both L_{10} and L_{50} lives. Primary interest is, however, in the magnitude of the experimental L_{10}, so considerable time savings can be achieved by curtailing the test runs after a finite operating period equal to at least three times the achieved experimental L_{10} life. Also, Andersson [12] demonstrated that savings in test time accompanied by increases in precision of test results can be achieved by using a sudden death test strategy. In this approach, the entire test lot of bearings is divided into subgroups of equal sizes. Each subgroup is then run as a unit until one bearing fails, at which time the testing of the subgroup is terminated. Figure 11.3 illustrates the effect of both lot size and test strategy on the precision of life test estimates obtainable from an endurance testing series.

11.4.4 MANUFACTURING ACCURACY OF TEST SAMPLES

To provide an accurate life estimate for the variable under evaluation, the experimenter must be sure that the test bearings are free from material and manufacturing defects and that all parts conform to established dimensional and form tolerances. This situation is not always easy to attain since experimental materials might respond differently to standard manufacturing processes, or they could require unique processing steps that are not yet totally defined. Experimental manufacturing processes require additional verification, or their use might produce unexpected variations in metallurgical or dimensional parameters. Therefore, adequate test control is achieved by detailed pretest auditing of the test parts to supplement the standard in-process evaluations. Table 11.1 and Table 11.2 contain lists of those metallurgical and dimensional parameters considered mandatory in a typical pretest audit, as well as an indication of the number of samples that need to be checked in each case. These lists are not to be construed as complete; other parameters could be evaluated beneficially if time and money permit.

11.5 TEST RIG DESIGN

Some specific characteristics are desired in an endurance test system to achieve the control requirements of a life test series. An individual test run takes a long time; therefore, the test machine must be capable of running unattended without experiencing variation in the applied test parameters such as load(s), speed, lubrication conditions, and operating temperature. The basic test system that could also be subject to fatigue, such as load–support bearings, shafts, and load linkages, must be many times stronger than the test bearings so that test runs can be completed with the fewest interruptions from extraneous causes. The assembly of the

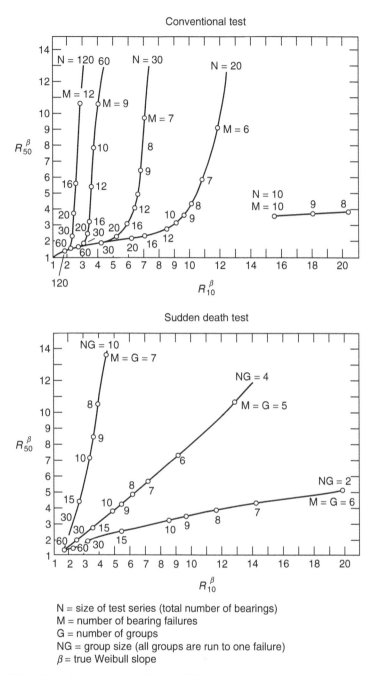

FIGURE 11.3 Effect of the lot size on confidence of life test results. (From Andersson, T., *Ball Bear. J.*, 217, 14–23, 1983. With permission.)

test machine should have only a minor influence on the test conditions to minimize variations between individual test runs. For example, the alignment of the test bearings should be automatically assured by the assembly of the test housing. If not, a simple direct means of monitoring and adjusting this parameter must be provided. Also, since a test series requires multiple setups, easy assembly and disassembly of the test system are desirable to minimize

TABLE 11.1
Typical Metallurgical Audit Parameters

100% Nondestructive Tests: Ring Raceways Only
Magnaflux for near-surface materials defects
Etch inspection for surface processing defects
Sample Destructive Tests: All Components
Microhardness to 0.1 mm (0.004 in.) depth below raceway surfaces
Microstructures to 0.3 mm (0.012 in.) depth below raceway surfaces
Retained austenite levels
Fracture grain size
Inclusions ratings

turnaround time and manpower requirements for test bearing changes. In addition, the test system must be easy to maintain and should be capable of operating reliably and efficiently for years to ensure long-term compatibility of test results. Design simplicity is a key ingredient in meeting all these demands. Sebok and Rimrott [13] presented a comprehensive discussion of the design philosophy for life test rigs; Figure 11.4 illustrates some typical endurance test configurations described.

The application of some of the design concepts of Figure 11.4 to actual endurance test systems will be briefly addressed. Figure 11.5 is a photograph of an SKF R2 rig for testing 35- to 50-mm bore ball and roller bearings under radial, axial, or combined radial and axial loads; Figure 11.6 is a schematic diagram of an SKF R3 rig, a similar design for testing larger bearings. The operating speed in these rigs may be varied within limits to achieve a given test condition and bearing lubrication can be provided by grease, sump oil, circulating oil, or air-oil mist.

Practical life test rig designs will vary, depending on the type of bearing to be tested and its normal operating mode. For example, Figure 11.7, according to Hacker [14], shows a four-bearing test rig concept used in the testing of tapered roller bearings. In this instance, while testing is conducted under an externally applied radial load, each bearing also sees an

TABLE 11.2
Typical Dimensional Audit Parameters

100% Assembled Rings
Radial Looseness
Average and peak vibration levels
Statistical Sample of Ring Grooves
Diameter and waviness
Radius and form
Cross-groove surface texture
Statistical Sample of Balls
Diameter and out-of-round
Set size variation
Waviness
Surface texture

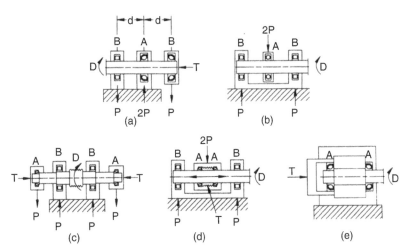

FIGURE 11.4 Typical bearing endurance test configurations discussed in Ref. [13]. A = test bearing, B = load bearing, P = applied radial load, T = applied thrust load, and D = drive.

FIGURE 11.5 SKF R2 endurance test rig. (Courtesy of SKF.)

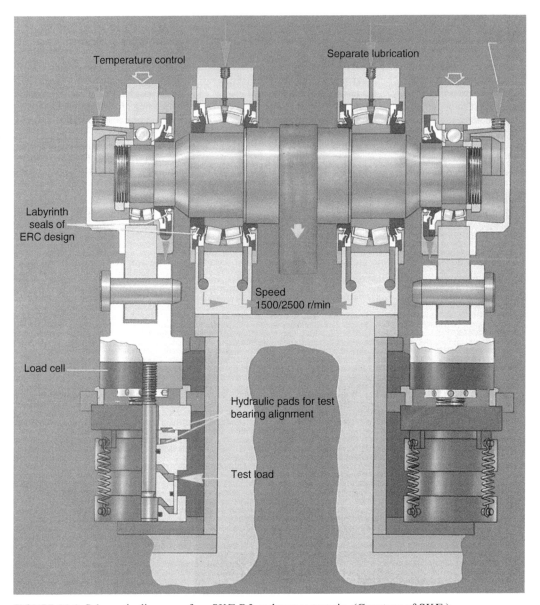

FIGURE 11.6 Schematic diagram of an SKF R3 endurance test rig. (Courtesy of SKF.)

internally induced thrust load. The size of the latter load is determined by the magnitude of the applied radial load, the fixed axial locations of the bearing cups and cones in the test housing, and the basic internal design of the test bearings. Figure 11.8 is a photograph of two such test rigs. The test rig design also accommodates testing of spherical roller and cylindrical roller bearings as indicated in Figure 11.9. The test rig design is also applicable to large bearing sizes as shown in Figure 11.10.

Tests are often conducted to define the life of bearings in special applications. They are frequently called life or endurance tests, but, more correctly, they are extended duration performance tests. The same basic test practices and test rig configurations are required for

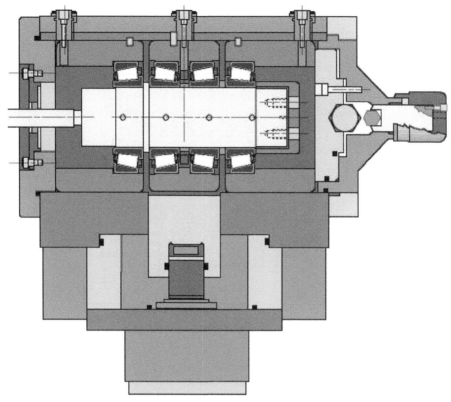

FIGURE 11.7 Schematic diagram of a tapered roller bearing test configuration. Four bearings are tested simultaneously under a sudden death test strategy. (Courtesy of the Timken Company.)

FIGURE 11.8 Photograph of a four-bearing test rig. The test housing can be used to test spherical roller and cylinder roller bearings as well as tapered roller bearings. (Courtesy of the Timken Company.)

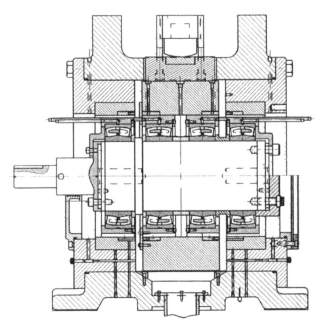

FIGURE 11.9 Schematic diagram of spherical roller bearing test configuration. Four bearings are tested simultaneously under a sudden death test strategy. (Courtesy of the Timken Company.)

FIGURE 11.10 Photograph of rigs for fatigue testing four large bearings simultaneously. These particular rigs accommodate bearings that have 460-mm (18 in.) outside diameter. (Courtesy of the Timken Company.)

FIGURE 11.11 A-frame automotive wheel hub bearing tester. (Courtesy of SKF.)

these tests, but some modifications of philosophy are required to simulate the major operating parameters of the application while achieving realistic test acceleration. An example of this type of tester is the SKF "A-frame" tester developed for evaluating automotive wheel hub bearing assemblies, see Figure 11.11. This tester simulates an automotive wheel bearing environment by using actual automobile wheel bearing mounting hardware, combined radial and axial loads applied at the tire periphery to produce moment loads on the bearing assembly, grease lubrication, and forced air cooling. Dynamic wheel loading cycles equivalent to those produced by vehicle lateral loading are applied cyclically to simulate a critical driving sequence. Testing is conducted in the sudden death mode so that wheel hub bearing unit life can be calculated using standard life test statistics. This test provides a way to compare the relative performance of automotive wheel support designs using life data generated under conditions similar to those of actual applications.

11.6 STATISTICAL ANALYSIS OF ENDURANCE TEST DATA

11.6.1 STATISTICAL DATA DISTRIBUTIONS

Many statistical distributions have been used to describe the random variability of the life of manufactured products. Such choices can be variously justified. For example, if a product has a reservoir of a substance that is consumed at a uniform rate throughout the product's life, and if the initial supply of the substance varies from item to item, according to a normal (Gaussian) distribution, then the product life will be normally distributed. Correspondingly, if the initial amount of the substance follows a gamma distribution, item life will be gamma distributed.

The Weibull distribution is a popular product-life model, justified by its property of describing, under fairly general circumstances, the way the smallest values in large samples vary among sets of large samples. Therefore, if item life is determined by the smallest life among many potential failure sites, it is reasonable to expect that life will vary from item to item according to a Weibull distribution.

Another property that makes the Weibull distribution a reasonable choice for some products is that it can account for a steadily increasing failure rate characteristic of wear-out failures, a steadily decreasing failure rate characteristic of a product that benefits from "burn-in," or a constant failure rate typical of products that fail due to the occurrence of a random shock.

The two-parameter Weibull distribution was adopted by Lundberg and Palmgren [2] to describe rolling bearing fatigue life on the strength of the excellence of the empirical fit to bearing fatigue life data. As shown in Chapter 8, when operating under moderate load and optimum lubrication conditions, a well-designed, manufactured, and applied bearing can endure indefinitely without experiencing fatigue failure. The Weibull model cannot describe this aspect of fatigue life. Nevertheless, under the relatively high loads common in fatigue testing practice, the Weibull distribution will closely approximate the observed fatigue behavior of rolling bearings.

11.6.2   THE TWO-PARAMETER WEIBULL DISTRIBUTION

11.6.2.1   Probability Functions

When it is said that a random variable, for example bearing life, follows the two-parameter Weibull distribution, it is implied that the probability that an observed value of that variable is less than some arbitrary value x can be expressed by

$$\text{Prob}\,(\text{life} < x) = F(x) = 1 - e^{-(x/\eta)^\beta} \qquad x,\, \eta,\, \beta > 0 \qquad (11.1)$$

$F(x)$ is known as the cumulative distribution function (CDF), η is the scale parameter, and β is the shape parameter. $F(x)$ may be considered the area under a curve $f(x)$ between 0 and the arbitrary value x. This curve is known as the probability density function (pdf) and has the form

$$f(x) = \frac{x^{\beta-1}}{\eta^\beta}\, e^{-(x/\eta)^\beta} \qquad (11.2)$$

Figure 11.12 is a plot of the Weibull pdf for various values of β. It is noted that a wide diversity of distribution forms are encompassed by the Weibull family, depending on the value of β. For $\beta = 1$, the Weibull distribution reduces to the exponential distribution. For β in the range of 3.0–3.5, the Weibull distribution is nearly symmetrical and approximates the normal pdf. The ability to assume such a range of shapes accounts for the extraordinary applicability of the Weibull distribution to many types of data.

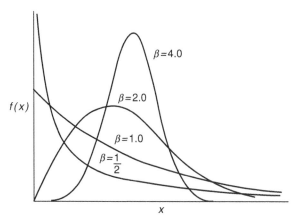

FIGURE 11.12 The two-parameter Weibull distribution for various values of the shape parameter β.

11.6.2.2 Mean Time between Failures

The average or expected value of a random variable is a useful measure of its "central tendency"; it is a single numerical value that can be considered to typify the random variable. It is defined as

$$E(x) = \int_0^\infty x f(x)\,\mathrm{d}x = \int_0^\infty x \left(\frac{x^{\beta-1}}{\eta^\beta}\right) \mathrm{e}^{-(x/\eta)^\beta}\,\mathrm{d}x \tag{11.3}$$

The value of the integral the above equation is

$$E(x) = \eta \Gamma\left(\frac{1}{\beta}+1\right) \tag{11.4}$$

$\Gamma(\ )$ is the widely tabulated gamma function. Table CD11.1 gives values of $\Gamma(1/\beta+1)$ for β ranging from 1.0 to 5.0.

In reliability theory, $E(x)$ is known as the mean time between failures (MBTF). It represents the average time between the failures of two consecutively run bearings; that is, the time between the failure of a bearing and the failure of its replacement. It does not represent the mean time between consecutive failures in a group of simultaneously running bearings. For this latter situation, provided $\beta \leq 1$, MBTF will vary with the failure order number. For example, the mean time between the first and second failures in samples of size 20 is different from the mean time between the 19th and 20th failures.

The scatter of a random variable is often characterized by a quantity called variance, defined as the average or expected value of the squared deviation of the variable from its expected value. Variance is given by

$$\sigma^2 = \int [x - E(x)]^2 f(x)\,\mathrm{d}x = \int \left[x - \eta\Gamma\left(\frac{1}{\beta}+1\right)\right]^2 \frac{x^{\beta-1}}{\eta^\beta} \mathrm{e}^{-(x/\eta)^\beta}\,\mathrm{d}x \tag{11.5}$$

The value of this integral is

$$\sigma^2 = \eta^2 \left[\Gamma\left(\frac{2}{\beta}+1\right) - \Gamma^2\left(\frac{1}{\beta}+1\right)\right] \tag{11.6}$$

Values of the quantity $[\Gamma(2/\beta+1) - \Gamma^2(1/\beta+1)]$ are given in Table CD11.1.

The units of variance σ^2 are the square of the units in which life is measured; for example, (revolutions)^2 or (hr)^2. The standard deviation or square root of σ^2 is often preferred as a measure of scatter because it is expressed in the same units as the variable itself. Neither the variance nor the standard deviation is cited much for the Weibull distribution; it is more usual to convey the magnitude of scatter by citing the values of a low percentile and a high percentile.

11.6.2.3 Percentiles

Equation 11.1 gives the probability that the observed value of a Weibull random variable is less than an arbitrary value. The inverse problem is to find a value of x, say x_p, for which the probability is a specified value p such that life will not exceed it. The term x_p is defined implicitly as

$$F(x_p) = 1 - e^{-(x_p/\eta)^\beta} = p \tag{11.7}$$

The solution of the above equation is

$$x_p = \eta \left[\ln \left(\frac{1}{1-p} \right) \right]^{1/\beta} \tag{11.8}$$

An important special case in rolling bearing engineering is the 10th percentile $x_{0.10}$, because it is historically customary that bearings are rated by the value of their 10th percentile life. In bearing literature, $x_{0.10}$ is called L_{10}. For consistency with the statistical literature on the Weibull distribution, $x_{0.10}$ is used in this discussion. It is expressible as

$$x_{0.10} = \eta \left[\ln \left(\frac{1}{1-0.10} \right) \right]^{1/\beta} = \eta(0.1054)^{1/\beta} \tag{11.9}$$

The median life $x_{0.50}$ is also of some interest:

$$x_{0.50} = \eta \left[\ln \left(\frac{1}{1-0.50} \right) \right]^{1/\beta} = \eta(0.6931)^{1/\beta} \tag{11.10}$$

Using Equation 11.8, the ratio of two percentiles, say x_p and x_q, is

$$\frac{x_q}{x_p} = \left[\frac{\ln(1-q)^{-1}}{\ln(1-p)^{-1}} \right]^{1/\beta} \tag{11.11}$$

Thus,

$$\frac{x_{0.50}}{x_{0.10}} = \left[\frac{\ln(1-0.50)^{-1}}{\ln(1-0.10)^{-1}} \right]^{1/\beta} = \left(\frac{0.6931}{0.1054} \right)^{1/\beta}$$

For $\beta = 10/9$, therefore, $x_{0.50} = 5.45$. This supports the rule, often quoted in the bearing industry, that median life $L_{50} \approx 5 \times L_{10}$, the rating life.

11.6.2.4 Graphical Representation of the Weibull Distribution

From Equation 11.1, the probability that a bearing survives a life x denoted $\mathcal{S}(x)$ is given by

$$\mathcal{S}(x) = 1 - F(x) = e^{-(x/\eta)^\beta} \tag{11.12}$$

Taking natural logarithms twice on both sides of the above equation leads to

$$\ln \ln \left(\frac{1}{\mathcal{S}} \right) = \beta[\ln(x) - \ln(\eta)] \tag{11.13}$$

The right-hand side is a linear function of $\ln(x)$. On special graph paper, called Weibull probability paper, for which the ordinate is ruled proportionately to $\ln[\ln(1/\mathcal{S})]$ and the abscissa is logarithmically scaled, the values of $\mathcal{S}$ vs. the associated values of x plot as straight line. If in the design of the paper the same cycle lengths are used for the logarithmic scale on both coordinate axes, the slope of the straight line representation will be numerically equal to β. In any case, the Weibull shape parameter or Weibull slope will be related to the slope of the straight line representation, and in some designs of probability paper an auxiliary scale is provided to relate the shape parameter and the slope.

Figure 11.13 is a plot on Weibull probability coordinates on which the distribution with $\beta = 1.0$ and $x_{0.10} = 15.0$ is represented. It was constructed by passing a 45° line through the point corresponding to the failure probability value $F = 0.1$ ($\mathcal{S} = 0.9$) and the life value $x_{0.10} = 15.0$. From this plot, the 20th may be read as the abscissa value at which a horizontal line at the ordinate value $F = 0.2$ intersects the straight line. Within graphical accuracy, $x_{0.20} = 32.0$. Inversely, the probability of failing before the life $x = 52.0$ is read to be roughly 30%. Representing a Weibull population on probability thus offers a graphical alternative to the use of Equation 11.1 and Equation 11.8 for the calculation of probabilities and percentiles. The graphical approach is sufficiently accurate for most purposes. The primary use of probability paper is not, however, for representing known Weibull distributions, but for estimating the Weibull parameters from life test results.

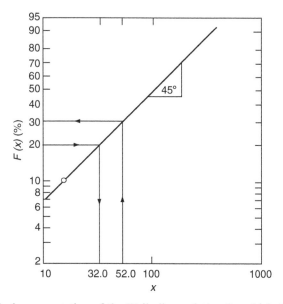

FIGURE 11.13 Graphical representation of the Weibull population for which $\beta = 1$ and $x_{0.10} = 15.0$.

11.6.3 ESTIMATION IN SINGLE SAMPLES

11.6.3.1 Application of the Weibull Distribution

Thus far it has been assumed that the Weibull parameters are known, and additionally required quantities such as probabilities, percentiles, expected values, variances, and standard deviations have been calculated in terms of these known parameters. This is a common situation in bearing application engineering, in which, given a catalog calculation of $x_{0.10}$ (L_{10}) and the standard Weibull slope of $\beta = 10/9$, it is required to calculate the median life, the MBTF, and so on. In developmental work involving new variables such as materials, lubricants, or component finishing processes, the focus is on determining the effect of these factors on the Weibull parameters. Accordingly, a sample of bearings modified from the standard in some way is subjected to testing under standardized conditions of load and speed until some or all fail. When all fail, the sample is said to be uncensored. In a censored sample, some bearings are removed from test before failure. Given the lives to failure or to test suspension of the unfailed bearings, the aim is to deduce the underlying Weibull parameters. This process is called estimation because it is recognized that, since life is a random variable, identical samples will result in different test lives. The Weibull parameter values estimated in any single sample must themselves be regarded as observed values of random variables that will vary from sample to sample according to a probability distribution known as the sampling distribution of the estimate. The scatter in the sampling distribution will decrease as the sample size is increased. The sample size therefore affects the degree of precision with which the parameters are determined by a life test. The precision is expressed by an uncertainty or confidence interval within which the parameter value is likely to lie. An estimation procedure that results in the calculation of a confidence interval is called interval estimation. A procedure that results in a single numerical value for the parameter is called point estimation. Point estimates in themselves are virtually useless, because, without some qualification, there is no way of judging how precise they are.

Accordingly, an analytical technique is given in the sequel for computing interval estimates of Weibull parameters. It is recommended that this technique be supplemented, however, with a point estimate obtained graphically. The graphical approach to estimation gives a synoptic view of the entire distribution and offers the opportunity to detect anomalies in the data that could easily be overlooked if reliance is placed entirely on the analytical technique.

11.6.3.2 Point Estimation in Single Samples: Graphical Methods

Assuming that a sample of n bearings is tested until all fail, the ordered times to failure are denoted $x_1 < x_2 < \cdots < x_n$. If the CDF of the Weibull population from which the sample was drawn were known, it would follow that lives x_i and the values $F(x_i)$, $i = 1, \ldots, n$, would plot as a straight line on Weibull probability paper. It has been shown that even if function $F(x)$ is not known, nevertheless, $F(x_i)$ will vary in repeated samples according to a known pdf. The mean or expected value of $F(x_i)$ has been shown to equal $i/(n+1)$. The median value of $F(x_i)$, also known as the median rank, has been shown by Johnson [15] to be approximately $(i - 0.3)/(n + 0.4)$. The procedure then is to plot the mean or median value of $F(x_i)$ against x_i for $i = 1, 2, \ldots, n$. The tradition in the bearing industry is to use the median rather than the mean as a plotting position choice, but the difference is small compared with the sampling variability.

Table 11.3 lists the ordered lives at failure for a sample of size $n = 10$, along with the actual and approximated values of the median ranks. Hence, the approximation is adequate within the limits of graphical accuracy. The median ranks are shown plotted against the lives in Figure 11.14.

TABLE 11.3
Random Uncensored Sample Size of $n = 10$

Failure Order Number (i)	Life	Median Rank	$(i - 0.3)/(n + 0.4)$
1	14.01	0.06697	0.06731
2	15.38	0.16226	0.16346
3	20.94	0.25857	0.25962
4	29.44	0.35510	0.35577
5	31.15	0.45169	0.45192
6	36.72	0.54831	0.54808
7	40.32	0.64490	0.64423
8	48.61	0.74142	0.74038
9	56.42	0.83774	0.83654
10	56.97	0.93303	0.93269

The straight line fitted to the plotted points represents the graphical estimate of the entire $F(x)$ curve. Estimates of the percentiles of interest are then read from the fitted straight line. For example, within graphical accuracy, the $x_{0.10}$ value is estimated as 15.3. The Weibull shape parameter, estimated simply as the slope of the straight line, is roughly 2.2.

The same graphical approach applies to right-censored data in which the censored observations achieve a longer running time than do the failures. The full sample size n is used to compute the plotting positions, but only the failures are plotted. When there is mixed censoring, that is, there are suspended tests among the failures, the plotting positions are no longer calculable by the method given because the suspensions cause ambiguity in determining the order numbers of the failures. Several alternative approaches are available for this situation, with generally negligible difference among them. Nelson's [16] method, known as

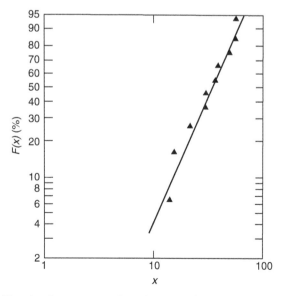

FIGURE 11.14 Probability plot for uncensored random sample of size $n = 10$.

TABLE 11.4
Calculation of Plotting Positions for a Hazard Plot

Life	Reverse Rank	Hazard (h)	Cumulative Hazard (H)	$F = 1 - e^{-H}$
0.569 S	10	—	—	—
8.910 F	9	0.1111	0.1111	0.1052
21.410 S	8	—	—	—
21.960 F	7	0.1429	0.2540	0.2243
32.620 S	6	—	—	—
39.290 F	5	0.2000	0.4540	0.3649
42.990 S	4	—	—	—
50.400 F	3	0.3333	0.7873	0.5449
53.270 S	2	—	—	—
102.600 S	1	—	—	—

hazard plotting, is recommended because it is easy to use. Column 1 of Table 11.4 gives the lives of failure or test suspension in a sample of size $n = 10$. Of the ten bearings, $r = 4$ have failed, and the lives of failure are marked with an "F" in Table 11.4. Similarly, the lives at test suspension are marked "S". The lives in column 1 are in ascending order of time on test irrespective of whether the bearing failed. Column 2, termed the reverse rank by Nelson [16], assigns the value n to the lowest time on test, the value $n - 1$ to the next lowest, and so on. Column 3, called the hazard, is the reciprocal of the reverse rank, but is calculated only for the failed bearings. Column 4 is the cumulative hazard and contains for each failure the sum of the hazard values in column 3 for that failure and each failure that occurred at an earlier running time. Thus, for the second failure, the cumulative hazard is $0.2540 = 0.1111 + 0.1429$. The cumulative hazard can then be plotted against life on probability that has been designed with an extra "hazard" scale. If graph paper of this type is not available, it is only necessary to calculate an estimate of the plotting position applicable to ordinary paper by transforming the cumulative hazard H to $F = 1 - e^{-H}$. This computation is shown in column 5 of Table 11.4. Figure 11.15 shows the resultant plot. It is noted that, as in the right-censored case, only the failures are plotted. The suspended tests have played a role, however, in determining the plotting positions of the failures.

11.6.3.3 Point Estimation in Single Samples: Method of Maximum Likelihood

The method of maximum likelihood (ML) is a general approach to the estimation of the parameters of probability distributions. The central idea is to estimate the parameters as the values for which the last observed test sample would most likely have occurred.

Considering an uncensored sample of size n, the likelihood is the product of the pdf $f(x) = x^{\beta-1}/\eta^{\beta} \exp[-(x/\eta)^{\beta}]$ evaluated at each observed life value. The ML estimates of η and β are the values that maximize this product. For a censored sample with $r < n$ failures, the likelihood function contains, in lieu of the density function $f(x)$, the term $1 - F(x) = \exp[-(x/\eta)^{\beta}]$ evaluated at each suspended life value. It can be shown that the ML estimate of β, denoted by a caret ($\wedge$), is the solution of the following nonlinear equation:

$$\frac{1}{\hat{\beta}} = \frac{\sum_{i=1}^{i=r} \ln x_i}{r} - \frac{\sum_{i=1}^{i=n} x_i^{\hat{\beta}} \ln x_i}{\sum_{i=1}^{i=n} x_i^{\hat{\beta}}} \tag{11.14}$$

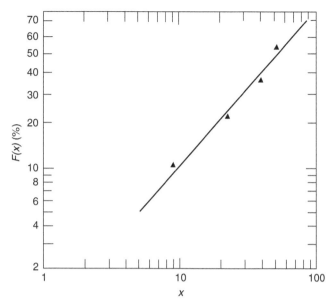

FIGURE 11.15 Probability plot for mixed censoring. Plotting positions are calculated based on cumulative hazard.

According to McCool [17], this equation has only a single positive solution. That solution is found using the Newton–Raphson method, although in highly censored cases the guess value used to start the method might need to be modified to avoid convergence to a negative value for β.

Having determined β from Equation 11.14, the ML estimate of η is obtained as follows:

$$\hat{\eta} = \left(\sum_{i=1}^{i=n} \frac{x_i^{\hat{\beta}}}{r} \right)^{1/\hat{\beta}} \tag{11.15}$$

The ML estimate of a general percentile is

$$\hat{x}_p = \hat{\eta} k_p^{1/\hat{\beta}} \tag{11.16}$$

where k_p is defined as

$$k_p = -\ln(1 - p) \tag{11.17}$$

Confidence limits can be set if the censoring mode corresponds to the suspension of testing when the rth earliest failure occurs. This type of censoring is customarily called type II censoring as contrasted to type I censoring in which testing is suspended at a predetermined testing time. In the latter, the number of failures is predetermined by the experimenter.

The basis for confidence intervals for β is that the random function $v(r, n) = \hat{\beta}/\beta$ follows a sampling distribution that depends on the sample size n and censoring number r, not on the underlying values of η and β. Functions with this property are known as pivotal functions. The sampling distribution of $v(r, n)$ cannot be found analytically, but may be determined empirically to whatever precision necessary by Monte Carlo sampling. In the Monte Carlo method, repeated samples are drawn by computer simulation from a Weibull distribution with arbitrary parameter values; for example, $\beta = 1.0$ and $\eta = 1.0$. The ML estimate $\hat{\beta}$ is formed for each sample and divided by the underlying value of β to yield a value of $v(r, n)$. With typically 10,000 such values, the percentiles may be computed from the sorted set and then equated to the expected value of the distribution.

Denoting the 5th and 95th percentiles of $v(r, n)$ as $v_{0.05}(r, n)$ and $v_{0.95}(r, n)$ leads to the following 90% confidence interval for β:

$$\frac{\hat{\beta}}{v_{0.95}(r, n)} < \beta < \frac{\hat{\beta}}{v_{0.05}(r, n)} \tag{11.18}$$

The raw ML estimates of the Weibull parameters are biased; that is, both the average and median of the β estimates in an indefinitely large number of samples will differ somewhat from the true β value for the population from which the samples were drawn. It is possible to correct the raw ML estimate so that either its average or its median will coincide with the underlying population value of β. Because the distribution of $v(r, n)$ is not symmetrical, it is necessary to choose whether the adjusted estimator should be median or mean unbiased. Median unbiasedness is recommended because then the ML point estimate will have the reasonable property that it is just as likely to be larger than the underlying true value as to be smaller. McCool [17] demonstrated that the median unbiased estimate of β, denoted by $\hat{\beta}'$, is expressible as

$$\hat{\beta}' = \frac{\hat{\beta}}{v_{0.50}(r, n)} \tag{11.19}$$

Table CD11.2 gives values of $v_{0.05}(r, n)$, $v_{0.50}(r, n)$, and $v_{0.95}(r, n)$ for $5 \leq n \leq 30$ and various values of r.

Correcting the bias of the estimate and setting confidence limits for a general percentile x_p depend on the pivotality of the random function $u(r, n, p) = \hat{\beta} \ln (\hat{x}_p / x_p)$. Given percentiles of $u(r, n, 0.10)$ determined by Monte Carlo sampling, a 90% confidence interval on $x_{0.10}$ can be set up:

$$\hat{x}_{0.10} e^{-u_{0.95}(r, n, 0.10)/\hat{\beta}} < x_{0.10} < \hat{x}_{0.10} e^{-u_{0.05}(r, n, 0.10)/\hat{\beta}} \tag{11.20}$$

A median unbiased estimate of $x_{0.10}$ can be calculated as

$$\hat{x}'_{0.10} = \hat{x}_{0.10} e^{-u_{0.50}(r, n, 0.10)/\hat{\beta}} \tag{11.21}$$

Values of the 5th, 50th, and 95th percentiles of $u(r, n, 0.10)$ are also given in Table CD11.2. *See Example 11.1.*

11.6.3.4 Sudden Death Tests

A popular test strategy in the bearing industry is the "sudden death" test. In sudden death testing, a test sample of size n is divided into l subgroups, each of size m ($n = l \times m$). When the first failure occurs in each subgroup, testing is suspended on that subgroup. When the test is over, there are l failures, the first failures in each of the subgroups. To estimate β, these first failures are substituted directly into Equation 11.14. Confidence limits for β are then calculated using Equation 11.18 with $r = n = l$. That is, the first failures are treated as members of an uncensored sample whose size is equal to the number of subgroups. Table CD11.3 gives the percentiles of $v(l, l)$ for $2 \leq l \leq 6$. The value of $\hat{x}_{0.10}$, determined by using the sample of first failures and Equation 11.16, is denoted as $\hat{x}_{0.10s}$. According to McCool [18], the ML estimate applicable to the complete sample is then calculated as follows:

$$\hat{x}_{0.10} = \hat{x}_{0.10s} m^{1/\hat{\beta}} \tag{11.22}$$

90% confidence limits for $x_{0.10}$ may be computed as

$$\hat{x}_{0.10} e^{-u_{0.95}(l, m, 0.10)/\hat{\beta}} < x_{0.10} < \hat{x}_{0.10} e^{-u_{0.05}(l, m, 0.10)/\hat{\beta}} \tag{11.23}$$

A median unbiased estimate of $x_{0.10}$ is calculated from

$$\hat{x}'_{0.10} = \hat{x}_{0.10} e^{-q_{0.50}/\hat{\beta}} \tag{11.24}$$

Table CD11.3 gives values of the percentiles of the random function $q(l, m, p)$ required for these calculations.

See Example 11.2.

11.6.3.5 Precision of Estimation: Sample Size Selection

A confidence interval reflects the uncertainty in the value of the estimated parameter due to the finite size of the life test sample. As the sample size increases, the two ends of the confidence interval approach each other; that is, the ratio of the upper to lower ends of the confidence interval approaches 1. For finite sample sizes, McCool [19] suggested this ratio as a useful measure of the precision of estimation. From Equation 11.18, the confidence limit ratio R for β estimation is

$$R = \frac{v_{0.95}(r, n)}{v_{0.05}(r, n)} \tag{11.25}$$

Values of R for various n and r are given in Table CD11.2 for conventional tests and Table CD11.3 for sudden death tests. It is noted that for a given sample size n, the precision improves (R decreases) as the number of failures r increases.

For $x_{0.10}$, the ratio of the upper to lower confidence limits contains the random variable $\hat{\beta}$. The approach taken by McCool [19] in this case was to use as a precision measure the median value of this ratio, denoted $R_{0.50}$. The expression for this median ratio contains the unknown value of the true shape parameter β. For planning purposes, one may use an historical value such as 10/9 or, alternatively, the value $R_{0.50}^{\beta}$ as the precision measure. Values of $R_{0.50}^{\beta}$ are given in Table CD11.2 for conventional testing and Table CD11.3 for sudden death testing.

11.6.4 Estimation in Sets of Weibull Data

11.6.4.1 Methods

Very often an experimental study of bearing fatigue life will include the testing of several samples, differing from each other with respect to the level of some qualitative factor under study. A qualitative factor is distinct from a quantitative factor such as temperature or load, which can be assigned a numerical value. Examples of qualitative factors are lubricants, cage designs, and bearing materials.

McCool [20] showed that more precise estimates can be made if the data in the samples making up the complete investigation are analyzed as a set. This is possible if it can be assumed that samples are drawn from Weibull populations, which, although they might differ in their scale parameter values, nonetheless have a common value of β.

Applicable tabular values for carrying out the analyses presuppose that the sample size n and the number of failures r are the same for each sample in the set; henceforth, this is assumed to be the case. It is thus assumed that k groups of size n have been tested until the rth failure occurred in each group. The first step is to determine whether it is plausible that the groups have a common value of β. This is done by analyzing each group individually to determine the values of $x_{0.10}$ and β. The largest and smallest of the $k\beta$ estimates are then determined, and their ratio formed. If the β values differ among the groups, this ratio would tend to be large. Table CD11.4 gives the values of the 90th percentile of the ratio $w = \hat{\beta}_{max} / \hat{\beta}_{min}$ for various r, n, and k. These values were determined by Monte Carlo sampling from k Weibull populations that had a common value of β. Therefore, values of the ratio of the largest to the smallest shape parameter estimates exceeding those in Table CD11.4 will occur only 10% of the time if the groups do have a common value of β. These values may be used as the critical values in deciding whether a common β assumption is justified.

Having determined that a common β assumption is reasonable, this common β value can be estimated using the data in each group, by solving the nonlinear equation

$$\frac{1}{\hat{\beta}_1} + \frac{1}{rk} \sum_{i=1}^{i=k} \sum_{j=1}^{j=r} \ln x_{i(j)} - \sum_{i=1}^{i=k} \frac{\sum_{j=1}^{j=n} x_{i(j)}^{\hat{\beta}_1} \ln x_{i(j)}}{k \sum_{j=1}^{j=n} x_{i(j)}^{\hat{\beta}_1}} \tag{11.26}$$

where $\hat{\beta}_1$ denotes the ML estimate of the common β value and $x_{i(j)}$ denotes the jth order failure time within the ith group. Confidence limits for β may be set analogously to Equation 11.17 as follows:

$$\frac{\hat{\beta}_1}{(v_1)_{0.95}} < \beta < \frac{\hat{\beta}_1}{(v_1)_{0.05}} \tag{11.27}$$

where $v_1 (r, n, k) = \hat{\beta}_1 / \beta$. A median unbiased estimate of β may be calculated from

$$\hat{\beta}' = \frac{\hat{\beta}_1}{(v_1)_{0.50}} \tag{11.28}$$

Table CD11.5 gives percentiles of $v_1(r, n, k)$ needed for setting 90% confidence limits and for bias correction of various values of n, r, and k. The scale parameter for the ith group may be re-estimated with β_1 as follows:

$$\hat{\eta}_i = \left(\frac{\sum x_{i(j)}^{\hat{\beta}_1}}{r} \right)^{1/\hat{\beta}_1} \tag{11.29}$$

The value of x_{pi} may be estimated from

$$\hat{x}_{pi} = \hat{\eta}_i k^{1/\hat{\beta}_1} \tag{11.30}$$

Confidence limits for $x_{0.10}$ may be computed as follows:

$$\hat{x}_{0.10}e^{-(u_1)_{0.95}/\hat{\beta}_1} < x_{0.10} < \hat{x}_{0.10}e^{-(u_1)_{0.05}/\hat{\beta}_1} \tag{11.31}$$

where $u_1 = \hat{\beta}_1 \ln(\hat{x}_{0.10}/x_{0.10})$ is the k sample generalization of $u(r,n,0.10)$. The median unbiased estimate of $x_{0.10}$ may be computed as

$$\hat{x}'_{0.10} = \hat{x}_{0.10}e^{-(u_1)_{0.50}/\hat{\beta}_1} \tag{11.32}$$

Now that $x_{0.10}$ has been estimated for each group using the ML estimate $\hat{\beta}_1$ of the common shape parameter, the next question of interest is whether these $x_{0.10}$ values differ significantly. That is, are the apparent differences among the $x_{0.10}$ estimates real, or could they be due to chance? To test whether the underlying true $x_{0.10}$ values are all equal, the magnitude of variation that could occur in the estimated values due to chance alone must be assessed. This can be done by using the random function $t_1(r,n,k)$ defined by

$$t_1(r,n,k) = \hat{\beta}_1 \ln \left[\frac{(\hat{x}_{0.10})_{max}}{(\hat{x}_{0.10})_{min}} \right] \tag{11.33}$$

where $(\hat{x}_{0.10})_{max}$ and $(\hat{x}_{0.10})_{min}$ are the largest and smallest values of $\hat{x}_{0.10}$ calculated among the k samples. The 90th and 95th percentiles of $t_1(r,n,k)$ may be used to assess the observed difference in the $x_{0.10}$ values. Any two samples, for example, sample i and sample j, for which values the quantity $\hat{\beta}_1 \ln [(\hat{x}_{0.10})_i / (\hat{x}_{0.10})_j]$ exceeds $(t_1)_{0.90}$, may be declared to differ from each other at the 10% level of significance. Correspondingly, using the 95th percentile of $t_1(r,n,k)$ results in a 5% significance level test for the equality of the $x_{0.10}$ values.

See Example 11.3.

11.7   ELEMENT TESTING

11.7.1   Rolling Component Endurance Testers

Conducting an endurance test series on full-scale bearings is expensive because numerous test samples are required to obtain a useful experimental life estimate. The identification of simpler, less costly, life testing methods has therefore been a longstanding objective. The use of elemental test configurations offers a potential solution to this need. In this approach, a test specimen that has a simplified geometry (e.g., flat washer, rod, or ball) is used, and RC is developed at multiple test locations. The aim is to extrapolate the life data generated in an element test to a real bearing application, thus saving calendar time and cost as compared with life data generated using full-scale bearing tests. This objective has historically not been achieved, generally because all of the operating parameters influencing fatigue life of rolling–sliding contacts were not reduced to stresses; rather, as is shown in Chapter 8, they were evaluated as life factors. The only stress directly evaluated in both element and full-scale bearing endurance testing has been the Hertz or normal stress acting on the contact. Lubrication, contamination, surface topography, and material effects have been evaluated as life factors. To be able to extrapolate the life data derived from element testing to full-scale bearing life data, it is necessary to evaluate both data sets from the standpoint of applied and induced stresses as compared with material strength. The methods to accomplish this for full bearings are developed in Chapter 8; Harris [21] developed a similar method for balls endurance tested in v-ring test rigs.

Even without direct correlation of life test data between elements and full-scale bearings, element testing has proven useful in the ability to rank the performance of various materials in initial screening sequences or in adverse environments, such as extremely low or high temperature, oxidizing atmospheres, and vacuum. Therefore, a discussion of element life testing techniques is warranted, even when the test data evaluation techniques do not permit direct correlation with full-scale bearing life test data. Caution must always be used, however, because the precision of the ranking process is open to question. Performance reversals have sometimes been experienced when comparing the screening element test results when the materials have been retested in actual bearings. Such reversals can be avoided if both the element test data and actual bearing test data evaluations are based on the total stress consideration.

The oldest and perhaps the most widely used element test configuration is the rolling four-ball machine or Barwell [22] tester developed in the 1950s. This system uses four 12.7-mm (0.50 in.) diameter balls to simulate an angular-contact ball bearing operating with a vertical axis under a pure thrust load. One ball is the primary test element serving as the inner ring of the bearing assembly. It is supported in pyramid fashion on the remaining three balls, which rotate freely in a conforming cup at a predetermined contact angle. A modification of this test method, the rolling five-ball tester, was subsequently developed at NASA Lewis Research Center (now NASA Glenn Research Center) [23]. To generate more stress cycles, the test ball is supported by a group of four balls. This system, illustrated in Figure 11.16 and Figure 11.17, has been used to generate an extensive amount of life test data on standard and experimental bearing steels. Accelerated life testing is typically conducted at Hertz stresses of 4,138 MPa (600,000 psi). This loading involves some plastic deformations of the materials, making extrapolation of the life test data to complete bearings unreliable. The tester may be used to compare RC materials and lubricants.

Another widely used element test system, as illustrated in Figure 11.18, is the RC tester developed by General Electric [24]. The test element in this configuration is a 4.76-mm (0.1875 in.) diameter rod rotating under load between two 95.25-mm (3.75 in.) diameter disks. The rod can be axially repositioned to achieve a number of RC tracks on a single bar. Unfortunately, this configuration is not as cost effective as it first appears. Stress concentrations occur at the edges of the rod contact unless the disks are profiled (crowned) in the axial direction. This significantly increases the cost of manufacturing the disks. During operation, fatigue failures on the rod also tend to damage the disk surfaces, requiring these to be refinished at regular intervals. To achieve accelerated testing, Hertz stresses as great as 5,517 MPa (800,000 psi) are frequently employed. This loading is substantially in the regime of plastic deformations; hence, extrapolation of data for prediction of bearing fatigue endurance is unreliable. The tester is mainly used to compare RC materials.

A variation of an RC endurance test rig using a cylindrical rod as the test element was described by Glover [25]. In this rig developed by Federal–Mogul–Bower and illustrated in Figure 11.19, the load is applied to the rod through three balls supported in tapered roller bearing outer rings (cups). The typical applied Hertz stress in each contact is 4,138 MPa (600,000 psi), which, as explained above, involves some plastic deformation.

Another element test configuration is the single ball tester developed by Pratt and Whitney, United Technologies Corporation, for evaluating balls used in aircraft gas turbine engine bearings [26]. This system, shown in Figure 11.20 and Figure 11.21, tests balls from approxi-

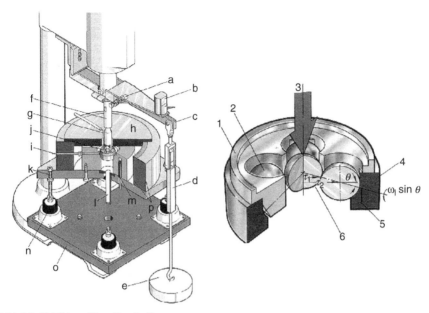

FIGURE 11.16 NASA rolling five-ball test system.

mately 19–65 mm (0.75–2.50 in.) in two v-ring raceways with lubrication to simulate the application. Ball/v-ring contact angles are typically 25° or 30° inducing spinning components of angular velocities in the contacts, and substantial sliding. Hertz stresses in the contacts are typically 4,000 MPa (580,000 psi), thus involving some plastic deformations. Harris [21] developed a stress-based ball life prediction method for this system. Subsequently, endurance

FIGURE 11.17 Group of five-ball test rigs.

FIGURE 11.18 General Electric Polymet RC disk machine.

test data accumulated using this rig were used in the development of fatigue limit stress values for several bearing RC component materials [27].

11.7.2 ROLLING–SLIDING FRICTION TESTERS

11.7.2.1 Purpose

The main purpose of the element test rigs described above is to accumulate RC fatigue endurance data in an economical, efficient, and rapid manner. The relative influence on bearing fatigue endurance of materials, material processing, lubricants, and so on can be investigated thereby. Some of the most significant stresses that determine the extent of bearing life are the surface shear stresses occurring in the rolling element–raceway contacts. Element test rigs can be designed to investigate the influence of friction on bearing endurance and also to help quantify the magnitude of traction in the rolling–sliding contacts that occur in many rolling bearing applications.

11.7.2.2 Rolling–Sliding Disk Test Rig

To experimentally determine the magnitude of the frictional stresses occurring in EHL contacts, rolling–sliding disk machines have been developed. The device developed by Nélias et al. [28] is illustrated in Figure 11.22. The disks are contoured to produce elliptical contact areas as illustrated in Figure 11.23. The motors in Figure 11.22 may turn at different speeds to

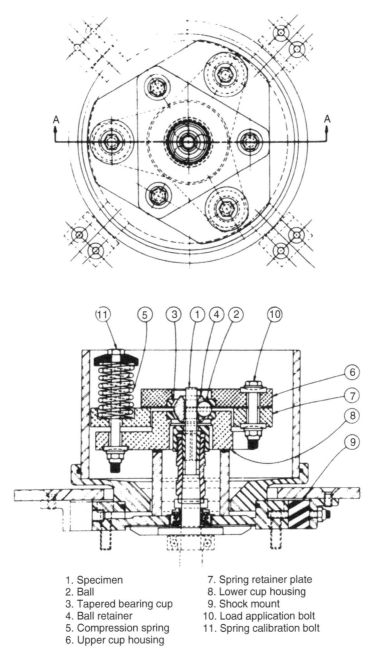

1. Specimen
2. Ball
3. Tapered bearing cup
4. Ball retainer
5. Compression spring
6. Upper cup housing
7. Spring retainer plate
8. Lower cup housing
9. Shock mount
10. Load application bolt
11. Spring calibration bolt

FIGURE 11.19 Ball–rod RC fatigue test rig.

achieve the desired rolling–sliding motion. Motor 2 is mounted in hydrostatic cylindrical bearings to permit friction torque, and hence, friction force measurement. The friction force F_f to applied force W ratio is called the traction coefficient. Using the analytical methods of Chapter 5, the effective local (x, y) friction coefficients can be estimated from the test results. In Chapter 8, it is shown how the test device in Figure 11.22 has been used to determine the characteristics of the effect of friction on fatigue of the rolling–sliding contacts in ball and

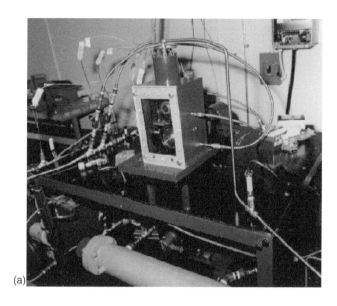

(a)

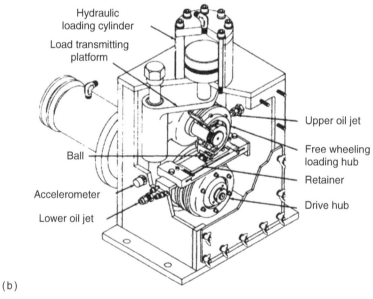

(b)

FIGURE 11.20 Photograph (a) and drawing (b) of a Pratt and Whitney single ball/v-ring test rig.

roller bearings. As discussed in Chapter 8, by equipping the test rig with the contaminated lubrication system of Figure 11.24, Ville and Nélias [29] investigated the effects of particulate contamination on rolling–sliding contact fatigue.

11.7.2.3 Ball–Disk Test Rig

A ball–disk test rig, initially developed by Wedeven [30] and shown in Figure 11.25, was designed to determine the nature of lubricant films in point contacts. Rolling velocity may be varied by varying the ball drive spindle angle and the radius at which the ball contacts

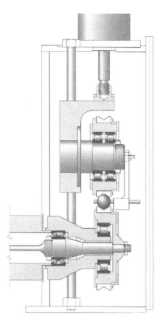

FIGURE 11.21 Schematic diagram of Pratt and Whitney single ball/v-ring test rig.

the disk. Using a disk element of a clear material such as sapphire or glass and optical interferometry, the pressure distributions in Hertz point contacts could be displayed (see Figure 4.11). The rig has been further developed by Wedeven [31] with separately powered ball drive and disk drive shafts and with an air bearing support of the disk drive. It is thereby possible to determine contact traction force vs. slide–roll ratio; the graphical display in Figure 11.26 was the output from the test rig. A mathematical model of the EHL circular point contact may also be developed, and by the matching analytical and experimental data, the localized friction components that comprise the traction may be determined.

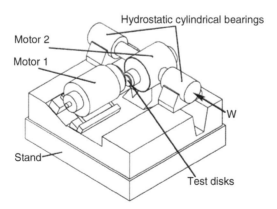

FIGURE 11.22 Schematic drawing of rolling–sliding disk testing device. (From Nélias, D., et al., *ASME Trans., J. Tribol.*, 120, 184–190, April 1998. With permission.)

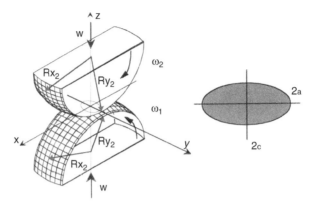

FIGURE 11.23 Illustration of elliptical contact area generated by the rolling–sliding disk test device. (From Nélias, D., et al., *ASME Trans., J. Tribol.*, 120, 184–190, April 1998. With permission.)

The rig can be equipped with an environment chamber to allow evaluation of the traction coefficient under conditions of high and low temperature, and high vacuum. It further permits optical examination of the circular point contact under the effects of lubricant particulate contamination; the photographs in Figure 10.34 were obtained using such a test rig.

11.8 CLOSURE

In Chapter 11 in the first volume of this handbook, it was demonstrated that although ball and roller bearing fatigue life rating and endurance formulas are founded in theory, they are

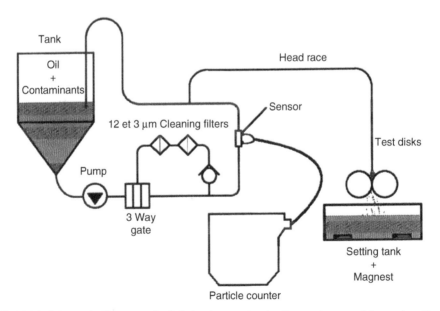

FIGURE 11.24 Schematic diagram of a lubrication contamination system used in conjunction with a rolling–sliding test rig. (From Ville, F. and Nélias, D., Early fatigue failure due to dents in EHL contacts, Presented at the STLE Annual Meeting, Detroit, May 17–21, 1998. With permission.)

FIGURE 11.25 Ball–disk traction test rig. (From Wedeven Associates, Inc., *Bridging Technology and Application through Testing*, Brochure, 1997. With permission.)

semiempirical relationships requiring the establishment of various constants to enable their use. These constants, which depend on the bearing raceway and rolling element materials, can be established only by appropriate testing. Because of the stochastic nature of rolling bearing fatigue endurance, testing procedures necessarily require bearing or material populations of sufficient size to render the test results meaningful. Sample sizing effects were discussed in detail herein.

Historically, to establish sufficiently accurate rating formula constants, it has been necessary to test complete bearings. With the development of stress-based life factors as

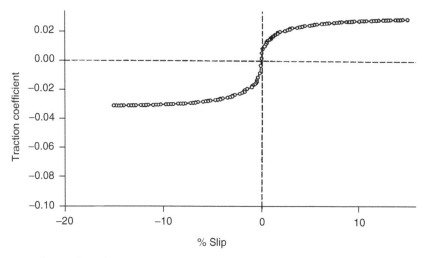

FIGURE 11.26 Curve of traction coefficient vs. percent sliding obtained from Wedeven ball–disk test rig.

shown in Chapter 8, however, it is now possible to use element testing methods to determine many of these constants. For example, endurance testing of balls in v-ring test rigs may be used to determine the basic material fatigue strengths of various materials. On the other hand, some of the stresses that influence bearing life depend on the raceway forming and surface finishing methods. To duplicate these effects, the exact component may need to be endurance tested.

REFERENCES

1. Weibull, W., A statistical theory of the strength of materials, *Proc. R. Swed. Inst. Eng. Res.*, 151, Stockholm, 1939.
2. Lundberg, G. and Palmgren, A., Dynamic capacity of rolling bearings, *Acta Polytech. Mech. Eng.*, *Ser. 1*, 3(7), *R. Swed. Acad. Eng.*, 1947.
3. Lundberg, G. and Palmgren, A., Dynamic capacity of roller bearings, *Acta Polytech. Mech. Eng.*, *Ser. 2*, 4(96), *R. Swed. Acad. Eng.*, 1952.
4. Ioannides, E. and Harris, T., A new fatigue life model for rolling bearings, *ASME Trans., J. Tribol.*, 107, 367–378, 1985.
5. International Organization for Standards, International Standard ISO 281, *Rolling Bearings—Dynamic Load Ratings and Rating Life*, 2006.
6. Tallian, T., On competing failure modes in rolling contact, *ASLE Trans.*, 10, 418–439, 1967.
7. Valori, R., Tallian, T., and Sibley, L., Elastohydrodynamic film effects on the load life behavior of rolling contacts, ASME Paper 65-LUBS-11, 1965.
8. Johnston, G., et al., Experience of element and full bearing testing over several years, *Rolling Contact Fatigue Testing of Bearing Steels, ASTM STP 771*, J. Hoo, Ed., 1982.
9. Sayles, R. and MacPherson, P., Influence of wear debris on rolling contact fatigue, *ASTM STP 771*, J. Hoo, Ed., 1982, pp. 255–274.
10. Fitch, E., *An Encyclopedia of Fluid Contamination Control*, Fluid Power Research Center, Oklahoma State University, 1980.
11. Tallian, T., *Failure Atlas for Hertz Contact Machine Elements*, 2nd Ed., ASME Press, 1999.
12. Andersson, T., Endurance testing in theory, *Ball Bear. J.*, 217, 14–23, 1983.
13. Sebok, G. and Rimrott, U., Design of rolling element endurance testers, ASME Paper 69-DE-24, 1964.
14. Hacker, R., Trials and tribulations of fatigue testing of bearings, SAE Technical Paper 831372, 1983.
15. Johnson, L., *Theory and Technique of Variation Research*, Elsevier, New York, 1970.
16. Nelson, W., Theory and application of hazard plotting for censored failure data, *Technometrics*, 14, 945–966, 1972.
17. McCool, J., Inference on Weibull percentiles and shape parameter for maximum likelihood estimates, *IEEE Trans. Reliab.*, R-19, 2–9, 1970.
18. McCool, J., Analysis of sudden death tests of bearing endurance, *ASLE Trans.*, 17, 8–13, 1974.
19. McCool, J., Censored sample size selection for life tests, *Proc. 1973 Ann. Reliab. Maintainab. Symp.*, IEEE Cat. No. 73CH0714–64, 1973.
20. McCool, J., Analysis of sets of two-parameter Weibull data arising in rolling contact endurance testing, *Rolling Contact Fatigue Testing of Bearing Steels, ASTM STP 771*, J.J. Hoo Ed., American Society for Testing and Materials, Philadelphia, 1982, pp. 293–319.
21. Harris, T., Prediction of ball fatigue life in a ball/v-ring test rig, *ASME Trans., J. Tribol.*, 119, 365–374, July 1997.
22. Barwell, F. and Scott, D., *Engineering*, 182, 9–12, 1956.
23. Zaretsky, E., Parker, R., and Anderson, W., NASA five-ball tester—over 20 years of research, *Rolling Contact Fatigue Testing of Bearing Steels, ASTM STP 771*, J. Hoo, Ed., 1982.
24. Bamberger, E. and Clark, J., Development and application of the rolling contact fatigue test rig, *Rolling Contact Fatigue Testing of Bearing Steels, ASTM STP 771*, J. Hoo, Ed., 1982.

25. Glover, G., A ball–rod rolling contact fatigue tester, *Rolling Contact Fatigue Testing of Bearing Steels, ASTM STP 771*, J. Hoo, Ed., 1982.
26. Brown, P., et al., Evaluation of powder-processed metals for turbine engine ball bearings, *Rolling Contact Fatigue Testing of Bearing Steels, ASTM STP 771*, J. Hoo, Ed., 1982.
27. Harris, T., Establishment of a new rolling bearing fatigue life calculation model, Final Report U.S. Navy Contract N00421-97-C-1069, February 23, 2002.
28. Nélias, D., et al., Experimental and theoretical investigation of rolling contact fatigue of 52100 and M50 steels under EHL or Micro-EHL conditions, *ASME Trans., J. Tribol*, 120, 184–190, April 1998.
29. Ville, F. and Nélias, D., Early fatigue failure due to dents in EHL contacts, Presented at the STLE Annual Meeting, Detroit, May 17–21, 1998.
30. Wedeven, L., *Optical Measurements in Elastohydrodynamic Rolling Contact Bearings*, Ph.D. Thesis, University of London, 1971.
31. Wedeven Associates, Inc., *Bridging Technology and Application through Testing*, Brochure, 1997.

Appendix

All equations in the text are written in metric or standard international system units. In this appendix, Table A.1 gives factors for conversion of Standard International system units to English system units. Note that for the former, only millimeters are used for length and square millimeters for area. Furthermore, the basic unit of power used herein is the watt (as opposed to kilowatt). To be consistent with this, Table A.2 provides the appropriate English system units constant for each equation in the text that has a Standard International system units constant.

TABLE A.1
Unit Conversion Factors[a]

Unit	Standard International System	Conversion Factor	English System
Length	mm	0.03937, 0.003281	in., ft
Force	N	0.2247	lb
Torque	mm · N	0.00885	in. · lb
Temperature difference	°C,K	1.8	°F, °R
Kinematic viscosity	mm^2/sec (centistokes)	0.001076	ft^2/sec
Heat flow, power	W	3.412	Btu/hr
Thermal conductivity	W/mm · °C	577.7	Btu/hr · ft · °F
Heat convection coefficient	W/mm^2 · °C	176,100	Btu/hr · ft^2 · °F
Pressure, stress	N/mm^2 (MPa)	144.98	psi

[a] English system units equal Standard International system units multiplied by conversion factor.

TABLE A.2
Equation Constants for SI and English System Units

Chapter Number	Equation Number	SI System Constant	English System Constant
1	33	3.84×10^{-5}	4.36×10^{-7}
	34	3.84×10^{-5}	4.36×10^{-7}
	35	1.24×10^{-5}	8.55×10^{-8}
	36	1.24×10^{-5}	8.55×10^{-8}
	37	1.24×10^{-5}	8.55×10^{-8}
	39	0.62×10^{-5}	4.28×10^{-8}
	42	0.62×10^{-5}	4.28×10^{-8}
	52	1.24×10^{-5}	8.55×10^{-8}
	54	0.62×10^{-5}	4.28×10^{-8}
	58	0.31×10^{-5}	2.14×10^{-8}
	65	3.84×10^{-5}	4.36×10^{-7}
	74	0.62×10^{-5}	4.28×10^{-8}
	75	0.31×10^{-5}	2.14×10^{-8}
3	27	2.26×10^{-11}	2.11×10^{-6}
	59	2.26×10^{-11}	2.11×10^{-6}
	60	3.39×10^{-11}	3.17×10^{-6}
	62	2.26×10^{-11}	2.11×10^{-6}
	63	3.39×10^{-11}	3.17×10^{-6}
	67	4.47×10^{-12}	4.18×10^{-7}
	68	8.37×10^{-12}	7.83×10^{-7}
	106	2.15×10^{5}	3.12×10^{7}
	107	3.39×10^{-11}	3.17×10^{-6}
4	64	4.597×10^{-12}	1.509×10^{-18}
	65	8.543×10^{-9}	4.066×10^{-13}
7	6	10^{3}	30.2
	7	9.551×10^{3}	288.4
	18	0.0332	0.332
	19	0.060	0.60
	20	2.30×10^{-5}	0.30
	21	0.030	0.30
	28	5.73	0.173
	30	5.73×10^{-8}	0.173×10^{-8}
8	4	77.9	5.914×10^{3}
	11	464	4.166×10^{4}

Index

isotropic, 144
oblique roughness, 117
pitting, 245
real, 139
shear stress, 134, 138, 193, 222
tangential stress, 131
topography, 116
Surface-initiated fatigue, 295

T
Tangential profile, 11
Tapered roller bearings
asymmetrical rollers, 56
combine radial, thrust and moment loading, 25
high speed operation, 71, 84, 87
roller end-flange loading, 25
roller end-flange sliding motion, 56
Temperature, 191
control in endurance testing, 309
excursion, 274
hardness, effect on, 274
influence coefficients, 200
level, 192
Lundberg-Palmgren theory and, 234
node system, 202, 207
steady state, 192
Thermal conductivity, 196
Thermal elastohydrodynamic lubrication, 114
Thermal imbalance, 274
Traction coefficient, 137, 339

V
Variance of Weibull distribution, 321
Velocities in fluid lubrication, 99, 145
VIMVAR M50 steel, 213, 223
VIMVAR M50NiL steel, 223

Viscosity
base oil in grease, 120
high pressure, 103
pressure effect, 102
temperature dependence, 168
Viscous forces in lubrication, 98
drag on rolling elements, 155
Volume under stress
critical, 252
Lundberg-Palmgren, 233
Ioannides-Harris, 234
Von Mises stress, 232, 237

W
Wear
definition, 293
ring cracking due to, 276
roller end-flange sliding, 56
types, 294
Weibull
analysis, 304
cumulative distribution function, 320
distribution, 319
graphical representation of distribution, 323
hazard, 326
mean time between failures, 321
percentiles, 322
probability density function, 320
shape parameter, 325
slope, 211, 324–325
two-parameter distribution, 320
variance, 321
Work-hardening of steel, 250

Y
Yaw angle, 46